Communications and Control Engineering Series

Editors: A. Fettweis · J. L. Massey · J.W. Modestino · M. Thoma

JACOPO CO. RICCATI

Sergio Bittanti · Alan J. Laub
Jan C. Willems (Eds.)

The Riccati Equation

Springer-Verlag
Berlin Heidelberg NewYork
London Paris Tokyo
Hong Kong Barcelona Budapest

Editors:

Sergio Bittanti
Politecnico di Milano, Dipartimento di Elettronica,
Piazza Leonardo da Vinci 32, I-20133 Milano, Italy

Alan J. Laub
Department of Electrical and Computer Engineering,
University of California, Santa Barbara, CA 93106, USA

Jan C. Willems
Department of Mathematics, University of Groningen,
P.O. Box 800, NL-9700 AV Groningen, The Netherlands

Frontispiece: Portrait of Count Jacopo Riccati reproduced by courtesy of the municipal library of Castelfranco Veneto (TV), Italy

---

ISBN 3-540-53099-1 Springer-Verlag Berlin Heidelberg NewYork
ISBN 0-387-53099-1 Springer-Verlag NewYork Berlin Heidelberg

Library of Congress Cataloging-in-Publication Data
The Riccati equation / S. Bittanti, A. J. Laub, J. C. Willems, editors.
(Communications and control engineering series)
Includes index.
ISBN 3-540-53099-1
ISBN 0-387-53099-1
1. Riccati equation. I. Bittanti, Sergio. II. Laub, Alan J.
III. Willems, Jan C. IV. Series.
1948- .
QA371.R36 1991

Typesetting: Text supplied by authors, formatted by Danny L. Lewis, Buchproduktion Berlin, with T_EX;
Printing: Color-Druck Dorfi GmbH, Berlin; Binding: Lüderitz & Bauer, Berlin
61/3020-543210

Preface

Conceived by Count Jacopo Francesco Riccati more than a quarter of a millennium ago, the Riccati equation has been studied in subsequent centuries by a number of distinguished scientists, including several members of the Bernoulli family, L. Euler and J. Liouville, to name but a few. The last thirty years have witnessed a considerable renaissance of the research around this equation, stemming from the widespread developments in and diffusion of systems and control concepts and applications. Indeed, since its introduction in control theory at the beginning of the sixties, the Riccati equation has known an impressive range of applications, such as linear quadratic optimal control, stability theory, stochastic filtering and stochastic control, synthesis of linear passive networks, differential games and, more recently, H_∞-control and robust stabilization.

The state of the art in this area was reviewed at a recent Workshop on the Riccati Equation in Control, Systems, and Signals held in Como (Italy) June 26-28, 1989. This workshop constituted the foundation from which the idea of this book germinated.

The purpose of this book is to present a self-contained treatment of the main issues evolving around the Riccati equation, in particular theory, applications, and numerical algorithms. The book, which consists of coordinated tutorial chapters written by different authors, is intended as a graduate text as well as a reference for scientists, especially engineers, and mathematicians.

The organization is as follows. Chapter 1 is devoted to the history and pre-history of the Riccati equation. Chapters 2 and 3 supply a comprehensive view of the algebraic Riccati equation, mainly based on a linear algebra approach. A geometrical analysis of the equation is carried out in Chapters 4 and 5. Chapters 2 to 5 deal with the constant coefficient case. The periodically time-varying Riccati equation is the subject of Chapter 6. The leading numerical techniques for the solution of the Riccati equation are overviewed in Chapter 7. The remaining four chapters address connections between the Riccati equation and some important problems in systems and control. More precisely, in Chapter 8, the role of the Riccati equation in the study of dissipative systems is elucidated. The linear quadratic optimal control problem in its various facets is the subject of Chapters 9 and 10. Finally, a unified survey on generalized Riccati equations in dynamic games is presented in Chapter 11.

The book is reasonably self-contained, but the reader should have some familiarity with basic concepts in system theory and linear algebra. Section 2 of

Chapter 2 presents a concise overview of some of these concepts. Some knowledge of geometric concepts is also useful, mainly for Chapters 4 and 5.

The editors would like to express their sincere thanks and acknowledgment to all authors for care and diligence in providing their contribution.

Sergio Bittanti, Alan J. Laub and Jan C. Willems

Contents

List of Contributors

Gregory Ammar
Department of Mathematics, Northern Illinois University, DeKalb, IL 60115, USA

Tamer Basar
University of Illinois at Urbana-Champaign, Decision and Control Laboratory, 1101 West Springfield Avenue, Urbana, IL 61801, USA

Robert Bitmead
Department of Systems Engineering, Australian National University, Canberra ACT 2601, Australia

Sergio Bittanti
Politecnico di Milano, Dipartimento di Elettronica, Piazza Leonardo da Vinci 32, 20133 Milano, Italy

Frank M. Callier
Department of Mathematics, FNDP, Rempart de la Vierge 8, B-5000 Namur, Belgium

Patrizio Colaneri
Centro di Teoria dei Sistemi, Dipartimento di Elettronica, Piazza Leonardo da Vinci 32, 20133 Milano, Italy

Giuseppe De Nicolao
Centro di Teoria dei Sistemi, Dipartimento di Elettronica, Piazza Leonardo da Vinci 32, 20133 Milano, Italy

Michel Gevers
Université Catholique de Louvain, Bâtiment Maxwell, B-1348 Louvain-la-Neuve, Belgium

Vladimir Kucera
Institute of Information Theory and Automation, Czechoslovak Academy of Sciences, P.O. Box 18, vodárenskou věži 4, 18208 Prague, Czechoslovakia

Peter Lancaster
University of Calgary, Department of Mathematics and Statistcs, 2500 University Drive NW, Calgary, Alberta, Canada T2N 1N4

Alan J. Laub
Department of Electrical and Computer Engineering, University of California, Santa Barbara CA 93106, USA

Clyde F. Martin
Department of Mathematics, Texas Tech University, Box U319, Lubbock, Texas 79409, USA

Leiba Rodman
Department of Mathematics, College of William and Mary, Williamsburg, Virginia, 23185, USA

Mark A. Shayman
Department of Electrical Engineering, University of Maryland, College Park MD 20724, USA

H. L. Trentelman
Mathematics Institute, P.O. Box 800, NL-9700 AV Groningen, The Netherlands

Jacques L. Willems
Engineering Faculty, University of Gent, Grotesteenweg-Noord 2, B-9710 Gent, Belgium

Jan C. Willems
University of Groningen, Department of Mathematics, P.O. Box 800, NL-9700 AV Groningen, The Netherlands

1 Count Riccati and the Early Days of the Riccati Equation

Sergio Bittanti

1.1 Introduction

Towards the turn of the seventeenth century, when the baroque was giving way to the enlightenment, there lived in the Republic of Venice a gentleman, the father of nine children, by the name of Jacopo Franceso Riccati. On the cold New Year's Eve of 1720, he wrote a letter to his friend Giovanni Rizzetti, where he proposed two new differential equations. In modern symbols, these equations can be written as follows:

$$\dot{x} = \alpha x^2 + \beta t^m \tag{1.1}$$

$$\dot{x} = \alpha x^2 + \beta t + \gamma t^2 \tag{1.2}$$

where m is a constant. This is probably the first document witnessing the early days of the Riccati Equation, an equation which was to become of paramount importance in the centuries to come.

1.2 A Glimpse into Riccati's Life

Count Jacopo Riccati was born in Venice on May 28, 1676. His father, a nobleman, died when he was only ten years old. The boy was raised by his mother, who did not marry again, and by a paternal uncle, who recognized unusual abilities in his nephew and persuaded Jacopo Francesco's mother to have him enter a Jesuit college in Brescia. Young Riccati enrolled at this college in 1687, probably with no intention of ever becoming a scientist. Indeed, at the end of his studies at the college, in 1693, he enrolled at the University of Padua as a student of law. However, following his natural inclination, he also attended classes of astronomy given by Father Stefano degli Angeli, a former pupil of Bonaventura Cavalieri. Father Stefano was fond of Isaac Newton's *Philosophiae Naturalis Principia*, which he passed on to young Riccati around 1695. This is probably the event which caused Riccati to turn from law to science.

After graduating on June 7, 1696, he married Elisabetta dei Conti d'Onigo on October 15, 1696. She bore him 18 children, of whom 9 survived childhood. Amongst them, Vincenzo (b.1707, d.1775), a mathematical physicist, and Gior-

Bittanti, Laub, Willems (Eds.)
The Riccati Equation

dano (b.1709, d.1790), a scholar with many talents but with a special interest for architecture and music, are worth mentioning.

Riccati spent most of his life in Castelfranco Veneto, a little town located in the beautiful country region surrounding Venice. Besides taking care of his family and his large estate, he was in charge of the administration of Castelfranco Veneto, as Provveditore (Mayor) of that town, for nine years during the period 1698-1729. He also owned a house in the nearby town of Treviso, where he moved after the death of his wife (1749), and where his children had been used to spending a good part of each year after 1747.

1.3 The "letterato" and the Man

Notwithstanding all his responsibilities, Count Riccati always found time for his beloved studies. He did not follow any lecture courses in mathematics or other scientific disciplines. Basically, the profound knowledge of the self-taught man was acquired by reading. Among the journals of the period, it is worth mentioning the most important academic reports, in particular the *Commentari dell'Accademia delle Scienze di Bologna*, the *Acta Eruditorum Lipsiae* and the reports of St. Petersburg's Imperial Academy. Besides these scientific journals, Count Riccati was a reader of many Italian journals of general interest, e.g., the *Galleria di Minerva* and *Giornale de' Letterati d'Italia*, both printed in Venice, the former from 1696 to 1717 and the latter from 1710 to 1740. These journals had diverse interests; a typical issue would contain poems, short novels, philosophic essays and occasionally some mathematics. They reflected the spirit of the period prevailing in Northern Italy, according to which what we now consider a scientist was supposed to be a "*letterato*", a person of profound knowledge and vast interests. In line with this, Riccati had far-reaching interests, ranging from mathematics to poetry, from physics to religion, as witnessed by his works and his rich library. In conclusion, although he agreed with Galileo Galilei that "*la pietra lavagna e' la pietra di paragone degli ingegni*" ("the blackboard is the appropriate field of comparison of talents"), he also believed that the brain should be better exercised in a variety of fields. As he wrote: "*L'intelletto d'ogni uomo dovrebbe essere educato fin dalla sua adolescenza a far tesoro delle scienze piu' eccellenti e delle arti piu' belle. Non dico gia' che qualunque materia si debba scandagliare fino al fondo. Secondo il genio e il temperamento, una almeno se ne scelga e sopra di essa di proposito si metta studio. Nelle altre si faccia come la pecchia, che da ogni fiore va succhiando una qualche stilla, onde le cose e le voci per cui si esprimono non riescano nuove, ne' ci convenga restar su due piedi in molte occasioni, e guardare un silenzio poco onorevole, per non dir una decina di spropositi*" ("Since adolescence, the mind should be educated to treasure the most eminent of sciences and the finest of arts. I do not want to claim that every topic should be probed in detail. Following one's own talent and inclination, one should select at least one topic, and study it in depth. In the others, one should follow the example of the bee which sucks a drop of nectar from each flower ...") (from *Opere*, Vol 1, p. 164).

All through his life, an important complement of reading was direct exchange of ideas, through correspondence and conversation. Riccati was in contact with Domenica Maria Gaetana Agnesi, Gabriele Manfredi, Giovanni Poleni, Giovanni Rizzetti, Giuseppe Suzzi, Antonio Vallisneri, Bernardino Zendrini, and many other Italians. He was also in contact with various European mathematicians, such as Jacob Hermann and some members of the influential Bernoulli family, mainly Nicolaus III (b.1695, d.1726). Most of Riccati's correspondence can be found in Castelfranco Veneto, with the exception of the letters exchanged with members of the Bernoulli family, which are kept in the Basel University Library.

Riccati was an undemonstrative, kind man who preferred his home to academies and universities. His way of life was a very simple one, and he travelled very little. Probably, the only extended period he spent away from home was the summer of 1719, when, following the recommendation of his physician, he moved to Val di Sole to take advantage of the healthy water of that valley. He turned down many notable invitations, including the most appealing one of becoming president of St. Petersburg's Academy (c. 1725). He also refused the chair of Mathematics at the University of Padua and the invitation to the Court of Wien as Aulic Adviser. He was a member of the *Academy of Science of Bologna*, but he was informed of the appointment after his nomination.

Count Riccati was a strong and hard-working person, with an active and fertile mind throughout the years of his life. On April 2, 1754, he had a sudden bout of fever. A fortnight later, on April 15, he passed away.

1.4 Riccati and the Academic World

While in various European Countries most scientists were already inclined to publish their results in short contributions as soon as they achieved them, the general attitude in Italy was to wait until a consistent amount of new results was achieved, and publish then a comprehensive book, an "*Opera*". There is no doubt that Count Riccati was influenced by this attitude, which perfectly matched his pacific nature. Another distinctive feature of Count Riccati's personality, so peculiar when compared with the general attitude of many academic scientists of his time, was his natural tendency to discuss freely the results of his achievements with friends and colleagues as soon as he obtained them, prior to any publication.

In this connection, the case of Maria Gaetana Agnesi (b.1718, d.1799), a mathematician of Milan, is worth elaborating. The oldest of 21 children, Maria Gaetana Agnesi was the daughter of a professor of mathematics, Pietro, who occupied a chair at the University of Bologna. Very early on, she was recognized as an infant prodigy, and she set out on her most important work, *Istituzioni Analitiche ad Uso della Gioventu' Italiana* (Textbook of Analysis for the Use of Young Italians), at the age of 20. She spent many years on this two volume treatise on differential and integral calculus. The *Istituzioni* were eventually published in 1748, with an immediate impact on the academic world. Appreciated for its clarity and synthesis, Agnesi's book was used as a basic textbook in Mathematics in several Italian Universities for several decades. It was also translated into French, German and English, and used in the corresponding countries by many freshmen in scientific disciplines. In the preface to her Istituzioni, Agnesi thanks Count Riccati for bringing to her knowledge an effective computational method of integrals which he had explained to her when the method was still unpublished. ("*Nel tomo secondo per entro il calcolo Integrale ritrovera' il lettore un Metodo affatto nuovo per i polinomi, ne' in luogo alcuno prodotto; questo e' del celebre, e mai abbastanza lodato Sig. Co. Jacopo Riccati Cavaliere di singolarissimo merito nelle scienze tutte, e ben noto al mondo letterario. Ha egli voluto fare a me questa grazia, che io non meritava, ed io rendo a lui, ed al pubblico quella giustizia, che si conviene*" – "In the second volume, when dealing with Integral Calculus, the reader will find a new method for polynomials; this is due to the famous Count Jacopo Riccati, a personality of unique merit in all sciences, and well known to the literate world. He was so kind as to favour me with such a gift, which I did not deserve, and I now do justice to him, and to the public, as it should be.") Unfortunately, not all the scientists met by Riccati were as faithful in recognizing his merits as Maria Gaetana Agnesi was.

1.5 Riccati and Differential Equations

An appropriate way of appreciating Riccati's contribution to differential calculus is to consult his *Opere*, a work in 4 volumes published in Lucca by G. Rocchi in 1765, after Riccati's death. The editor was his son Giordano, to whom we are indebted for the care he took in collecting most of his father's works. However, if one is willing to follow the true sequence of Riccati's discoveries over the years of his life, one should complement the reading of his publications with the correspondence he wrote and received.

Riccati's main interest in the area of differential equations focused on the methods of separation of variables. Probably, such an interest originated in the reading of Gabriele Manfredi's book *De constructione aequationum differentialium primi gradus* printed in Bologna in 1707 (Manfredi occupied the Chair of Mathematics at Bologna University for many years). Riccati developed various methods, such as the method of *dimezzata separazione* (about 1715), and of *coefficienti ed esponenti indeterminati* (about 1717). Both methods were conceived to solve problems having a physical basis; in particular, the second one was motivated by a differential equation arising in a pendulum-type problem.

A compendium of Riccati's methods can be found in the lecture notes which he prepared for his private classes to Giuseppe Suzzi and Ludovico da Riva, who studied mathematics with him during 1722 and 1723. Subsequently, Suzzi and da

Riva became professors of, respectively, physics and astronomy at the University of Padua. The lecture notes, which can be found in the *Opere*, are entitled *Della separazione delle indeterminate nelle equazioni differenziali di primo e di secondo grado, e della riduzione delle equazioni differenziali del secondo grado e d'altri gradi ulteriori* (On the separation of variables in differential equations of first and second order, and on the reduction of differential equations of second order and higher orders). The notes, actually 154 pages, are comprised of three parts and two appendices. The first and second parts are devoted to first order differential equations. Specifically, the first part (*Dei metodi inventati da vari celebri Autori per separare le indeterminate nelle equazioni differenziali di primo grado*) is devoted to methods for the separation of variables invented by other celebrated mathematicians. Here, special reference is made to the work of Gabriele Manfredi. In the second part (*Dei metodi inventati dall'autore per separare le indeterminate nelle equazioni differenziali di primo grado*), the methods of solution due to Riccati are discussed with reference to different equations which we would now call "Riccati equations." Finally, the third part deals with second order differential equations.

As for the specific equations he studied, in these lecture notes and elsewhere, Riccati's interest evolved around scalar equations of the type

$$\dot{x} = ax^2 + bx + c$$

with time varying or constant parameters. In particular we should mention, besides (1.1) and (1.2), a number of further equations of first order, such as

$$\dot{x} = \alpha t^p x^2 + \beta t^m \tag{1.3}$$

$$\dot{x} = \alpha t^p x^q + \beta t^m \tag{1.4}$$

where m, p, and q are constants. Obviously, (1.1) is a particular case of (1.3) (with p = 0) and (1.3) is a particular case of (1.4) (with q = 2). One of the problems which led Riccati to the analysis of these equations was to study the behaviour of the slope of a line connecting the origin of a plane to a point the coordinates of which satisfy linear differential equations. This aspect is further elaborated in Chapter 4. As for the second order equations, he was particularly attracted by the equation

$$\ddot{x} = \alpha t^m,$$

which he called "*equazione ingannatrice*" ("misleading equation"). A similar appreciation for such an equation was also expressed by Euler some years later.

1.6 History and Prehistory of the Riccati Equation

Most probably, Riccati began to conceive and study eqs. (1.1)–(1.4) in 1715. Unfortunately, in agreement with his general attitude, he did not immediately publish the fruit of his work. To the best of our knowledge, the first evidence of these studies goes back to 1719, when Riccati met Nicolaus II Bernoulli (1687-1759), who lived in Padua then (Nicolaus II held the Chair of Mathematics at the University of Padua from 1716 to 1719). In this meeting, he brought to the attention of this member of the Bernoulli family "his" differential equations, and the methods of solutions he knew. In particular, he raised the question of finding those triples (m, p, q) for which it was possible to separate the variables.

On April 1, 1719, Nicolaus II wrote an interesting letter to Pierre Remond de Monfort, where he discusses the cases of separation of variables of which he became aware during this private communication, in particular the case when $q = 2$ and $m + 3p = -4$. A second reference to such a conversation can be found in a letter, dated February 5, 1721, written by Riccati to another member of the Bernoulli family, Nicolaus III, cousin of Nicolaus II.

During the years 1720 and 1721, Nicolaus III was in Venice as a tutor in a noble family. He and Riccati met a few times, and then exchanged a conspicuous correspondence during that period, and following it. The contact between the two was probably produced by the letter we referred to at the beginning of this paper, the letter Riccati wrote on January 1, 1721, to Giovanni Rizzetti. In this important document, Riccati makes reference to a meeting he had in Bologna with Gabriele Manfredi, with whom he discussed various questions, in particular the problem of the separation of variables in differential equations. Then he writes, "*When I came back home, I began to write all that I had thought of on this subject, but I was delayed by two reasons, namely the difficulty in computation and the suspicion that what I believed to be new was already known to analysts of great experience, such as the members of the Bernoulli family. This is why I would like to ask you to be so kind as to communicate the two subsequent formulas, which are among the simplest ones encompassed by my method, to the Eruditissimo Signor Niccolo' Bernullj ...*" After writing eqs. (1.1) and (1.2), Riccati continues as follows, "*Until now, I have not been able to separate the variables in the above formulas in general, and I don't know whether it will be possible. However, I have found infinite values of the integer m for which the variables can be separated ...*".

From the subsequent letters written by Nicolaus III to Riccati, it is apparent that Nicolaus III encountered many difficulties in proving that the variables were separable for Riccati's sequence of integers m. Furthermore, it appears that, sometime in 1721, Riccati communicated verbally this famous sequence to Nicolaus III. Indeed, in Nicolaus III's letter to Riccati dated August 26, 1721, one can read, "*I have eventually found the same cases of separation of variables you referred to when we met in Val di Sole*". Then, he presents his line of reasoning to prove that in equation (1.3) the separation was possible for $m = -3p - 4$ and $m = (-p - 4)/3$. This sequence is also reported on in the second part of Riccati's lecture notes on the separation of variables mentioned above.

In the same letter, Nicolaus III invites Riccati to write a paper for the *Acta Eruditorum Lipsiae* to pose his problem to all mathematicians, "*who, after the death of Leibnitz and others, had begun to languish.*"

Riccati accepted the suggestion of Nicolaus III and, in the Supplements of that Journal, Vol. 8, Sect. 2, pages 67-73, 1724, the paper *Animadversiones in Aequa-*

tiones differentiales secundi gradus was published. In this seven-page paper written in Latin, Riccati reduces a second order differential equation to equation (1.3). Then he poses the problem of finding all exponents m and p for which the separation of variables was possible. More precisely, the last statement of the paper is:

In superiori formula $x^m\, dq = du + uudx : q$, dato ad libitum exponente m, statuatur quantitas $q = x^n$. Peto qua ratione determinandi sint valores alterius exponentis n, ut succedat indeterminatarum separatio, & aequationis constructio per solas quadraturas.

(In the previous formula $x^m\, dq = du + uudx : q$, given any exponent m, let $q = x^n$. I pose the question of finding those values of the exponent n such that the separation of variables is possible ...). Note that, in the terminology of the period, u^2 is indicated as uu and differentials are used in place of derivatives. Moreover, the symbol : indicates division.

This paper, published in 1724, is usually considered as the first official document on the Riccati equation. Strangely enough, however, in 1723, in the Supplements of the same Journal, pages 503-510, an *Appendix* to the above paper was published. In other words, the appendix was published prior to the paper! The reason for this remains a mystery. As an obvious consequence, however, we can conclude that Riccati's original paper, written on the suggestion of the letter of August 1721 of Nicolaus III, was probably submitted in 1721 or 1722, although its publication was postponed until 1724.

1.7 The Riccati Equation over the Centuries

As we have already said, Riccati knew infinitely many cases when the separation of variables was possible in his equations. Apparently, he did not know whether his solution was the most general one. In the paper of 1724 in the *Acta Eruditorum Lipsiae*, interested as he was in posing the problem in its full generality, Riccati did not even mention the infinite sequence he communicated to Nicolaus III in 1721. In the same issue of the *Acta Eruditorum Lipsiae*, Riccati's paper was immediately followed by a comment by Daniel I Bernoulli (Daniel I (b.1700, d.1782) was a brother of Nicolaus III). The comment ends as follows: "*I now add the solution (to Riccati's problem). However, in order to leave to others the possibility of attempting, I will supply it in a disguised form, the significance of which will be clarified in due time. The solution of the problem posed by Riccati Esq. in disguised form is:*

14a, 6b, 6c, 8d, 33e, 5f, 2g, 4b, 33i, 6l, 21m, 26n, 16o, 8p, 5q, 17r, 16s, 25t, 32u, 5x, 3y, +, −, ..., +, =, 4, 2, 1"

We don't know whether Bernoulli's riddle has ever been solved by anyone.

During the latter half of the second decade of the eighteenth century, various authors studied Riccati's problem in its full generality, mainly his pupil Giuseppe

Suzzi, Nicolaus III and, as already said, Daniel I Bernoulli. Moreover, in 1776, Lagrange proposed the scalar and time-varying Riccati equation in its full generality.

With reference to equation (1.1) Giuseppe Suzzi and Daniel I derived the formula $m = -4k/(2k \pm 1)$, supplying, for all integers k, infinitely many cases when the variables can be separated. Nicolaus III considered the more general equation (1.3) and derived the formula $m = (-2kp - 4k \pm p)/(2k \pm 1)$, where k is any positive integer. Here, letting $p = 0$, the sequence discovered by Suzzi and Daniel I is obtained. Moreover, by letting $k = 0$ and taking $-$ in the $\pm$, the Riccati sequence $m = -3p - 4$ follows. Some years later, in 1732, Euler obtained the same result by an ingenious procedure of integration by series applied to equation (1.1). He also applied the continued fraction method to the equation and its generalizations. As a matter of fact, the sequence $m = -4k/(2k \pm 1)$ provides all (and the only) integers for which the variables can be separated in equation (1.1), as proven in 1841 by Liouville. However, already in 1778, in the famous encyclopedia by D'Alembert and Diderot (*L'Enciclopédie ou dictionaire des arts et des métiers*), at the heading "Riccati equation", one can find a proof that $m = -4k/(2k + 1)$ is a sequence of integers for which the separation of variables is possible. By the way, this is one of the first places where the denomination "Riccati equation" was explicitly used. To be precise, in the encyclopedia, the heading is "Ricati equation". So, it was D'Alembert and Diderot who launched the term "Riccati equation". But it was also they who initiated a series of misspellings of Riccati's surname which would propagate along the centuries. In modern times, especially in English-language literature, one sometimes reads "Ricatti" in place of Riccati but this is generally simply an error.

The importance of the Riccati equation in modern times need hardly be emphasized. A generalization of the equation into a matrix form (the *matrix Riccati equation*) plays a major role in many design problems of modern engineering, especially filtering and control. Correspondingly, the last 30 years have witnessed a considerable development of the research activity around this equation. The emergence of system theory concepts in the second part of the twentieth century in the control science community has led to a breakthrough in the theoretical analysis. Notions such as controllability and observability, stabilizability and detectability, have been found to be important tools to study the equation, and many properties previously obscure have been eventually brought to light. Attention has also been paid to the discrete-time counterpart of the Riccati equation, a difference equation usually also named after Riccati. Efficient numerical algorithms to solve these equations, both in continuous and discrete time, are now available.

This very active research of the last 30 years will undoubtedly continue to be so for many decades to come, and we can only wonder what new developments will be brought by the forthcoming century on this venerable equation, conceived by Count Riccati more than a quarter of a millennium ago.

Acknowledgement

During the preparation of this paper, various colleagues and friends helped me in collecting and exploring historical papers. Other colleagues read and commented on an early draft of the paper. For their kind contribution, I would like to thank A. Bazzeo, P. Bernhard, A. Campagner, M. Campi, G. De Nicolao, L. Grugnetti, G. and M. Guardabassi, E. Knobloch, A.J. Laub, A. Locatelli, M. Maranzana-Figini, C. Truesdell, J.C. Willems and H.K. Wimmer. The views expressed, needless to say, are my own.

1.8 References

A very detailed report of 67 pages on Riccati's life and work, written soon after Riccati's death, is given in

1. C. di Rovero. *Vita del Conte Jacopo Riccati.* in *Opere del Conte Jacopo Riccati nobile trevigiano.* G. Rocchi. Lucca 1765.

A very good report is also

2. A. A. Michieli. *Una famiglia di matematici e di poligrafi trivigiani: i Riccati.* in *Atti del Reale Istituto Veneto di Scienze, Lettere ed Arti.* Tomo 102, Parte ii, 1942-43.

Further useful references of general interest are (in chronological order):

3. G.B. Marzari. *Elogio di Jacopo Riccati pronunciato nella grand' aula del regio liceo del Tagliamento per il riaprimento degli studi il 15 Novembre 1812.* G. Trento e figli, Treviso, 1812.
4. Cantor M. *Vorlesungen über Geschichte der Mathematik*, Band iii, Leipzig Teubner, 1901.
5. I. Szabo. *Die Familie der Mathematiker Riccati.* I Mitteilung. in *Humanismus und Technik*, Berlin 1974, p. 37-75; II Mitteilung in *Humanismus und Technik*, Berlin 1974, p. 109-131.
6. L. Grugnetti. *Sulla vecchia e attuale equazione di Riccati.* in *Rendiconti del Seminario Facolta' di Scienze dell' Universita' di Cagliari*, Vol. 55, Fasc. 1, 1985, p. 7-23.

One of most interesting papers by Euler on the Riccati equation is the following one, written in Latin.

7. L. Euler. *De risolutione aequationis* $dy + a\,yydx = bx\,dx$. Novi Commentari academiae scientiarum Petropolitanae 9 (1762/3, 1764), p. 154-169.

Euler's work on the Riccati equation and other equations can be appreciated from the recent translation from Latin due to M.F. Wyman and B.F. Wyman:

8. L. Euler. *An Essay on Continued Fractions.* Mathematical Systems Theory, Vol. 18, p. 295-328, 1985 (translation of the original paper by Euler published in 1744).

Liouville's proof that $m = -4k/(2k \pm 1)$ is the only sequence for which the integration of equation (1.1) is possible can be found in:

9. J. Liouville. *L'equation de Riccati.* Journal de Mathematiques pures et appliques. Tome VI, p. 1-13, 1841.

The famous letter written by Riccati to Giovanni Rizzetti dated January 1st, 1721, as well as five letters to Nicolaus III Bernoulli, can be found in

10. L. Grugnetti. *L'Equazione di Riccati.* Bollettino di Storia delle Matematiche, Vol VI, fasc. 1, p. 45-82, 1986.

The above letters witness the very early days of the Riccati equation. In Grugnetti's paper, one can also find 6 letters written by Nicolaus III to Riccati in 1721, during the period of his stay in Italy, plus two letters written by Nicolaus III from St. Petersburg in 1726, one to Riccati and one to Gabriele Manfredi. Most correspondence is in Italian, except one letter in Latin.

The remarkable life of Domenica Maria Gaetana Agnesi is reported in many papers and books, such as

11. A. Masotti. *Maria Gaetana Agnesi*. Rendiconti del seminario Matematico e Fisico di Milano, 14, 1940
12. L.M. Osen. *Women in Mathematics*. The MIT Press, Cambridge, 1974
13. C. Truesdell. *Maria Gaetana Agnesi*. Archive for History of Exact Sciences, 40, 2, p. 113-142, 1989.

As general reference for the history of mathematics, the author used

14. C.B. Boyer. *The History of Mathematics*. J. Wiley and Sons, 1974.

In particular, the terminology for the members of the Bernoulli family conforms to that of Boyer.

The town of Castelfranco Veneto organized in 1990, April 5 and 6, a conference on *I Riccati e la cultura della Marca nel settecento europeo* (The Riccatis' and the culture of the Castelfranco region in the eighteenth century). The conference book (in Italian) is expected to appear in 1991.

2 Solutions of the Continuous and Discrete Time Algebraic Riccati Equations: A Review

Peter Lancaster and Leiba Rodman

2.1 Introduction

This review is concerned with two algebraic Riccati equations. The first is a quadratic matrix equation for an unknown $n \times n$ matrix X of the form

$$XDX + XA + A^*X - C = 0, \tag{2.1}$$

where A, D, C are $n \times n$ complex matrices with C and D hermitian. Further hypotheses are imposed as required, although Section 2.3 contains some discussion of more general non-symmetric quadratic equations. The second equation has the fractional form

$$X = A^*XA + Q - (C + B^*XA)^*(R + B^*XB)^{-1}(C + B^*XA), \tag{2.2}$$

where R and Q are hermitian $m \times m$ and $n \times n$ matrices, respectively, and A, B, C are complex matrices with respective sizes $n \times n$, $n \times m$, and $m \times n$. The two equations are frequently referred to as the "continuous" and "discrete" Riccati equations, respectively, because they arise in physical optimal control problems in which the time is treated as a continuous variable, or a discrete variable.

Both equations arise in now classical problems of systems theory, differential equations, and filter design, as well as in differential games. References to these motivating problems are strewn throughout the literature and particularly in several earlier reviews. So they will not be reproduced here. The earlier reviews are, perhaps, less wide-ranging than this exposition. References [1], [2], [3, Chapter II.6], [4] and [5] (in chronological order) deal in varying detail with equations of type (2.1), and [6] is concerned with equations of type (2.2).

The authors have been impressed by both the parallels and the discrepancies between the structure of the two equations. They are treated here side by side in the belief that the reader will also appreciate their contrasts and comparisons.

Although this paper contains numerous references, it cannot be claimed that we have been exhaustive in our search of the literature. There have been innumerable contributions on the theory, application and numerical solution of Riccati equations, and a complete review would far exceed the authors' stamina. We discuss the following topics:

Bittanti, Laub, Willems (Eds.)
The Riccati Equation

(1) Geometric description of the solutions under weak conditions on the coefficients (Sections 2.3, 2.4, 2.5, 2.6, 2.12, and 2.16).
(2) Existence and comparison theorems generated by the recursive method (Sections 2.7 and 2.17).
(3) The connection with factorization problems for certain rational matrix functions (Sections 2.9, 2.10, 2.11, 2.14, and 2.15).
(4) Special properties of equations with real coefficient matrices (Sections 2.12, 2.13, 2.14 and 2.18).

It will be seen that our knowledge of (2.1) is better developed than that of (2.2), and in the case of the former we discuss:

(5) Dependence of the solution on the coefficients (Sections 2.8 and 2.13).

In the case of item (1) above, we examine the connection between two descriptions of the solution set of equation (2.1). The first originates with Willems [7] and Coppel [8] (see also the recent treatment of Ando [5]), and the second evolved later from the study of the spectral structure of an associated Hamiltonian matrix. (This began with work by Potter [9] and Kučera [10]; see [11], [2], and [12] for more recent developments.)

It is also noteworthy that, in our view, there is an advantage to be gained in separating the symmetries of the complex Riccati equation itself from the additional symmetry that occurs when the matrix coefficients are real. This philosophy is reflected in our presentation.

In addition to the five topics listed above, we touch on some recent developments. In Sections 2.10 and 2.14, we present a fascinating connection of the Riccati equation (2.1) with the construction of rectangular rational matrix functions that are contractions on the real line (and with their unitary completions) which was discussed in detail by Gohberg and Rubinstein [13] (see [14] for related results in this direction).

The latter problem area, as in the bulk of this review, involves the case in which matrix D of equation (2.1) is semidefinite. In the case of equation (2.2), we naturally seek solutions X for which $R + B^*XB$ is invertible, or even positive definite. Riccati equations in which these conditions are violated are also of current interest. An introduction to an equation of type (2.1) with D indefinite, and which arises in the problem of J-spectral factorization, is given in Section 2.11. Another case arises for equations of type (2.2) with $R+B^*XB$ not invertible (the inverse is replaced by a generalized inverse) in the classical linear-quadratic optimal control problem with *singular* differential or difference constraints. These are known as descriptor systems. We do not discuss these problems, but refer the reader to works of Bender and Laub [15], [16] and Mehrmann [17], [18] for more information and earlier references.

Our plan in this review is to present enough ideas and definitions for the reader to be able to appreciate the statement of the theorems which are mostly presented without proof. The proofs have been included in many cases (but not exclusively) where there is believed to be some original contribution or extension of existing results.

2.2 Preliminaries and Notations

In this section we summarize some notations, ideas, and results required to understand the body of this review. More details on all of these topics can be found in references [19] and [20]. Let us begin with some notations concerning linear spaces, subspaces, and matrices. By $\mathbf{C}^n$ (or $\mathbf{R}^n$) is meant the complex (real) linear space of column vectors of n complex (real) numbers. Then $\mathbf{C}^{m\times n}$ (or $\mathbf{R}^{m\times n}$) is the linear space of $m \times n$ matrices with complex entries. The operations in these spaces are, of course, the entry-wise definitions of addition and scalar multiplication. We use I to denote the $n \times n$ identity matrix. The size of I is generally clear from the context.

A *subspace* is a subset of $\mathbf{C}^n$ or $\mathbf{R}^n$ that is also closed under the operations of addition and scalar multiplication. For arbitrary sets or subspaces $\mathcal{S}_1$ and $\mathcal{S}_2$, $\mathcal{S}_1 \subset \mathcal{S}_2$ denotes either inclusion or equality.

If $x \in \mathbf{C}^n$ or $x \in \mathbf{R}^n$, x^* is the transposed (row) vector of conjugated entries of x. The *standard inner product* on either $\mathbf{C}^n$ or $\mathbf{R}^n$ is defined by

$$(x, y) = y^* x,$$

and we write $\|x\| = (x, x)^{1/2}$ (the euclidean norm) and say x, y are *orthogonal* if $(x, y) = 0$. In the rest of this section, we refer only to $\mathbf{C}^n$, although $\mathbf{C}^n$ can be replaced throughout by $\mathbf{R}^n$.

If $\mathcal{S}$ is a subset of $\mathbf{C}^n$, then $\mathcal{S}^\perp$ denotes the subspace of $\mathbf{C}^n$ consisting of the set of vectors that are orthogonal to all members of $\mathcal{S}$. If $\mathcal{S}$ is a subspace, then $\mathcal{S}$, $\mathcal{S}^\perp$ are *orthogonal complements* and this is denoted by $\mathbf{C}^n = \mathcal{S} \oplus \mathcal{S}^\perp$. If $\mathcal{S}_1, \mathcal{S}_2$ are complementary subspaces, but not necessarily mutually orthogonal, then we write $\mathbf{C}^n = \mathcal{S}_1 \dot{+} \mathcal{S}_2$. If $x_1, ..., x_k \in \mathbf{C}^n$, then span $\{x_1, x_2, ..., x_k\}$ is the subspace of $\mathbf{C}^n$ of all linear combinations of $x_1, x_2, ..., x_k$.

If $A \in \mathbf{C}^{m\times n}$, then A^* denotes the $n \times m$ matrix obtained by transposition and conjugation of the entries of A. Define the image and kernel of A by

$$Im\ A = \{x \in \mathbf{C}^m | x = Ay \text{ for some } y \in \mathbf{C}^n\};$$

$$\text{Ker } A = \{x \in \mathbf{C}^n | Ax = 0\}.$$

Recall the important facts that

$$\mathbf{C}^n = \text{Ker } A \oplus Im\ A^*, \mathbf{C}^m = \text{Ker } A^* \oplus Im\ A. \tag{2.3}$$

The set of all eigenvalues of a matrix $A \in \mathbf{C}^{n\times m}$ is called the *spectrum* of A and written $\sigma(A)$. The *partial multiplicities* of an eigenvalue λ_0 are the sizes of Jordan blocks in a Jordan normal form for A having eigenvalue λ_0. The *algebraic multiplicity* of eigenvalues λ_0 is the sum of the partial multiplicities.

There are two notions of *stability* for a square matrix. The first is associated with differential systems and is defined by having the real part of λ ($Re\lambda$) negative whenever $\lambda \in \sigma(A)$. The second is associated with discrete, or difference systems, and defined by $|\lambda| < 1$ whenever $\lambda \in \sigma(A)$.

A *Jordan chain* of a matrix A is a set of vectors $x_1, x_2, ..., x_r$ such that $x_1 \neq 0$ and for some $\lambda \in \sigma(A)$

$$(A - \lambda I)x_1 = 0$$

$$(A - \lambda I)x_j = x_{j-1},\ j = 2, 3, ..., r.$$

In particular, x_1 is a (right) *eigenvector* of A, and the members of a Jordan chain are necessarily linearly independent.

A subspace $\mathcal{S}$ of $\mathbf{C}^n$ is said to be A-*invariant*, where $A \in \mathbf{C}^{n\times n}$, if $A\mathcal{S} \subset \mathcal{S}$. Here $A\mathcal{S} = \{x \in \mathbf{C}^n | x = Ay \text{ for some } y \in \mathcal{S}\}$. Note that the span of a Jordan chain for A is A-invariant. If $\mathcal{S}_1$ is A-invariant and has dimension k, then A is similar to a blocktriangular matrix

$$\begin{bmatrix} A_1 & A_3 \\ 0 & A_2 \end{bmatrix}, \tag{2.4}$$

with $A_1 \in \mathbf{C}^{k\times k}$. If $\mathbf{C}^n = \mathcal{S}_1 \dot{+} \mathcal{S}_2$ and both $\mathcal{S}_1$ and $\mathcal{S}_2$ are A-invariant, then the pair $\mathcal{S}_1, \mathcal{S}_2$ is said to be A-*reducing*, and A is similar to a block-diagonal matrix diag$[A_1, A_2]$, and the size of A_j is the dimension of $\mathcal{S}_j$(dim $\mathcal{S}_j$) for $j = 1$ and 2. If $\mathcal{S}$ is A-invariant, the *restriction* of A to $\mathcal{S}$ is the natural transformation on $\mathcal{S}$ denoted by $A|_{\mathcal{S}}$. Then $\sigma(A|_{\mathcal{S}}) \subset \sigma(A)$ always holds. If there is no larger A-invariant subspace for which this inclusion holds, then $\mathcal{S}$ is called a *spectral subspace* of A. If $\sigma_1 \subset \sigma(A)$, and Γ is a smooth closed contour with σ_1 inside Γ and the rest of $\sigma(A)$ outside Γ, then the matrix

$$P = \frac{1}{2\pi i}\int_\Gamma (I\lambda - A)^{-1} d\lambda \tag{2.5}$$

is a *projection* (*i.e.*, $P^2 = P$), and Im P is A-invariant. Furthermore, Im P is a spectral subspace of A and $\sigma(A|_{Im\ P}) = \sigma_1$. Also, $I - P$ is a projection onto the complementary spectral subspace associated with $\sigma(A)\backslash\sigma_1$.

Suppose that $\mathcal{S}$ is a subspace of $\mathbf{C}^n$ and P is any projection onto $\mathcal{S}$, *i.e.*, Im $P = \mathcal{S}$. The invariance of $\mathcal{S}$ under A can be characterized in terms of P. Thus, $\mathcal{S}$ is A-invariant if and only if $PAP = AP$ for any projector P onto $\mathcal{S}$. This extends naturally to reducing pairs: the pair of subspaces $\mathcal{S}_1, \mathcal{S}_2$ is A-reducing if and only if $AP = PA$, where P is the unique projector onto $\mathcal{S}_1$ along $\mathcal{S}_2$, (*i.e.*, Im $P - \mathcal{S}_1$ and Ker $P = \mathcal{S}_2$).

For a pair of matrices $A \in \mathbf{C}^{n\times n}$, $B \in \mathbf{C}^{n\times r}$, the subspace $\sum_{j=0}^{n-1} Im(A^jB)$ of $\mathbf{C}^n$ is called the *controllable subspace* of the pair. It is the smallest A-invariant subspace containing Im B. If the controllable subspace fills out $\mathbf{C}^n$, then the pair (A, B) is said to be *controllable*. An equivalent statement is that the $n \times nr$ matrix $[B \;\; AB \cdots A^{n-1}B]$ has linearly independent rows. We will use the Hautus criterion (see [21, 22]) later on; according to this criterion, (A, B) is controllable if and only if the $n \times (n + r)$ matrix $[\lambda I - A, B]$ has linearly independent rows for every $\lambda \in \mathbf{C}$. If the controllable subspace of (A, B) has dimension k, there is a nonsingular matrix S such that (compare with (2.4)),

$$SAS^{-1} = \begin{bmatrix} A_1 & A_3 \\ 0 & A_2 \end{bmatrix}; \quad SB = \begin{bmatrix} B_1 \\ 0 \end{bmatrix}, \tag{2.6}$$

where A_1 is $k \times k$, B_1 is $k \times r$, and (A_1, B_1) is a controllable pair. This is sometimes referred to as a *Kalman form* for A and B. The pair (A, B) is said to be *stabilizable* if the matrix A_2 in a Kalman form (2.6) is stable. This is equivalent to asserting the existence of a matrix F such that $A + BF$ is stable. Note that, as with stability itself, this admits two notions of stabilizable pairs.

Observability and detectability are concepts that are duals for controllability and stabilizability, respectively. Thus, the pair (C, A) with C of size $r \times n$ and A of size $n \times n$ is *observable* if and only if (A^*, C^*) is controllable. This can also be expressed in the form: (C, A) is an observable pair if the $nr \times n$ matrix

$$\begin{bmatrix} C \\ CA \\ \vdots \\ CA^{n-1} \end{bmatrix}$$

has linearly independent columns. The pair (C, A) is said to be *detectable* if and only if (A^*, C^*) is stabilizable.

2.3 Solutions and Invariant Subspaces

Consider the general Riccati equation

$$XBX + XA - DX - C = 0, \tag{2.7}$$

where A, B, C, D have sizes $n \times n$, $n \times m$, $m \times n$, and $m \times m$, respectively, and $m \times n$ matrix solutions X are to be found.

For any $m \times n$ matrix X and $n \times n$ identity I, we call the subspace

$$G(X) = Im \begin{bmatrix} I \\ X \end{bmatrix} \subset \mathbf{C}^{m+n}$$

the *graph* of X. Also a subspace of $\mathbf{C}^{m+n}$ is a *graph subspace* if it has the form $G(X)$ for some X.

The first simple proposition connects solutions of (2.7) with invariant subspaces of the $(m + n) \times (m + n)$ matrix

$$T = \begin{bmatrix} A & B \\ C & D \end{bmatrix}. \tag{2.8}$$

Proposition 2.1. *For any $m \times n$ matrix X, the graph of X is T-invariant if and only if X is a solution of* (2.7).

This proposition, and the following theorem are the natural generalizations of an idea first introduced by Potter [9].

Theorem 2.2. *Equation* (2.7) *has a solution $X \in \mathbf{C}^{m\times n}$ if and only if there is a set of vectors $v_1, ..., v_n$ in $\mathbf{C}^{m+n}$ forming a set of Jordan chains for T (of equation* (2.8)), *and if*

$$v_j = \begin{bmatrix} y_j \\ z_j \end{bmatrix}, \quad j = 1, 2, \ldots,$$

where $y_j \in \mathbf{C}^n$, then $y_1, y_2, \ldots, y_n$ form a basis for $\mathbf{C}^n$.

Furthermore, if

$$Y = [y_1 y_2 \cdots y_n] \in \mathbf{C}^{n\times n}, \; Z = [z_1 z_2 \cdots z_m] \in \mathbf{C}^{m\times n},$$

every solution of (2.7) *has the form $X = ZY^{-1}$ for some set of Jordan chains $v_1, v_2, ..., v_n$ for T.*

The invertibility of Y is, of course, the condition ensuring that span$\{v_1 v_2 \cdots v_n\}$ is a graph subspace of $\mathbf{C}^{m+n}$. This result suggests that, generically, the number of solutions of (2.7) to be expected will not exceed the binomial coefficient $\binom{m+n}{n}$, the number of ways in which the vectors $v_1, v_2, ..., v_n$ can be chosen from a basis of $m + n$ eigenvectors for T.

2.4 Geometric Theory for the Continuous Algebraic Riccati

The equation we consider now has the form

$$\mathcal{R}(X) := XDX + XA + A^*X - C = 0, \tag{2.9}$$

where $D \geq 0$, $C^* = C$ and all matrices are in $\mathbf{C}^{n\times n}$. The existence of nonsymmetric or hermitian solutions are both questions of interest. Define $2n \times 2n$ matrices M and H by

$$M = i\begin{bmatrix} A & D \\ C & -A^* \end{bmatrix}; \quad H = \begin{bmatrix} -C & A^* \\ A & D \end{bmatrix}, \tag{2.10}$$

and we say that this is the matrix pair *associated* with equation (2.9). Note that if T is constructed as indicated in equations (2.8) and (2.7), then we have $M = iT$ and so Proposition 2.1 and Theorem 2.2 apply verbatim if T is replaced by M.

Note also that $H^* = H$ and, without loss of generality, we can assume that H is invertible and has signature (sig H) equal to zero (otherwise replace A by $A + i\alpha I$ where $\alpha \in \mathbf{R}$ and is large enough). Furthermore, $HM = M^*H$, which is interpreted as saying that M is self-adjoint in the indefinite scalar product on $\mathbf{C}^{2n}$ determined by H.

Another, apparently similar choice of indefinite scalar product is frequently made, namely, using

$$\hat{H} \rightarrow \hat{H} = \begin{bmatrix} 0 & i\alpha I \\ -i\alpha I & 0 \end{bmatrix}, \tag{2.11}$$

($0 \neq \alpha \in \mathbf{R}$). We again have $\hat{H}^* = \hat{H}$, $\hat{H}$ nonsingular, and $\hat{H}M = M^*\hat{H}$. Most of the theory goes through with only minor modifications, but there are subtle differences in the analysis of the stability of solutions of (2.9) under perturbation that persuade us to use H rather than $\hat{H}$ (see [23] and Section 2.7 below).

The matrix H is also related to the Riccati function in the following way, as is easily verified:

$$\begin{bmatrix} I \\ X \end{bmatrix}^* H \begin{bmatrix} I \\ X \end{bmatrix} = \mathcal{R}(X) + (X^* - X)(A + DX), \tag{2.12}$$

for any $X \in \mathbf{C}^{n \times n}$. In particular, if the domain of $\mathcal{R}$ is restricted to *hermitian* matrices, then

$$\mathcal{R}(X) = \begin{bmatrix} I \\ X \end{bmatrix}^* H \begin{bmatrix} I \\ X \end{bmatrix}. \tag{2.13}$$

Now we know from Section 2.3 that, to solve the equation $\mathcal{R}(X) = 0$, we must focus on M-invariant subspaces that are also graph subspaces associated with solution matrices X. Equation (2.13) tells us that, for hermitian solutions X, the corresponding graph subspace must be *H-neutral.* That is, if $x \in G(X)$, then $x^*Hx = 0$.

More generally, equation (2.12) says that, if X is a solution of $\mathcal{R}(X) = 0$ for which

$$(X^* - X)(A + DX) \leq 0, \tag{2.14}$$

then the graph of X is H-nonpositive. Such solutions of $\mathcal{R}(X) = 0$ will be called *special* solutions.

Observe also that for the scalar product defined by $\hat{H}$ of (2.11),

$$\begin{bmatrix} I \\ X \end{bmatrix}^* \hat{H} \begin{bmatrix} I \\ X \end{bmatrix} = i\alpha(X - X^*),$$

where X is any $n \times n$ matrix. It follows that *the graph of X is $\hat{H}$-neutral if and only if X is hermitian*, a property that is generally not enjoyed by H.

Whichever scalar product is chosen, our attention is drawn to H-neutral, or possibly H-nonpositive M-invariant subspaces. Such subspaces can be identified using the fact that M is H-self-adjoint. But first it is important to recognize the symmetries of the spectrum of M, say $\sigma(A)$. The following proposition is vital, and elementary (a proof can be found, *e.g.*, in [3], Section I.2.2).

Proposition 2.3. *Let* $H, A \in \mathbf{C}^{n\times n}$ *with* $H^* = H$, $\det\, H \neq 0$, *and* $HA = A^*H$. *Then*

(a)$\sigma(A)$ *is symmetric with respect to the real axis, i.e.,* $\lambda \in \sigma(A)$ *implies* $\bar{\lambda} \in \sigma(A)$ *and* $\lambda, \bar{\lambda}$ *have the same partial multiplicities as eigenvalues of* A.

(b)*If* iA *is a real matrix, then* $\sigma(A)$ *is symmetric with respect to both the real and imaginary axes.*

Note that when the coefficients of $\mathcal{R}(X)$ are real matrices, then iM is real, so that the double symmetry of part (b) of the proposition arises in this case. There is some advantage to be gained from separating these symmetries, although many authors (see [2], for example) prefer to work directly with case (b) (the *symplectic* case) which is, indeed, of great practical importance. As long as $\sigma(M) \not\subset \mathbf{R}$, the construction of H-neutral subspaces is easy once one makes the observation that *if* $\lambda \in \sigma(M)$ *and* $\lambda \neq \bar{\lambda}$, *then the spectral subspace of* M *corresponding to* λ *is* H-*neutral.* More generally, introduce the notion of a $c-set$ of eigenvalues; a subset of $\sigma(A)$ is called a $c-set$ if it contains no real eigenvalues and no conjugate pairs of eigenvalues.

Proposition 2.4. (Theorem I.2.5 in [3]) *With the hypotheses of Proposition* 2.3, *let* $\mathcal{N}$ *be a direct sum of spectral subspaces of* A *associated with the eigenvalues in a* $c-set$ S *of* $\sigma(A)$, *then* $\mathcal{N}$ *is* H-*neutral.*

Applying this proposition to M, we see immediately that if $\sigma(M)$ contains no real eigenvalues, then there exists an n-dimensional, M-invariant, H-neutral subspace $\mathcal{S}$, and to link this with the existence of a solution of $\mathcal{R}(X) = 0$, it only remains to show that $\mathcal{S}$ is a graph subspace. The following proposition shows that these hypotheses are realized in some cases of interest.

Proposition 2.5. *If* $D \geq 0$, $C \geq 0$, (A, D) *is stabilizable and* (C, A) *detectable, then* M *has no real eigenvalues.*

We provide a proof of this well-known result (the proof itself uses known ideas as well, but is not easily located in the literature in a concise form; see, however, [24]).

Proof. Suppose $\lambda \in \mathbf{R}$ and $M \begin{bmatrix} x_1 \\ x_2 \end{bmatrix} = \lambda \begin{bmatrix} x_1 \\ x_2 \end{bmatrix}$, $\begin{bmatrix} x_1 \\ x_2 \end{bmatrix} \neq \begin{bmatrix} 0 \\ 0 \end{bmatrix}$. Thus

$$-Ax_1 + Dx_2 = -i\lambda x_1, \quad Cx_1 + A^*x_2 = -i\lambda x_2. \tag{2.15}$$

The first equation implies

$$x_2^* D x_2 = -i\lambda x_2^* x_1 + x_2^* A x_1 = -i\lambda x_1^* x_2 + x_1^* A^* x_2,$$

and using the second equation

$$x_2^* D x_2 = -x_1^* C x_1.$$

The hypotheses $D \geq 0$ and $C \geq 0$ imply $Cx_1 = Dx_2 = 0$.

But then equations (4.7) reduce to give

$$x_1^*(A^* + i\lambda I) = 0 \text{ and } x_1^* C = 0.$$

$$x_2^*(A - i\lambda I) = 0 \text{ and } x_2^* D = 0.$$

Using the Hautus criterion (see Section 2.2), since $-i\lambda$ is not a stable eigenvalue of A, stabilizability of (A^*, C) and (A, D) implies that

$$\text{rank}[A^* + i\lambda I, C] = n;$$

$$\text{rank}[A - i\lambda I, D] = n.$$

Consequently, $x_1 = x_2 = 0$, and we have a contradiction. □

If A is H-self-adjoint (as in Proposition 2.3) with no real eigenvalues, the identification of H-neutral A-invariant subspaces of dimension n is relatively easy. Such subspaces may also exist when A has real eigenvalues and their identification requires more detailed analysis of the spectral subspaces of the real eigenvalues. Some indication of this appears in Theorems 2.8 and 2.9 below. Let us only remark here that this task is facilitated by the simultaneous reduction of (A, H) to an appropriate canonical form, as described in some detail in Part I of [3], for example.

It turns out that, in addition to the symmetry properties imposed on $\mathcal{R}(X)$ in (2.9), the *controllability* of the pair (A, D) is sufficient to ensure the (so-far) missing condition that an n-dimensional, M-invariant, H-neutral subspace should be a graph subspace. With this additional hypothesis, the following major results can be proved (ref. [8], [11], [2]).

Theorem 2.6. *If $D \geq 0$, $C^* = C$, and (A, D) is controllable, then the following statements are equivalent:*

(i) *There exists a hermitian solution of equation* (2.9);
(ii) *there exists an n-dimensional, M-invariant, H-neutral subspace;*
(iii) *the partial multiplicities of the real eigenvalues of M (if any) are all even.*

It has been shown by Faibusovich [25] that, when A, D, C are real matrices, the controllability of (A, D) in this theorem can be replaced by a weaker condition of "sign-controllability."

Theorem 2.7. *If $D \geq 0$, $C^* = C$, (A, D) is controllable, and $\mathcal{N}$ is an n-dimensional, M-invariant, H-nonpositive subspace, then $\mathcal{N}$ is a graph subspace determined by a special solution X of $\mathcal{R}(X) = 0$ (i.e., X satisfies inequality (2.14)).*

Conversely, if X is a special solution of $\mathcal{R}(X) = 0$, then the graph of X, $G(X)$, is n-dimensional, M-invariant, and H-nonpositive.

The existence of special solutions is now guaranteed by combining Theorem 2.7 with a well-known theorem of Pontryagin (see [3], for example), which asserts that, given an invertible $2n \times 2n$ hermitian matrix H with signature zero, and a $2n \times 2n$ matrix M such that $HM = M^*H$, there exists an n-dimensional, M-invariant, H-nonpositive subspace.

The role played by the real spectrum of M in the generation of H-neutral subspaces is clarified by the next result (see [26]). We denote by Γ the set of hermitian solutions of (2.9).

Theorem 2.8. *Suppose $D \geq 0$, $C^* = C$, (A, D) is controllable, and Γ is not empty. Let $\mathcal{M}_0$ be the spectral subspace of M corresponding to its real eigenvalues ($\mathcal{M}_0 = \{0\}$ if $\sigma(M) \cap \mathbf{R} = \emptyset$) and $m = \dim\ \mathcal{M}_0$. Then there is a unique subspace $\mathcal{S}_0 \subset \mathcal{M}_0$ of dimension $m/2$ which is M-invariant and H-neutral.*

Furthermore, $\sigma(M|_{\mathcal{S}_0}) = \sigma(M) \cap \mathbf{R}$ and the partial multiplicities of an eigenvalue $\lambda \in \sigma(M|_{\mathcal{S}_0})$ are those of λ as an eigenvalue of M divided by two.

Let $\mathcal{M}_+$ be the spectral subspace of M corresponding to the eigenvalues of M in the open upper half-plane. One n-dimensional, M-invariant, H-neutral subspace is now represented as the (H-orthogonal) direct sum of $\mathcal{M}_+$ and $\mathcal{S}_0$ of Theorem 2.8. More generally, one can add $\mathcal{S}_0$ to a spectral subspace of M associated with a *maximal* $c\!-\!set$ (see Proposition 2.4) to characterize the so-called *spectral solutions* in the set Γ. Better still, we have (Theorem II.4.9 in [3]):

Theorem 2.9. *Suppose $D \geq 0$, $C^* = C$, (A, D) is controllable, and that Γ is not empty. For every M-invariant subspace $\mathcal{M} \subset \mathcal{M}_+$, there is a unique $X \in \Gamma$, say $X = \Phi(\mathcal{M})$, such that $G(X) \cap \mathcal{M}_+ = \mathcal{M}$. Conversely, for each $X \in \Gamma$, there is a unique M-invariant subspace $\mathcal{M} \subset \mathcal{M}_+$ such that $X = \Phi(\mathcal{M})$.*

Further analysis in this direction allows one to *count* the distinct hermitian solutions of $\mathcal{R}(X) = 0$ in some cases. Thus, let m_i be the algebraic multiplicity of an eigenvalue λ of M, and assume that $\mathrm{Ker}(\lambda I - M)$ has dimension one for every non-real eigenvalue λ. Then the number of distinct hermitian solutions of $\mathcal{R}(X) = 0$ is

$$\sqcap_{i=1}^{\alpha}(1 + m_i),$$

where α is the number of distinct eigenvalues λ of M in the open upper half-plane. If there is a non-real λ for which $\dim\ \mathrm{Ker}(\lambda I - M) > 1$, then there is a continuum of hermitian solutions. In particular, if M has all its 2n eigenvalues distinct and no real eigenvalues, then there are 2^n distinct hermitian solutions.

2.5 Partial Order

In the "linear-quadratic" optimal control problem, the spectrum of the matrix $A + DX$, where X satisfies $\mathcal{R}(X) = 0$ is of great physical significance. This spectrum is easily identified by taking advantage of the simple similarity relation:

$$T^{-1}MT = i\begin{bmatrix} A + DX & D \\ 0 & -(A + DX)^* \end{bmatrix} \tag{2.16}$$

where $T = \begin{bmatrix} I & 0 \\ X & I \end{bmatrix}$. In particular, we have

$$M \begin{bmatrix} I \\ X \end{bmatrix} = i \begin{bmatrix} I \\ X \end{bmatrix} (A + DX), \tag{2.17}$$

and $\sigma(i(A + DX)) \subset \sigma(M)$. Thus that part of the spectrum of M determining the solution $X \in \Gamma$ (as in Section 2.4) reappears, after rotation through a right angle, as the spectrum of the state matrix $A + DX$ associated with the feedback matrix X. For example, when M has no real eigenvalues, there are solutions X_+ and X_- of $\mathcal{R}(X) = 0$ for which Re $\lambda > 0$ for $\lambda \in \sigma(A + DX_+)$ and $Re(\lambda) < 0$ for $\lambda \in \sigma(A + DX_-)$. Furthermore, it is not difficult to see that for *any* hermitian solution X of $\mathcal{R}(X) = 0$, we have $X_- \leq X \leq X_+$ (where $A \geq B$ means that $A - B \geq 0$, *i.e.*, $A - B$ is positive semi-definite). Thus, X_+ and X_- can be described as maximal and minimal hermitian solutions of $\mathcal{R}(X) = 0$, respectively. More generally, admitting real eigenvalues in M, the following result can be proved (Theorem II.4.12 of [3]).

Theorem 2.10. *Suppose $D \geq 0$, $C^* = C$, (A, D) is controllable, and Γ is not empty. Then there exists a maximal solution X_+, and a minimal solution X_-, in Γ. The solution X_+ (resp. X_-) is the unique solution in Γ for which $\sigma(A+DX)$ is in the closed left (resp. right) half-plane, and is obtained by taking $\mathcal{M} = \mathcal{M}_+$ (resp. $\mathcal{M} = \{0\}$) in Theorem 2.9.*

We now have a natural partial order on Γ. Furthermore, we have a connection between the members of Γ and invariant subspaces of M, which have the natural order determined by inclusion. This connection can be made more precise as follows (see [2], [23], [8], [4]):

Theorem 2.11.

(a) *The map Φ introduced in Theorem 2.9 is a homeomorphism between Γ and the set of all M-invariant subspaces in $\mathcal{M}_+$.*
(b) *If $X, Y \in \Gamma$, then $X \leq Y$ if and only if $\Phi^{-1}(X) \subset \Phi^{-1}(Y)$.*

The homeomorphism of part (a) requires the introduction of two topologies; the natural choices are made of the metric induced by any matrix norm on Γ and the gap-metric on the M-invariant subspaces.

A parallel to the latter results is provided by results of Willems and Coppel ([7], [8]). In these references, the results are obtained for the case of *real* A, D, and C; we recast them here for the complex case. Interesting comments on the relative merits of these two descriptions are given by Shayman [2]. First define the hermitian matrix $X_+ - X_-$, the "gap" between the extremal solutions, and let $\mathcal{N} = \text{Ker}(X_+ - X_-)$.

Theorem 2.12. *Make the hypotheses of Theorem* 2.10. *Then for any* $x \in \mathcal{N}$ *and any* $X \in \Gamma$,

$$(A + DX)x = (A + DX_+)x = (A + DX_-)x. \tag{2.18}$$

Moreover, for each $X \in \Gamma$, $\mathcal{N}$ *is the spectral subspace of* $A + DX$ *corresponding to all of its pure-imaginary eigenvalues.*

Proof. Let $X \in \Gamma$ and write $\mathcal{N}_1 = \operatorname{Ker}(X - X_-)$, $\mathcal{N}_2 = \operatorname{Ker}(X_+ - X)$. Observe first that $\mathcal{N} = \mathcal{N}_1 \cap \mathcal{N}_2$. The inclusion $\mathcal{N}_1 \cap \mathcal{N}_2 \subset \mathcal{N}$ is clear. Also, if $x \in \mathcal{N}$, then $(X - X_+)x = (X - X_-)x$ and we have

$$((X - X_+)x, x) = ((X - X_-)x, x).$$

But $(X - X_+) \leq 0$ and $X - X_- \geq 0$, so $(X - X_+)x = (X - X_-)x = 0$, and so $\mathcal{N} \subset \mathcal{N}_1 \cap \mathcal{N}_2$.

We obviously have

$$(A + DX_-)x = (A + DX)x, \quad x \in \mathcal{N}_1;$$

$$(A + DX_+)x = (A + DX)x, \quad x \in \mathcal{N}_2,$$

and so (2.18) follows from $\mathcal{N} = \mathcal{N}_1 \cap \mathcal{N}_2$.

A direct computation using $\mathcal{R}(X_+) = \mathcal{R}(X_-) = 0$ shows that

$$\begin{aligned} &-(X_+ - X_-)D(X_+ - X_-)+ \\ &(X_+ - X_-)(A + DX_+) + (A^* + X_+D)(X_+ - X_-) = 0. \end{aligned} \tag{2.19}$$

Using this, it is easily verified that $\mathcal{N}$ is $(A + DX_+)$-invariant and hence, using (2.18), that $\mathcal{N}$ is $(A+DX)$-invariant for any $X \in \Gamma$. Furthermore, $\sigma((A+DX)|_{\mathcal{N}})$ does not depend on X. As the non-pure imaginary parts of $\sigma(A + DX_-)$ and $\sigma(A + DX_+)$ are disjoint, it follows that all the eigenvalues of the restriction $(A+DX_+)|_{\mathcal{N}}$ are pure imaginary, and similarly for $\sigma(A+DX)|_{\mathcal{N}}$ for any $X \in \Gamma$.

Next, it is shown that the subspace $\mathcal{N}$ contains the spectral subspace of $A + DX_+$ corresponding to *all* of its pure imaginary eigenvalues. Suppose not. Then there is a pure imaginary eigenvalue $-i\omega$ of $A + DX_+$ with eigenvector $z \in \mathcal{N}$. Let $x = (X_+ - X_-)z$, then $x \neq 0$ and (2.19) implies that

$$x^*Dx = x^*(A + DX_+)z + z^*(A^* + X_+D)x$$

$$= -i\omega x^*z + i\omega z^*x = 0,$$

since $x^*z \in \mathbf{R}$; hence, $Dx = 0$. Using (2.19) once more, it is now verified that

$$(A + DX_+)^*x = i\omega x.$$

Consequently, for $m = 1, 2, \ldots$,

$$D(A + DX_+)^{*m}x = (i\omega)^m Dx = 0, \quad x \neq 0.$$

This contradicts the controllability of $(A + DX_+, D)$ (which follows from the controllability of (A, D)).

It has been proved that $\mathcal{N}$ is the spectral subspace of $A+DX_+$ corresponding to all of its pure imaginary eigenvalues. It follows from (2.16) that, for any $X \in \Gamma$, the set of pure imaginary eigenvalues of $A+DX$ is just the set of real eigenvalues of M, and also that the dimension of the corresponding spectral subspace of $A+DX$ does not depend on X. Since (2.18) implies that $\mathcal{N}$ is also $(A+DX)$-invariant for any $X \in \Gamma$, it follows that $\mathcal{N}$ is, indeed, the spectral subspace of $A+DX$ corresponding to all of its pure imaginary eigenvalues. □

The next theorem establishes a one-to-one correspondence between the members of Γ and certain invariant subspaces of $A+DX_+$. First, with $\mathcal{N}$ as defined in Theorem 2.12, let $\mathcal{N}_-$ be the spectral subspace of $A+DX_+$ that is complementary to $\mathcal{N}$. Since $\sigma(A+DX_+)$ is in the closed left half-plane, $\mathcal{N}_-$ corresponds to all the eigenvalues of $A+DX_+$ with negative real parts. Then we have:

Theorem 2.13. *Suppose $D \geq 0$, $C^* = C$, (A, D) is controllable, and that Γ is not empty. Let $\mathcal{S}$ be an $(A+DX_+)$-invariant subspace with $\mathcal{S} \subset \mathcal{N}_-$ and define*

$$\mathcal{T} = ((X_+ - X_-)\mathcal{S})^{\perp}. \tag{2.20}$$

Then $\mathcal{T}$ is $(A+DX_-)$-invariant, $\mathbf{C}^n = \mathcal{S} \dot{+} \mathcal{T}$, and if P is the projection onto $\mathcal{S}$ along $\mathcal{T}$, then

$$X = X_+ P + X_-(I-P) \tag{2.21}$$

is in Γ. Conversely, if $X \in \Gamma$, there is an $(A+DX_+)$-invariant subspace $\mathcal{S} \subset \mathcal{N}_-$ for which X is given by (2.21), *and this correspondence is one-to-one.*

Let us compare Theorem 2.13 with the results of Sections 2.3 and 2.4. Theorem 2.9 associates each solution X of (2.9) in Γ with a unique M-invariant subspace $\mathcal{M}$ of $\mathcal{M}_+$ (the spectral subspace of M corresponding to all of its eigenvalues in the open upper half-plane). The subspace $\mathcal{M} \dot{+} \mathcal{S}_0$ (with $\mathcal{S}_0$ defined in Theorem 2.9) can then be complemented to generate an n-dimensional, M-invariant, H-nonpositive subspace $\mathcal{N}$, as in Theorem 2.7, and, on introducting a Jordan basis for $\mathcal{N}$, the solution X is determined as in Theorem 2.2. In contrast, $(A+DX_+)$-invariant subspaces $\mathcal{S}$ of $\mathcal{N}_-$ are used in Theorem 2.13 to characterize matrices $X \in \Gamma$ and the $(A+DX)$-invariant subspace $\mathcal{N}$ plays a role analogous to that of $\mathcal{S}_0$ in Section 2.4. The connection between the two is clarified by the observation (which follows from Theorems 2.12 and 2.8) that $x \in \mathcal{N}$ if and only if, for any $X \in \Gamma$,

$$\begin{bmatrix} x \\ Xx \end{bmatrix} \in \mathcal{S}_0.$$

The proof of Theorem 2.13 follows the argument of Willems [7] and Coppel [8] closely and will not be reproduced here. We remark that equivalent results have also been obtained by Ando [5].

2.6 Relaxing the Controllability Condition

The structural properties of the set Γ of hermitian solutions of $\mathbf{R}(X) = 0$ (when $\Gamma \neq \emptyset$) discussed in the two preceding sections depend on the hypothesis that (A, D) is a controllable pair. It is natural to ask whether this can be relaxed to admit pairs (A, D) that are merely stabilizable. This is so, but at some expense; notably the possibility that Γ has no minimal solution. An example of this (taken from [27]) is:

$$D = \begin{bmatrix} 1 & 0 \\ 0 & 0 \end{bmatrix}, \quad A = \begin{bmatrix} 0 & 0 \\ 0 & -1 \end{bmatrix}, \quad C = \begin{bmatrix} 1 & 0 \\ 0 & 0 \end{bmatrix},$$

in which (A, D) is stabilizable but not controllable. The set Γ consists of:

$$\begin{bmatrix} 1 & 0 \\ 0 & 0 \end{bmatrix}, \quad \begin{bmatrix} -1 & b \\ \bar{b} & -|b|^2/2 \end{bmatrix}, \quad b \in \mathbf{C}.$$

The obvious first step in investigation of this case is to make a coordinate transformation consistent with the orthogonal decomposition $\mathbf{C}^n = \mathcal{C} \oplus \mathcal{C}^{\perp}$, where $\mathcal{C}$ is the controllable subspace of (A, D). Then we may assume that A, D, C have the Kalman form (cf. formula (2.6)):

$$A = \begin{bmatrix} A_1 & A_{12} \\ 0 & A_2 \end{bmatrix}, \quad D = \begin{bmatrix} D_1 & 0 \\ 0 & 0 \end{bmatrix}, \quad C = \begin{bmatrix} C_{11} & C_{12} \\ C_{12}^* & C_{22} \end{bmatrix}, \tag{2.22}$$

where (A_1, D_1) is controllable, A_2 is stable, and $D_1 \geq 0$. Partition any solutions X in Γ conformally with (2.22):

$$X = \begin{bmatrix} X_1 & X_{12} \\ X_{12}^* & X_2 \end{bmatrix}.$$

Then $\mathcal{R}(X) = 0$ is found to be equivalent to the three equations

$$X_1 D_1 X_1 + X_1 A_1 + A_1^* X_1 - C_1 = 0, \tag{2.23}$$

$$(A_1^* + X_1 D_1) X_{12} + X_{12} A_2 = C_{12} - X_1 A_{12}, \tag{2.24}$$

$$A_2^* X_2 + X_2 A_2 = -X_{12}^* D_1 X_{12} - X_{12}^* A_{12} - A_{12}^* X_{12} + C_2. \tag{2.25}$$

Now introduce the transformations

$$H_1 = i \begin{bmatrix} 0 & I \\ -I & 0 \end{bmatrix}, \quad M_1 = i \begin{bmatrix} -A_1 & D_1 \\ C_1 & A_1^* \end{bmatrix}$$

on $\mathcal{C} \times \mathcal{C}$ (compare with equations (2.10) and (2.11)), and the results of Section 2.4 can be applied to the solutions of (2.23). Given a hermitian solutions X_1 of (2.23), the solvability of (2.24) and (2.25) follows from the fact that A_2 is stable (when we replace X_1 in (2.24) by the maximal solution of (2.23)). The next theorem is obtained on developing this argument (see [27] and Theorem 2.6 above).

Theorem 2.14. *If $D \geq 0$, $C^* = C$, and (A, D) is stabilizable, then the following statements are equivalent:*

(i) *Γ is not empty;*
(ii) *there exists an M_1-invariant, H_1-neutral subspace with the same dimension as $\mathcal{C}$ (the controllable subspace of (A, D));*
(iii) *the partial multiplicities of the real eigenvalues of M_1 (if any) are all even.*

Extensions of Theorems 2.9 and 2.11 to the case when (A, D) is stabilizable and particularly the definition of the map Φ, requires a further hypothesis guaranteeing the solubility of (2.24) for *all* $X \in \Gamma$. Thus, we call equation (2.9) *regular* if (keeping in mind the representations (2.22)),

$$\sigma(M_1) \cap \sigma(A_2) = \emptyset.$$

Let $\mathcal{L}_+$ be the maximal M-invariant subspace for which $\sigma(M_1|_{\mathcal{L}_+})$ lies in the open upper half-plane (in particular, $\mathcal{L}_+ \subset \mathcal{M}_+$).

Theorem 2.15. ([27]) *Suppose $D \geq 0$, $C^* = C$, (A, D) is stabilizable, Γ is not empty, and (2.9) is regular. Then*

(a) *for every M_1-invariant subspace $\mathcal{L} \subset \mathcal{L}_+$, there is a unique $X \in \Gamma$, say $X = \Phi_1(\Gamma)$ such that $G(PXP) \cap \mathcal{L}_+ = \mathcal{L}$, where P is the orthogonal projection onto $\mathcal{C}$. Conversely, if $X \in \Gamma$, there is a unique M_1-invariant subspace $\mathcal{L} \subset \mathcal{L}_+$ such that $X = \Phi_1(\mathcal{L})$.*
(b) *If $X, Y \in \Gamma$, then $X \leq Y$ if and only if $\Phi_1^{-1}(X) \leq \Phi_1^{-1}(Y)$.*

An interesting consequence of the regularity of the Riccati equation follows from this theorem.

Corollary 2.16. ([27]) *Under the hypotheses of Theorem* 2.15, *Γ contains both maximal and minimal solutions.*

Further investigation of the non-regular case can be found in the paper [27].

2.7 Constructive Existence and Comparison Theorems

It is not difficult to see that, if the Newton-Raphson iterative process for vector-valued functions is applied to the non-linear equation $\mathcal{R}(X) = 0$, then one is faced with the solution of a sequence of *linear* equations. These equations for the symmetric equation (2.9) are of Lyapunov type. It is remarkable that, on making a suitable initialization of the process (and with hypotheses that are now familiar), convergence of the process can be guaranteed and a useful constructive approach to analysis of the algebraic Riccati equation results. The idea was first developed by Kleinman [28] for controllable systems, and went through a sequence of refinements admitting stabilizable systems, as well as a weakening of the hypothesis that Γ is not empty, to the assumption that there is a hermitian solution for the Riccati *inequality*

$$\mathcal{R}(X) := XDX + XA + A^*X - C \leq 0. \tag{2.26}$$

Consideration of such inequalities also admits interesting comparison theorems (see Theorem 2.18 below) for the maximal solutions of different equations. Several authors have taken part in the evolution of this technique. See [28], [29], [8], [27], [30], for example.

Denote the set of all hermitian solutions of (2.26) by Λ and observe that $\Lambda \supset \Gamma$. The following theorems can be found (in a more general form) in [30].

Theorem 2.17. *Assume that $D \geq 0$, $C^* = C$, (A, D) is stabilizable, and that Λ is not empty. Then there is an $X_+ \in \Gamma$ such that $X_+ \geq X$ for all $X \in \Lambda$. Also, $\sigma(A + DX_+)$ is in the closed left half-plane.*

Note that X_+ is maximal for Λ, as well as for Γ.

To formulate a comparison theorem, consider the hermitian matrix (used to define a scalar product in Section 2.4):

$$H = \begin{bmatrix} -C & A^* \\ A & D \end{bmatrix},$$

and recall equation (2.13). Let

$$\tilde{\mathcal{R}}(X) = X\tilde{D}X + X\tilde{A} + \tilde{A}^*X - \tilde{C}$$

denote another Riccati function and write

$$\tilde{H} = \begin{bmatrix} -\tilde{C} & \tilde{A}^* \\ \tilde{A} & \tilde{D} \end{bmatrix}.$$

Theorem 2.18. *Assume that $D, \tilde{D} \geq 0$, C and $\tilde{C}$ are hermitian, (A, D) and $(\tilde{A}, \tilde{D})$ are stabilizable, and that $\tilde{\mathcal{R}}(X) \leq 0$ has an hermitian solution. If $H \leq \tilde{H}$, then $\mathcal{R}(X) = 0$, $\tilde{\mathcal{R}}(X) = 0$ have maximal hermitian solutions X_+ and $\tilde{X}_+$, respectively, and $X_+ \geq \tilde{X}_+$.*

The theorem has an important and possibly more familiar consequence.

Corollary 2.19. *If $D \geq 0$, $C \geq 0$, and (A, D) is stabilizable, then Γ is not empty, X_+ exists and $X_+ \geq 0$.*

To prove the Corollary, take $\tilde{A} = A$, $\tilde{D} = D$, and $\tilde{C} = 0$ for the comparison equation $\tilde{\mathcal{R}}(X) = 0$. This has the solution $\tilde{X} = 0$ and the Corollary is obtained on applying the theorem.

Note that if the stabilizable hypothesis is replaced by the stronger controllability condition, then minimal solutions exist and have analogous properties. In particular, it follows from the theorem that, as C and D decrease, X_+ will decrease, X_- will increase, and so the "gap" $X_+ - X_-$, introduced in Section 2.5, will also decrease (as observed in [7]).

2.8 Parameter Dependence and Perturbations

Consider again the equation

$$XDX + XA + A^*X - C = 0 \tag{2.27}$$

with $D \geq 0$ and $C^* = C$. In this section, in contrast with the previous material, we consider D, C, and A as variable matrices and study the behavior of hermitian solutions of (2.27) (if they exist) as functions of A, C, and D. We present only a sample of basic results and approaches, and refer the reader to the papers [31, 32, 33, 34, 26, 23, 35, 36, 9] and to Chapter III.4 in [3] for more information.

We start with continuity properties of the maximal and minimal hermitian solutions of (2.27). Recall that a hermitian solution X_+ (resp. X_-) of (2.27) is called *maximal* (resp. *minimal*) if $X_+ \geq X$ (resp. $X \geq X_-$) for every hermitian solution X of (2.27). If (A, D) is controllable and the set Γ of hermitian solutions of (2.27) is non-empty, then the existence of maximal and minimal solutions is guaranteed (cf. Theorem 2.10).

Theorem 2.20. ([31, 36]) *Let W be the set of all ordered triples (A, D, C) of $n \times n$ matrices with the following two properties:*

(i) $D \geq 0$, $C = C^*$, (A, D) *is controllable;*
(ii) *there is a hermitian solution of* (2.27).

Then the maximal X_+ and the minimal X_- solutions of (2.27) *are continuous functions of $(A, D, C) \in W$ (the set W is endowed with the induced topology from the set of all ordered triples of $n \times n$ matrices in the natural topology).*

Thus, the maximal and minimal solutions are well behaved as far as the continuity goes. One can extend this result to the so-called unmixed solutions (see [2, 23]).

Simple examples show that the maximal and minimal solutions are not analytic functions of (A, D, C). The following example (borrowed from [37]) illustrates this point.

Example 2.1. Let

$$A(t) = \begin{bmatrix} 0 & -1 \\ 0 & 0 \end{bmatrix}, \quad D(t) = \begin{bmatrix} 0 & 0 \\ 0 & 1 \end{bmatrix}, \quad \text{and } C(t) = \begin{bmatrix} t^2 & 0 \\ 0 & 0 \end{bmatrix}, \quad t \in \mathbf{R}.$$

Then the associated matrix $M = M(t)$ is

$$M(t) = i \begin{bmatrix} 0 & -1 & 0 & 0 \\ 0 & 0 & 0 & 1 \\ t^2 & 0 & 0 & 0 \\ 0 & 0 & 1 & 0 \end{bmatrix}.$$

It is not hard to check that the eigenvalues of $M(t)$ are ϵ_1, ϵ_2, ϵ_3, ϵ_4, where $\epsilon_k = |t|^{1/2}\exp(\frac{1}{4}\pi i + \frac{1}{2}k\pi i)$, $k = 1, 2, 3, 4$. The corresponding eigenvectors are

$$x_k(t) = [1 \;\; \epsilon_k \; -\epsilon_k^3 \;\; \epsilon_k^2]^T.$$

For $i \neq j$ we have

$$\operatorname{span}\{x_i(t), x_j(t)\} = Im \begin{bmatrix} I \\ x_{ij}(t) \end{bmatrix},$$

where

$$x_{ij}(t) = \begin{bmatrix} \epsilon_i\epsilon_j(\epsilon_i + \epsilon_j) & -(\epsilon_i^2 + \epsilon_i\epsilon_j + \epsilon_j^2) \\ \epsilon_i\epsilon_j & \epsilon_j + \epsilon_i \end{bmatrix}.$$

For any choice of i, j we have

$$\epsilon_j + \epsilon_i = |t|^{1/2} \text{ constant},$$

so $x_{ij}(t)$ cannot be analytic. It follows that there is no analytic solution to the Riccati equation (hermitian or not). Observe that in this example $(A(t), D(t))$ is controllable for all $t \in \mathbf{R}$, the matrix $M(t)$ has no real eigenvalues for $t \neq 0$ and has the zero eigenvalue with multiplicity 4 when $t = 0$. Thus, by Theorem 2.6 for every $t \in \mathbf{R}$, there is a hermitian solution of

$$XD(t)X + XA(t) + A(t)^*X - C(t) = 0. \qquad \square$$

The extra assumption needed to ensure the analyticity of maximal (or minimal) hermitian solutions turns out to be the invariance of the number of real eigenvalues of M (counting with multiplicities):

Theorem 2.21. *Let $A(t)$, $D(t)$, and $C(t)$ be analytic $n \times n$ matrix functions on t defined on a real interval (α, β), with $D(t)$ positive semidefinite hermitian, $C(t)$ hermitian, and $(A(t), D(t))$ stabilizable for every $t \in (\alpha, \beta)$. Assume that for all $t \in (\alpha, \beta)$, the Riccati equation*

$$X(t)D(t)X(t) + X(t)A(t) + A(t)^*X(t) - C(t) = 0 \tag{2.28}$$

has a hermitian solution. Further assume that the number of real eigenvalues (counting multiplicities) of

$$M(t) := i\begin{bmatrix} A(t) & D(t) \\ C(t) & -A(t)^* \end{bmatrix}$$

is constant. Then the maximal solution $X_+(t)$ of (2.28) is an analytic function of $t \in (\alpha, \beta)$. Conversely, if $X_+(t)$ is an analytic function of $t \in (\alpha, \beta)$, then the number of real eigenvalues of $M(t)$ is constant.

Proof. The first part of Theorem 2.21 is proved in [35] (for the case when $M(t)$ has no real eigenvalues, it can be proved using the implicit function theorem, see [38]).

Assume now that $X_+(t)$ is analytic. As the number of real eigenvalues of $M(t)$ is twice the number of pure imaginary eigenvalues of $A(t) + D(t)X_+(t)$, it is enough to prove that the number of pure imaginary eigenvalues of $A(t) + D(t)X_+(t)$ is constant. By Theorem 2.17, $\sigma(A(t) + D(t)X_+(t))$ is in the closed left half-plane. Let λ_0 be a pure imaginary eigenvalue of $A(t_0) + D(t_0)X_+(t_0)$ for some $t_0 \in (\alpha, \beta)$. Then the eigenvalues $\lambda(t)$ of $A(t)+D(t)X_+(t)$ that are close to λ_0 admit developments into fractional power series (Puiseux series, see, e.g., [39])

$$\lambda(t) = \lambda_0 + \sum_{j=1}^{\infty} \alpha_j (t - t_0)^{j/p}$$

for t sufficiently close to t_0. It is easy to see that the only way $\lambda(t)$ can be in the closed left half-plane for all choices of the branch of $(t - t_0)^{1/p}$ is when $\lambda(t)$ stays pure imaginary. This proves the second part of Theorem 2.21. □

Observe that only stabilizability is assumed in Theorem 2.21; so the minimal solution need not exist. However, if $(A(t), D(t))$ is controllable, then the minimal solution $X_-(t)$ of (2.28) exists and is an analytic function of $t \in (\alpha, \beta)$ as well. Again, Theorem 2.21 can be extended (under suitable hypotheses) to the unmixed solutions (see [35]). It is an open problem whether Theorem 2.21 holds in case of analytic dependence of several real variables.

Next, we consider the notion of stable (in the sense of robustness) hermitian solutions of (2.27). A hermitian solution X of (2.27) is called *conditionally stable* if for every $\epsilon > 0$ there exists $\delta > 0$ such that every Riccati equation

$$YD'Y + YA' + A'^*Y - C' = 0 \tag{2.29}$$

with $\|A - A'\| + \|C - C'\| + \|D - D'\| < \delta$ and $D' \geq 0$, $C = C'$ has a hermitian solution Y with $\|X - Y\| < \epsilon$ provided the set of hermitian solutions of (2.29) is not empty. In informal terms, it means that any Riccati equation sufficiently close to the original equation has a hermitian solution as close to X as we wish provided the new equation (2.29) has hermitian solutions at all. If the proviso that the set of hermitian solutions of (2.29) is non-empty is removed from the definition, then we obtain the notion of *unconditional stability*. Clearly, an unconditionally stable hermitian solution is also conditionally stable, but the converse is not true in general, as Theorems 2.22 and 2.23 below make apparent. In these theorems, it will be assumed that the pair (A, D) is controllable.

Theorem 2.22. ([23]) *Let X be a hermitian solution of* (2.27). *The following statements are equivalent*:

(i) *X is conditionally stable*;

(ii) *X is isolated, i.e., there is an $\epsilon_0 > 0$ such that the only hermitian solution X' of* (2.27) *satisfying $\|X' - X\| < \epsilon_0$ is X itself*;

(iii) *for every non-real eigenvalue λ of the associated matrix $M = i \begin{bmatrix} A & D \\ C & -A^* \end{bmatrix}$ such that* $\dim \operatorname{Ker}(\lambda I - M) > 1$ *the spectral subspace of M corresponding to*

λ either is contained in the graph space $G(X)$ of X or has zero intersection with $G(X)$;

(iv) *each common non-real eigenvalue of $i(A+DX)$ and of $-i(A^*+XD)$ is an eigenvalue of M of geometric multiplicity one.*

The idea of the proof of this theorem is to use the description of hermitian solutions in terms of invariant subspaces of M (as indicated in Section 2.4), together with the characterization of certain classes of stable invariant subspaces of M developed in [26].

Theorem 2.23. *There exists an unconditionally stable hermitian solution of* (2.27) *if and only if M has no real eigenvalues. In such cases a hermitian solution is unconditionally stable if and only if it is conditionally stable.*

The proof is based on the observation that if M has real eigenvalues, it can be perturbed (using admissible perturbations A', C', and D' of A, C, and D, respectively) in such a way that the perturbed matrix M' has *simple* real eigenvalues; it follows then from Theorem 2.6 that the equation with perturbed coefficients

$$YD'Y+YA'+A'^*Y-C'=0$$

has no hermitian solutions.

2.9 Hermitian Solutions via the Frequency Domain

Here we approach the characterization of hermitian solutions of

$$XDX+XA+A^*X-C=0 \tag{2.30}$$

from another direction, namely, in terms of rational matrix functions and their minimal factorizations. It will be assumed throughout this section that $D\geq 0$, $C=C^*$, and (A,D) is controllable.

The first result concerns the existence of hermitian solutions. We let D_0 be any $n\times n$ matrix such that $D=D_0D_0^*$.

Theorem 2.24. *The following statements are equivalent:*

(i) *the equation* (2.30) *has hermitian solutions;*

(ii) *the rational matrix function*

$$Z(\lambda)=I+D_0^*(\lambda I+iA^*)^{-1}C(\lambda I-iA)^{-1}D_0$$

is non-negative on the real axis, i.e.,

$$x^*Z(\lambda)x\geq 0$$

for every $x\in \mathbf{C}^n$ and every real λ which is not a pole of $Z(\lambda)$;

(iii) *the rational matrix function*

$$W(\lambda) = I + (\lambda I - iA)^{-1} D_0 Z(\lambda) D_0^* (\lambda I - iA^*)^{-1} \tag{2.31}$$

is non-negative on the real axis.

Note that, in the classical "linear-quadratic" optimal control problem $Z(\lambda)$ has physical meaning because it is, in essence, the matrix function of the quadratic form defining the cost functional in terms of the Laplace transform of the input vector.

The equivalence of (i) and (ii) was established in [7]; for the full proof of this theorem, see Section II.4.4 in [3]. Now we shall use the equivalence of (i) and (ii) of Theorem 2.24 to describe hermitian solutions of (2.30) in terms of certain factorizations of the function $W(\lambda)$. To this end, we recall some basic facts concerning rational matrix functions (see, e.g., the monographs [40, 3, 20, 41, 42] for more information).

Let $W(\lambda)$ be an $n \times n$ rational matrix function which is analytic (*i.e.*, with no pole) at infinity, and $W(\infty) = I$. A *realization* of $W(\lambda)$ is, by definition, a representation in the form

$$W(\lambda) = I + C(\lambda I - A)^{-1} B, \tag{2.32}$$

where the (complex) matrices C, A, B are of sizes $n \times m$, $m \times m$, and $m \times n$, respectively. The matrix A is called the main matrix of realization (2.32). Some simple algebra shows that for $W(\lambda)$ given by (2.32), the inverse matrix function $W(\lambda)^{-1}$ is given by the realization

$$W(\lambda)^{-1} = I - C(\lambda I - (A - BC))^{-1} B. \tag{2.33}$$

The realization (2.32) is called *minimal* if the size of the matrix A is the smallest possible among all realizations of $W(\lambda)$ or, equivalently, if the pair (A, B) is controllable and the pair (C, A) is observable. The size m of the matrix A in a minimal realization for $W(\lambda)$ is called the *McMillan degree* of $W(\lambda)$. Every rational matrix function with value I at infinity admits a minimal realization which is unique up to similarity; that is, up to transformations of the form $C \to CS$, $A \to S^{-1}AS$, $B \to S^{-1}B$ for some invertible matrix S. Also, (2.32) is minimal if and only if (2.33) is.

Now let a factorization

$$W(\lambda) = W_1(\lambda) W_2(\lambda), \tag{2.34}$$

be given where $W_1(\lambda)$ and $W_2(\lambda)$ are $n \times n$ rational matrix functions with $W_1(\infty) = W_2(\infty) = I$. The factorization (2.34) is called *minimal* if the McMillan degrees of W_1 and W_2 add up to the McMillan degree of W. The minimal factorizations can be conveniently described in terms of minimal realizations. Namely, let (2.32) be a minimal realization for $W(\lambda)$, and let $\pi : \mathbf{C}^m \to \mathbf{C}^m$ be a projector such that

$$A(\text{Ker } \pi) \subset \text{Ker } \pi; \quad (A - BC)(Im\ \pi) \subset Im\ \pi. \tag{2.35}$$

Thus, using the direct sum decomposition $\mathbf{C}^m = \text{Ker } \pi \dot{+} Im\ \pi$, we may write

$$A = \begin{bmatrix} A_{11} & A_{12} \\ 0 & A_{22} \end{bmatrix}; \quad B = \begin{bmatrix} B_1 \\ B_2 \end{bmatrix}; \quad C = [C_1\ C_2];$$

$$A - BC = \begin{bmatrix} A_{11} - B_1C_1 & 0 \\ -B_2C_1 & A_{22} - B_2C_2 \end{bmatrix}.$$

Then

$$W(\lambda) = (I + C_1(\lambda I - A_{11})^{-1}B_1)(I + C_2(\lambda I - A_{22})^{-1}B_2)$$

is a minimal factorization, and every minimal factorization of $W(\lambda)$ arises in this way.

We now return to Theorem 2.24.

It turns out that one can rewrite $W(\lambda)$ given by (2.31) in the form

$$W(\lambda) = I + R^*\hat{H}(\lambda I - M)^{-1}R, \tag{2.36}$$

where $R = \begin{bmatrix} 0 \\ I \end{bmatrix}$, $\hat{H} = i\begin{bmatrix} 0 & I \\ -I & 0 \end{bmatrix}$, and $M = i\begin{bmatrix} A & D \\ C & -A^* \end{bmatrix}$ are matrices associated with the Riccati equation (2.30) (as in Section 2.4). Moreover, the representation (2.36) is a minimal realization for $W(\lambda)$. Because $W(\lambda)$ is non-negative on the real axis, it admits minimal factorizations of the form $W(\lambda) = (L(\bar{\lambda}))^*L(\lambda)$ (see [43]; Chapter II.3 in [3]). It turns out that such factorizations can be used to describe the hermitian solutions of (2.30). To set up this description, consider the second Riccati equation

$$XDX + A^*X + XA - (C + I) = 0, \tag{2.37}$$

and observe that by the comparison theorem (Theorem 2.21), if equation (2.30) admits hermitian solutions, so does (2.37). Also note that if $X = X^*$ is a solution of (2.30) and $\tilde{X} = \tilde{X}^*$ is a solution of (2.37), then the difference $X - \tilde{X}$ is invertible. Indeed, let $x \in \mathbf{C}^n$ be such that $Xx = \tilde{X}x$. Then

$$x^*\text{XDXx} + x^*XAx + x^*A^*Xx = x^*\tilde{X}D\tilde{X}x + x^*\tilde{X}Ax + x^*A^*\tilde{X}x.$$

But the left-hand side of this equality is x^*Cx, while the righthand side is $x^*(C + I)x$. So $x = 0$.

Theorem 2.25. ([4]) *Assume that the rational matrix function $W(\lambda)$ given by* (2.31) *is non-negative on the real axis. Let X and $\tilde{X}$ be hermitian solutions of* (2.30) *and* (2.37), *respectively. Then $W(\lambda)$ admits a minimal factorization*

$$W(\lambda) = (L(\bar{\lambda}))^*L(\lambda) \tag{2.38}$$

with

$$L(\lambda) = I_n + R^*\hat{H}\begin{bmatrix} ZX & -Z \\ \tilde{X}ZX & -\tilde{X}Z \end{bmatrix}(\lambda I - M)^{-1}R, \tag{2.39}$$

where $Z = (X - \tilde{X})^{-1}$. *Conversely, any rational matrix function* $L(\lambda)$ *with* $L(\infty) = I$ *and for which* (2.38) *is a minimal factorization of* $W(\lambda)$, *has the form* (2.39) *for some pair* $X, \tilde{X}$ *of hermitian solutions of* (2.30) *and* (2.37), *respectively. Moreover, the zeros of* $L(\lambda)$ *are the eigenvalues of* $i(A + D\tilde{X})$, *and the poles of* $L(\lambda)$ *are the eigenvalues of* $-i(A + DX)^*$, *and in both cases multiplicities are counted. More exactly: the partial zero* (*resp. pole*) *multiplicities at zero* (*resp. pole*) λ_0 *of* $L(\lambda)$ *are precisely the partial multiplicities of* $i(A + D\tilde{X})$ (*resp.* $-i(A + DX)^*$) *corresponding to* λ_0.

Recall that λ_0 is called a zero of a rational matrix function $L(\lambda)$ if λ_0 is a pole of $L(\lambda)^{-1}$, *i.e.*, a pole for at least one entry in $L(\lambda)^{-1}$. The zero (resp. pole) multiplicities of $L(\lambda)$ at λ_0 can be identified as the partial multiplicities of $A_0 - B_0C_0$ (resp. of A_0) at λ_0, where A_0, B_0, and C_0 are taken from a minimal realization

$$L(\lambda) = I + C_0(\lambda I - A_0)^{-1}B_0.$$

Choosing a fixed hermitian solution $\tilde{X}$ of (2.37), one obtains from Theorem 2.25 a one-to-one correspondence between the set of hermitian solutions of (2.30) and a set of certain minimal factorizations of type (2.38). For instance:

Corollary 2.26. *There is a one-one correspondence between hermitian solutions* X *of* (2.30) *and minimal factorizations of* $W(\lambda)$ *of the form* $W(\lambda) = (L(\bar{\lambda}))^* L(\lambda)$ such that $L(\infty) = I_n$ *and* $L(\lambda)$ *has all its zeros in the closed upper half-plane. This correspondence is given by formula* (2.39), *where* $\tilde{X}$ *is the maximal hermitian solution of* (2.37).

2.10 Rational Matrix Functions that are Contractions on the Line

In this section, we present a short account of some of the results from [13] on rational matrix functions that are contractions, and on their unitary completions. The algebraic Riccati equation and its hermitian solutions play a prominent role here.

A rational matrix function $W(\lambda)$ (generally rectangular of size $p \times m$) will be called a *proper contraction* if it satisfies two properties:

(a) $I - W(\lambda)(W(\lambda))^* \geq 0$ for all real λ;
(b) $W(\lambda)$ has no pole at infinity, and the matrix $I - W(\infty)(W(\infty))^*$ is positive definite.

In particular, it follows from (a) that $W(\lambda)$ has no poles on the real line.

Every $p \times m$ rational matrix function $W(\lambda)$ satisfying (b) admits a realization

$$W(\lambda) = D + C(\lambda I - A)^{-1}B, \tag{2.40}$$

where $D = W(\infty)$. We assume in the sequel that the realization (2.40) is *minimal.* With the realization (2.40), we associate the *state characteristic matrix*

$$Q := \begin{bmatrix} \alpha & \beta \\ \gamma & \alpha^* \end{bmatrix} := \begin{bmatrix} A + BD^*(I - DD^*)^{-1}C & B(I - D^*D)^{-1}B^* \\ C^*(I - DD^*)^{-1}C & A^* + C^*(I - DD^*)^{-1}DB^* \end{bmatrix} \tag{2.41}$$

and the Riccati equation

$$X\gamma X - iX\alpha^* + i\alpha X + \beta = 0. \tag{2.42}$$

Observe that the matrices α, β, and γ are $n \times n$, where n is the *McMillan degree* of $W(\lambda)$($i.e.$, the size of the matrix A in the minimal realization (2.40)). Also, $\beta \geq 0$, $\gamma \geq 0$. One can also show, as a consequence of the minimality of the realization (2.40), that the pairs (α, β) and (α^*, γ) are controllable.

As a side remark, observe that the formula (2.41) can be rewritten in the form

$$\begin{bmatrix} \gamma & \alpha^* \\ \alpha & \beta \end{bmatrix} = \begin{bmatrix} C & D \\ A & B \end{bmatrix} \circ \begin{bmatrix} C & D \\ A & B \end{bmatrix}^*,$$

where " $\circ$ " denotes the star (or Redheffer) product (introduced in [44]).

Theorem 2.27. ([13]) *Let $W(\lambda)$ be a $p \times m$ rational matrix function satisfying the property* (b). *Then $W(\lambda)$ is a proper contraction if and only if the associated Riccati equation* (2.42) *has a hermitian solution.*

It is not difficult to verify that, because the minimal realization (2.40) is unique up to similarity, the existence of hermitian solutions of (2.42) does not depend on the choice of minimal realization (2.40).

There is a simple connection between the matrix M associated with the Riccati equation (2.42) and the state characteristic matrix Q. Indeed,

$$M = i \begin{bmatrix} -i\alpha^* & \gamma \\ -\beta & -i\alpha \end{bmatrix},$$

and a calculation shows that

$$Q = \begin{bmatrix} 0 & iI \\ I & 0 \end{bmatrix} M \begin{bmatrix} 0 & I \\ -iI & 0 \end{bmatrix}.$$

Thus, Q and M are similar, and Theorem 2.6, combined with Theorem 2.27, yields

Corollary 2.28. ([13]) *A rational matrix function $W(\lambda)$ satisfying (b) is a proper contraction if and only if the partial multiplicities of the real eigenvalues of the state characteristic matrix (if any) are all even.*

It turns out that for a proper contraction $W(\lambda)$, every hermitian solution of (2.42) is invertible, and the number of negative (resp. positive) eigenvalues of any hermitian solution of (2.42) is equal to the number of poles of $W(\lambda)$ in the lower (resp. upper) half-plane. These facts were also proved in [13].

We turn now to the study of *minimal unitary completions* of a proper contraction $W(\lambda)$. These completions are rational matrix functions $\tilde{W}(\lambda)$ defined by

$$\tilde{W}(\lambda) = \begin{bmatrix} W_{11}(\lambda) & W(\lambda) \\ W_{21}(\lambda) & W_{22}(\lambda) \end{bmatrix} : \mathbf{C}^p \oplus \mathbf{C}^m \to \mathbf{C}^p \oplus \mathbf{C}^m,$$

where $\tilde{W}(\lambda)(\tilde{W}(\lambda))^* = I$ for every $\lambda \in \mathbf{R}$ and the McMillan degree of $\tilde{W}(\lambda)$ coincides with the McMillan degree of $W(\lambda)$. This kind of completion is important in network theory (see, *e.g.*, [45]). The following result (proved in [13]) describes all minimal unitary completions in terms of hermitian solutions of the associated Riccati equation.

Theorem 2.29. *Let $W(\lambda)$ be a $p \times m$ proper contraction with minimal realization*

$$W(\lambda) = D + C(\lambda I - A)^{-1}B,$$

and with the associated Riccati equation

$$X\gamma X - iX\alpha^* + i\alpha X + \beta = 0, \tag{2.43}$$

where $\begin{bmatrix} \alpha & \beta \\ \gamma & \alpha^* \end{bmatrix}$ is the state characteristic matrix. Then for every hermitian solutions H of (2.43), *the function*

$$\tilde{W}_H(\lambda) = \begin{bmatrix} M & D \\ N & P \end{bmatrix} + \begin{bmatrix} C \\ Y \end{bmatrix} (\lambda I - A)^{-1}[X \;\; B], \tag{2.44}$$

where $N = -D^*$; $M = (I - DD^*)^{1/2}$; $P = (I - D^*D)^{1/2}$;

$$X = (iHC^* - BD^*)(I - DD^*)^{-1/2};$$

$$Y = (I - D^*D)^{-1/2}(iB^*H^{-1} - D^*C),$$

is a minimal unitary completion of $W(\lambda)$. Conversely, every minimal unitary completion $\tilde{W}(\lambda)$ of $W(\lambda)$ is of the form

$$\tilde{W}(\lambda) = \begin{bmatrix} I & 0 \\ 0 & T \end{bmatrix} \tilde{W}_H(\lambda) \begin{bmatrix} S & 0 \\ 0 & I \end{bmatrix}$$

for some unitary matrices S and T (which are uniquely determined by the value of $\tilde{W}(\lambda)$ at infinity) and some hermitian solution H of (2.43). *Furthermore, there is one-to-one correspondence between the set of hermitian solutions of* (2.43) *and the set of all minimal unitary completions of $W(\lambda)$ with fixed value at infinity.*

An interesting observation from Theorem 2.29 is that all minimal unitary completions of a given proper contraction $W(\lambda)$ have the same poles as $W(\lambda)$ with the same multiplicities, but their zeros are generally different from those of $W(\lambda)$.

2.11 *J*-Spectral Factorization

Recently, the H^∞ optimization problems attracted the attention of many mathematicians and engineers, and the continuous algebraic Riccati equation and its hermitian solutions play a significant role in development of the theory. Without even attempting to mention all major recent contributions to this theory where the algebraic Riccati equation is crucial, we quote [46–49] as a sample.

In this section, we indicate the role of algebraic Riccati equations in the study of J-spectral factorization. This type of factorization is crucial in the development of some approaches to H^∞ optimization, but is also important and interesting from the mathematical point of view. As we shall see, in contrast with the previous section, the algebraic Riccati equation which appears here has indefinite quadratic terms. In this exposition, we follow generally [48].

It is convenient to introduce the following definitions. A $2n \times 2n$ (complex) matrix

$$H = \begin{bmatrix} H_{11} & H_{12} \\ H_{21} & H_{22} \end{bmatrix} \tag{2.45}$$

will be called *Hamiltonian* if $H_{12} = H_{12}^*$, $H_{21} = H_{21}^*$, and $H_{22} = -H_{11}^*$. We associate with a Hamiltonian matrix H as in (2.45) the Riccati equation

$$QH_{12}Q + QH_{11} + H_{11}^*Q - H_{21} = 0. \tag{2.46}$$

We say that a Hamiltonian matrix $H \in dom(Ric)$ if (a) H has precisely n eigenvalues (counted with multiplicities) in both the open left and open right half-plane, and (b) the associated Riccati equation admits a hermitian solution Q for which the matrix $H_{11}+H_{12}Q$ is stable ($i.e.$, has all its eigenvalues in the open left half-plane).

It is easy to see that for $H \in dom(Ric)$, there is a unique hermitian solution Q of the associated Riccati equation with stable $H_{11} + H_{12}Q$. Indeed, for such a Q, we have

$$H\begin{bmatrix} I \\ Q \end{bmatrix} = \begin{bmatrix} I \\ Q \end{bmatrix}(H_{11} + H_{12}Q).$$

So, because of the condition (a) in the definition of dom(Ric), the subspace $Im\begin{bmatrix} I \\ Q \end{bmatrix}$ must be the spectral subspace of H corresponding to its eigenvalues in the open left half-plane, and the matrix Q is uniquely determined by this property. We use the notation $Q = Ric(H)$.

In the next theorem, we let

$$J = \begin{bmatrix} I_p & 0 \\ 0 & -I_q \end{bmatrix}; \quad p+q=n;$$

and denote by $\overline{\pi^+}$ the closed right half-plane (including infinity).

Theorem 2.30. *Let the $n \times n$ rational matrix function $G(\lambda)$ be given by the (not necessarily minimal) realization*

$$G(\lambda) = D + C(\lambda I - A)^{-1}B,$$

where D is invertible and A is stable (so $G(\lambda)$ is analytic and invertible at infinity and has no poles in $\overline{\pi^+}$). Then there exists an $n \times n$ rational matrix function $W(\lambda)$ with no zeros and poles in $\overline{\pi^+}$ and such that

$$(W(-\bar{\lambda}))^* JW(\lambda) = (G(-\bar{\lambda}))^* JG(\lambda) \tag{2.47}$$

for all $\lambda \in \mathbf{C}$ which are not poles of one of the functions $W(\lambda)$, $G(\lambda)$, $(W(-\bar{\lambda}))^*$, $(G(-\bar{\lambda}))^*$ if and only if

$$H \in dom(Ric),$$

where

$$H = \begin{bmatrix} A & 0 \\ -C^*JC & -A^* \end{bmatrix} - \begin{bmatrix} B \\ -C^*JD \end{bmatrix} (D^*JD)^{-1}[D^*JC \;\; B^*]. \tag{2.48}$$

In this case $W(\lambda)$ satisfies (2.47) if and only if $W(\lambda)$ is given by

$$W(\lambda) = W_\infty + L(\lambda I - A)^{-1}B,$$

where W_∞ is some (necessarily invertible) matrix such that $D^*JD = W_\infty^* JW_\infty$, and

$$L = JW_\infty^{*-1}(D^*JC + B^*Q); \quad Q = Ric(H).$$

A more general version of this result is proved in [48].

Proof. We shall indicate the proof of part "if" only, and refer the reader to the original paper [48] for the proof of the "only if" part. We remark that the approach to the proof of the "only if" part is based on the description of canonical Wiener-Hopf factorization in terms of realizations (see [40]).

So assume $H \in dom(Ric)$. Then $A - B(D^*JD)^{-1}(D^*JC + B^*Q)$ is stable. On the other hand, we have (cf. formula (2.33))

$$W(\lambda)^{-1} = W_\infty^{-1} - W_\infty^{-1}L(\lambda I - (A - BW_\infty^{-1}L))^{-1}BW_\infty^{-1},$$

and since

$$A - B(D^*JD)^{-1}(D^*JC + B^*Q) = A - BW_\infty^{-1}L,$$

it follows that all poles of $W(\lambda)^{-1}$, *i.e.*, zeros of $W(\lambda)$, are in the open left half-plane. Next, rewrite the Riccati equation associated with H in the form

$$QA + A^*Q + C^*JC - L^*JL = 0,$$

and use this equation to verify directly that

$$(W(-\bar{\lambda}))^* JW(\lambda) = (G(-\bar{\lambda}))^* JG(\lambda). \qquad \square$$

The representation (2.47) is called a *J-spectral factorization*, and note that this is a particular case of a canonical Wiener-Hopf factorization of the rational matrix function $V(\lambda) := (G(-\bar{\lambda}))^* JG(\lambda)$. Clearly, $V(\lambda)$ is a rational matrix function analytic and invertible at infinity and satisfying $(V(-\bar{\lambda}))^* = V(\lambda)$ for every $\lambda \in \mathbf{C}$ which is not a pole of $V(\lambda)$; moreover, $V(\lambda)$ has no poles on the imaginary axis. As $W(\lambda)$ is analytic and invertible on the imaginary axis, a necessary condition for the existence of a J-spectral factorization is that $G(\lambda)$ has no zeros on the imaginary axis (this condition is implicit in the containment $H \in dom(Ric)$). In this case, $V(\lambda)$ has no zeros on the imaginary axis as well. Thus, Theorem 2.30 can be viewed as a result on existence and construction of a J-spectral factorization of rational matrix functions $V(\lambda)$ which are hermitian on the imaginary line and have no zeros and poles on the imaginary line (including infinity).

2.12 Geometric Theory: The Real Case

In this section we briefly overview some of the material presented in Sections 2.4-2.7 for the particular case when the coefficients of the Riccati equation (2.9) are real. Thus, we consider the equation,

$$XDX + XA + A^T X - C = 0, \tag{2.49}$$

where A, C, and D are real $n \times n$ matrices with $D^T = D \geq 0$ and $C = C^T$. We will be interested mostly in real symmetric solutions of (2.49). In the real case, it is convenient to use, together with the associated matrices,

$$M = i\begin{bmatrix} A & D \\ C & -A^T \end{bmatrix}; \quad H = \begin{bmatrix} -C & A^T \\ A & D \end{bmatrix};$$

also the matrices

$$M_r = \begin{bmatrix} A & D \\ C & -A^T \end{bmatrix} \quad \text{and } \hat{H}_r = \begin{bmatrix} 0 & I \\ -I & 0 \end{bmatrix}.$$

We know already that M is H-self-adjoint, and, in particular, the spectrum of M is symmetric relative to the real axis. But, since A, D, C are real, the spectrum of M is also symmetric with respect to the imaginary axis. So, if $\lambda_0 \in \sigma(M)$, then $\bar{\lambda}_0, -\lambda_0, and -\bar{\lambda}_0$ are also in $\sigma(M)$, and the partial multiplicities of $\lambda I - M$ corresponding to $\lambda_0, \bar{\lambda}_0, -\lambda_0, and -\bar{\lambda}_0$ are the same. Obviously, this observation applies to M_r as well.

A necessary condition for existence of the real symmetric solutions of (2.49) is that (complex) hermitian solutions exist. This condition turns out to be sufficient also (at least under the controllability assumption). More precisely, we have the following description of all real symmetric solutions analogous to Theorem 2.9:

Theorem 2.31. ([3]) *Suppose the coefficients of* (2.49) *are real, with* $D \geq 0$, $C = C^T$ *and* (A, D) *controllable, and assume that equation* (2.49) *has a hermitian* (*not* *necessarily real) solution. Let* $\tilde{\mathcal{N}}_+$ *be the spectral subspace of* M *corresponding to the eigenvalues in the quadrant* $\{\lambda \in \mathbf{C} | Im\ \lambda > 0,\ \mathrm{Re}\ \lambda \geq 0\}$. *Then, for every* M*-invariant subspace* $\tilde{\mathcal{L}} \subset \tilde{\mathcal{N}}_+$, *there exists a unique real symmetric solution* X *of* (2.49) *such that*

$$G(X) \cap \tilde{\mathcal{N}}_+ = \tilde{\mathcal{L}}.$$

Conversely, if X *is a real symmetric solution of* (2.49), *then* $G(X) \cap \tilde{\mathcal{N}}_+$ *is* M*-invariant.*

The correspondence between the invariant subspaces $\tilde{\mathcal{L}}$ and real symmetric solutions X of (2.49) is order preserving, in the same sense as Theorem 2.11.

It is not difficult to see that the maximal hermitian solution X_+ and the minimal hermitian solution X_- (if they exist) of (2.49) are necessarily real. Indeed, by taking complex conjugates, we see that $\bar{X}_+$ is again a hermitian solution of (2.49), and by the maximality property we have $\bar{X}_+ \leq X_+$. This implies $\bar{X}_+ = X_+$, so X_+ is real. The same argument applies to X_- as well.

Consider now real special solutions X of (2.49), *i.e.*, such that

$$(X^T - X)(A + DX) \leq 0$$

(see Section 2.4). It is easy to see that X is a real special solution if and only if the graph $G(X)$ (considered as a subspace in $\mathbf{R}^{2n}$) is M_r-invariant and H-non-negative. As in the complex case, real special solutions always exist (under the controllability assumption):

Theorem 2.32. *Let* A, C, D *be real with* $C = C^T$, $D \geq 0$ *and* (A, D) *controllable. Assume further that the matrix* H *is invertible and has signature zero. Then there exists a real special solution of* (2.49).

Proof. In view of Theorem 2.7, we have to demonstrate existence of an n-dimensional M_r-invariant H-non-negative subspace in $\mathbf{R}^{2n}$. First, observe that M_r is invertible. Indeed,

$$M_r = \begin{bmatrix} 0 & I \\ -I & 0 \end{bmatrix} H.$$

Now use the canonical form of the pair of matrices (M_r, H) under the transformations

$$(M_r, H) \rightarrow (S^{-1} M_r S, S^T H S) \tag{2.50}$$

for invertible real matrices S. This canonical form can be found, for instance, in [37], and in another version in [50] (see also [51]). To describe this canonical form, introduce the notation

$$J_n(a) = \begin{bmatrix} a & 0 & \cdots & 0 & 0 \\ 1 & a & \cdots & 0 & 0 \\ \vdots & \ddots & & \vdots & \vdots \\ \vdots & & \ddots & a & 0 \\ 0 & \cdots & \cdots & 1 & a \end{bmatrix}, \quad a \in \mathbf{R},$$

(lower triangular Jordan block of size $n \times n$);

$$J_n \begin{bmatrix} a & b \\ -b & a \end{bmatrix} = \begin{bmatrix} \begin{matrix} a & b & 0 & 0 \\ -b & a & 0 & 0 \\ 1 & 0 & a & b \\ 0 & 1 & -b & a \end{matrix} & & 0 \\ & \ddots & \\ 0 & & \begin{matrix} a & b & 0 & 0 \\ -b & a & 0 & 0 \\ 1 & 0 & a & b \\ 0 & 1 & -b & a \end{matrix} \end{bmatrix}; \quad a, b \in \mathbf{R}, b \neq 0$$

(almost lower triangular real Jordan block of size $2n \times 2n$);

$$G_j = \begin{bmatrix} 0 & \cdots & 0 & F^{j-1} \\ \vdots & & -F^{j-1} & 0 \\ \vdots & & \vdots & \vdots \\ (-1)^{j-1}F^{j-1} & \cdots & 0 & 0 \end{bmatrix}, \quad \text{where } F = \begin{bmatrix} 0 & 1 \\ -1 & 0 \end{bmatrix}$$

(G_j is a $2j \times 2j$ real symmetric matrix). Then, using the transformations (2.50), one can reduce the pair (M_r, H) to a direct sum of pairs of matrices of the following types:

I. $\left(J_n(a) \oplus -J_n(a)^T, \begin{bmatrix} 0 & I_n \\ I_n & 0 \end{bmatrix} \right)$, $a > 0$;

II. $\left(J_n \begin{bmatrix} 0 & b \\ -b & 0 \end{bmatrix}, \pm G_n \right)$, $b > 0$;

III. $\left(J_n \begin{bmatrix} a & b \\ -b & \mathrm{a} \end{bmatrix} \oplus - \left(J_n \begin{bmatrix} a & b \\ b & \mathrm{a} \end{bmatrix} \right)^T, \begin{bmatrix} 0 & I_{2n} \\ I_{2n} & 0 \end{bmatrix} \right)$, $a, b > 0$.

By inspection of these pairs of matrices, the existence of an n-dimensional, M_r-invariant, H-non-negative subspace is easily deduced. □

2.13 Perturbations of Solutions of Real Riccati Equations

In this section, we present a sample of recent results (developed in [52]) concerning stability of real symmetric solutions of the Riccati equation

$$XDX + XA + A^T X - C = 0 \tag{2.51}$$

with real $n \times n$ matrices A, D, and C. The definitions of conditional and unconditional stability given in Section 2.8 apply verbatim to the real case (of course, with the additional requirement that the coefficients of the perturbed equation are real as well). We consider the associated matrix

$$M_r = \begin{bmatrix} A & D \\ C & -A^T \end{bmatrix},$$

and denote by $\mathcal{R}(M_r; \lambda)$ the (real) spectral subspace of M_r corresponding to its real eigenvalue λ, and by $\mathcal{R}(M_r; a \pm ib)$ the (real) spectral subspace of M_r corresponding to the pair $a \pm ib$ of its complex conjugate non-real eigenvalues.

Theorem 2.33. ([52]). *Assume that $C = C^T$, $D = D^T \geq 0$ and (A, D) is controllable. A real symmetric solution X of* (2.51) *is conditionally stable if and only if the following conditions are satisfied:*

(i) *The intersection of the (real) graph subspace $G(X)$ and of $\mathcal{R}(M_r; \lambda)$ is either zero or $\mathcal{R}(M_r; \lambda)$ whenever $\lambda \neq 0$ is a real eigenvalue of M_r with geometric multiplicity bigger than* 1;
(ii) *$G(X) \cap \mathcal{R}(M_r, \lambda)$ is an even dimensional subspace whenever $\lambda \neq 0$ is a real eigenvalue of M_r of geometric multiplicity one and even algebraic multiplicity;*
(iii) *$G(X) \cap \mathcal{R}(M_r, a \pm ib)$ is either zero or $\mathcal{R}(M_r, a \pm ib)$ whenever $a \pm ib$ is a conjugate pair of non-real, non-pure imaginary eigenvalues of M_r with geometric multiplicity at least* 2.

Theorem 2.34. ([52]) *In the notation and under the hypotheses of Theorem* 2.33, *there exists an unconditionally stable, real symmetric solution X of* (2.51), *if and only if M_r has no pure imaginary eigenvalues. If this condition is satisfied, then X is unconditionally stable if and only if it is conditionally stable.*

The proofs of these results are based on the correspondence between invariant subspaces of M_r and real symmetric solutions of (2.51), and on the results on (conditional and unconditional) of invariant Lagrangian subspaces developed in [53]. Other types of stability of real symmetric solutions of (2.51) are also studied in [52].

2.14 Real Contractive Rational Matrix Functions

We present here analogues of the results of Section 2.10. The definitions introduced there will be used in this section without further explanation.

Consider $p \times m$ proper contraction $W(\lambda)$ which is real, i.e., $W(\lambda)$ is a real matrix for every real λ. Any real rational matrix function $W(\lambda)$ which is analytic at infinity admits a minimal realization,

$$W(\lambda) = D + C(\lambda I - A)^{-1}B$$

with real matrices A, B, C, and D. In the sequel, only real minimal realizations of $W(\lambda)$ wil be used. According to Theorem 2.27, the associated Riccati equation,

$$X\gamma X - iX^*\alpha + i\alpha X + \beta = 0, \tag{2.52}$$

has hermitian solutions.

We study now minimal unitary completions of $W(\lambda)$ which are also real.

Theorem 2.35. *Let H be a hermitian solution of* (2.52). *Then the associated minimal unitary completion $\tilde{W}_H(\lambda)$ of $W(\lambda)$ (given by formula* (2.44)) *is real if and only if H is pure imaginary: $H = iH_0$ for some real skew-symmetric matrix H_0.*

Proof. The formula (2.44) shows that if H is pure imaginary, then $\tilde{W}_H(\lambda)$ is real. Conversely, let $\tilde{W}_H(\lambda)$ be real. Formula (2.44) shows that

$$C(\lambda I - A)^{-1} iHC^* \tag{2.53}$$

is real for real λ. This implies easily that $CA^j iHC^*$ is real for $j = 0, 1, ...$, and from the observability of (A, C), it follows that iHC^* is real. The equation

$$H\gamma H - iH^*\alpha + i\alpha H + \beta = 0$$

can be rewritten in the form

$$HC^*(I - DD^*)^{-1}CH - iH(A^* + C^*(I - DD^*)^{-1}DB^*)$$

$$+ i(A + BD^*(I - DD^*)^{-1}C)H + B(I - DD^*)^{-1}B^* = 0,$$

and we see now that $-iHA^* + iAH$ is real. Write $H = Z_1 + iZ_2$, where Z_1 and Z_2 are real matrices; then we have $CZ_1 = 0$ *and* $-Z_1A^* + AZ_1 = 0$. Premultiply the latter equation by C to obtain $CAZ_1 = 0$, and by induction, $CA^jZ_1 = 0$ for $j = 0, 1, ...$. Now by the observability of (A, C), $Z_1 = 0$, and H is pure imaginary. □

In view of Theorem 2.35, it is natural to ask if there exist any pure imaginary hermitian solutions of (2.52). The following example shows that the answer is generally no, and consequently, not every real proper contraction admits minimal real unitary completion.

Example 2.2. Let

$$\alpha = \begin{bmatrix} 0 & 1 \\ -1 & 0 \end{bmatrix}; \quad \gamma = \begin{bmatrix} a & 0 \\ 0 & 0 \end{bmatrix}; \quad \beta = \begin{bmatrix} 0 & 0 \\ 0 & a \end{bmatrix},$$

where $0 < a < 1$. The eigenvalues of the state characteristic matrix $Q = \begin{bmatrix} \alpha & \beta \\ \gamma & \alpha^T \end{bmatrix}$, or, what is the same, the eigenvalues of the matrix $M = \begin{bmatrix} \alpha^T & i\gamma \\ -i\beta & \alpha \end{bmatrix}$, turn out to be $\pm\sqrt{1-a}$; $\pm\sqrt{1+a}$. According to Proposition 2.1 and Theorem 2.2 of [13], there exists a real proper contraction $W(\lambda)$ with the state characteristic matrix Q. It is not difficult (but tedious) to check that the corresponding Riccati equation

$$X\gamma X - iX^T\alpha + i\alpha X + \beta = 0$$

has no pure imaginary hermitian solutions. By Theorem 2.35, there is no real minimal unitary completion of $W(\lambda)$. □

2.15 The Discrete Riccati Equation: The Frequency Domain Approach

The now classic linear-quadratic optimal control problem in differential form leads to the symmetric Riccati equation (2.9) for the steady-state solution. It is well known that the analogous problem for difference equations leads to the so-called discrete Riccati equation:

$$X = A^*XA + Q - (C + B^*XA)^*(R + B^*XB)^{-1}(C + B^*XA). \qquad (2.54)$$

Here R and Q are hermitian matrices of sizes $n \times n$ and $m \times m$, respectively. The coefficients A, B, C are $n \times n$, $n \times m$, and $m \times n$, respectively, and $n \times n$ solution matrices X are to be found for which, of course, $R + B^*XB$ is invertible. The probabilistic design of discrete Kalman filters gives rise to the same equation (see [54], for example).

This equation is apparently more complicated than the quadratic equation (2.9) and, not surprisingly, this is reflected in its analysis. As a result, the geometric theory in terms of invariant subspaces of an associated $2n \times 2n$ matrix has been relatively slow to evolve, although analogues of Potter's ideas for continuous systems (referred to in Section 2.3) first appeared about 1970 (see [55], [56], and [57] for some early references). The two other lines of attack, namely, using factorization of certain rational matrix functions (as in Section 2.9), and using recursive methods (as in Section 2.7) follow more readily. Following the historical development, the rational function approach is introduced first. In the next section, we introduce the geometric theory, and existence and comparison theorems obtained by the recursive method are presented in Section 2.17.

The nature of the main results can be seen if attention is confined to a relatively simple equation. Thus, for the purposes of Section 2.15 and 2.16, it is assumed that $C = 0$ in (2.54) and A is assumed to be stable in the sense that $\lambda \in \sigma(A)$ implies $|\lambda| < 1$, and, in addition, A is invertible. Thus, $0 < |\lambda| < 1$ for all $\lambda \in \sigma(A)$. There is little loss of generality here, as we also assume that (A, B) is controllable, in which case linear feedback (in the sense of systems theory) can be used to obtain the properties assumed for A with a new matrix $A - BK$, if A does not already enjoy these properties. Thus, we consider

$$X = A^*XA + Q - (B^*XA)^*(R + B^*XB)^{-1}(B^*XA). \tag{2.55}$$

As in the case of continous systems (see Section 2.9), and associated rational matrix function arises on applying simple transform theory to the underlying discrete linear system, namely (see [58]),

$$\Psi(z) = B^*(z^{-1}I - A^*)^{-1}Q(zI - A)^{-1}B + R. \tag{2.56}$$

Note that $\Psi(z)$ is an analytic hermitian matrix function when $|z| = 1$.

Let us define an admissible solution of the Riccati equation to be a hermitian matrix X for which $R + B^*XB$ is invertible and (2.55) (or (2.54)) is satisfied.

Theorem 2.36. *Assume that (A, B) is controllable and $\sigma(A) \subset \mathcal{D} := \{\lambda : 0 < |\lambda| < 1\}$. Then there is a one-to-one correspondence between the admissible solutions of* (2.55) *and the set of all rational matrix functions $\Psi(z)$ with all their poles in $\mathcal{D}$ and $\Psi(\infty) = I$, and with the property that*

$$\Psi(z) = (\Phi(z^{-1}))^* D \Phi(z) \tag{2.57}$$

is a minimal factorization of $\Psi(z)$ for some matrix D.

Furthermore, the correspondence is defined on the admissible solutions X of (2.55) *by*

$$\Phi(z) = I + (R + B^*XB)^{-1}(B^*XA)(zI - A)^{-1}B, \tag{2.58}$$

and D is determined by $D = R + B^*XB$.

This theorem was proved for the maximal solution of (2.54) (see Section 2.17) by Molinari [59], and was extended to the class of admissible solutions in [60]. We remark that, given an admissible solution X, the fact that $\Psi(z)$ has a factorization of the form (2.57) (with $D = R + B^*XB$) is just a computation.

An interesting deduction from the theorems is the fact that the signature (i.e., the difference between the number of positive eigenvalues and the number of negative eigenvalues, counted with multiplicities) of $R + B^*XB$ is invariant for all admissible solutions X. As solutions for which $R + B^*XB > 0$ are of particular interest, the following result gives more information on this case (see [60]).

Theorem 2.37. *Under the hypotheses of Theorem* 2.36, *the following statements are equivalent:*

(a) *There exists a hermitian solution X such that $R + B^*XB > 0$;*

(b) *for every admissible solution* X, $R + B^*XB > 0$;
(c) $\Psi(z) \geq 0$ *for all* z *on the unit circle.*

If the conditions (a), (b), (c) (or one of them) of Theorem 2.37 are valid, then there is a maximal admissible solution X_+ of (2.55), *i.e.*, such that $X_+ \geq X$ for any other admissible solution X. The proof of this statement can be completed by using an iterative technique (see [61, 30]).

2.16 Geometric Theory for the Discrete Riccati Equation

We are now to outline an approach to the solution of (2.55) that is analogous to the theory of Sections 2.3 and 2.4 for equation (2.9). As analogues of the matrix pair in (2.10) (see also (2.11)), we choose T and iS where

$$T = \begin{bmatrix} A & 0 \\ -A^{*-1}Q & A^{*-1} \end{bmatrix}, \quad S = \begin{bmatrix} 0 & -I \\ I & 0 \end{bmatrix}. \tag{2.59}$$

Now iS is nonsingular and hermitian and $T^*(iS)T = iS$. This is interpreted as saying that T is iS-unitary. Recall the symmetry properties of the spectrum of an H-self-adjoint matrix described in Proposition 2.3. The spectrum of T has similar properties but now the symmetry is with respect to the unit circle, and not the real line (unless T is also real). In other words, if $\lambda_0 \in \sigma(T)$ (necessarily $\lambda_0 \neq 0$), then $\bar{\lambda}_0^{-1} \in \sigma(T)$, and the partial multiplicities of $\bar{\lambda}_0^{-1}$ as an eigenvalue of T coincide with those of λ_0.

A little computation serves to show that T is the main matrix in a realization for $\Psi(z)$ of (2.56). In fact,

$$\Psi(z) = R + [-B^*A^{-1}Q,\ B^*A^{*-1}](zI - T)^{-1}\begin{bmatrix} B \\ 0 \end{bmatrix}.$$

Then it can be shown that $\Psi(z)^{-1}$ has a realization with main matrix $T^\times$ where

$$\begin{aligned} T^\times &= T - \begin{bmatrix} B \\ 0 \end{bmatrix} R^{-1}[-B^*A^{*-1}Q, B^*A^{*-1}] \\ &= \begin{bmatrix} A + BR^{-1}B^*A^{*-1}Q & -BR^{-1}B^*A^{*-1} \\ -A^{*-1}Q & A^{*-1} \end{bmatrix}, \end{aligned}$$

and is also iS-unitary.

The geometric theory now depends on the fact that, for any admissible solution X of (2.55), the graph subspace $Im \begin{bmatrix} I \\ X \end{bmatrix}$ (see Section 2.3) is $T^\times$-invariant and iS-neutral (cf. Proposition 2.1). The neutral property is obvious, and the $T^\times$-invariance is an easy calculation. These are the natural generalizations of the eigenvalue-eigenvector techniques for the construction of admissible solutions. An early paper using this approach is by Vaughan [55], and numerical implementation is considered in [62].

The construction of $T^{\times}$-invariant subspaces can now be accomplished using analogues of Propositions 2.3 and 2.4. The difference is that we must take advantage of the symmetry of $\sigma(T^{\times})$ with respect to the unit circle, rather than the real line. An unexpected feature of this development is the need to work with $T^{\times}$ rather than T.

Using some detailed information on factorization of rational matrix functions which are self-adjoint on the unit circle, the following theorem is proved in reference [60]. There is a marked contrast between this proof and that of the analogous Theorem 2.6. In the latter case, more direct arguments suffice and there is no appeal to the factorization of matrix functions.

Theorem 2.38. *Let (A, B) be controllable with $0 < |\lambda| < 1$ for every eigenvalue λ of* A. *Assume also that R is invertible and there is an $\eta \in \mathbf{C}$ such that $|\eta| = 1$ and $\Psi(\eta) > 0$. Then the following statements are equivalent.*

(a) *There is an admissible solution of* (2.55).
(b) *There exists an n-dimensional $T^{\times}$-invariant, iS-neutral subspace.*
(c) *The partial multiplicities of $T^{\times}$ at its eigenvalue on the unit circle* (*if any*) *are all even.*
(d) *$\Psi(z) \geq 0$ for all z on the unit circle.*

When (a)-(d) *hold, every n-dimensional, $T^{\times}$-invariant, iS-neutral subspace $\mathcal{M}^{\times}$ is a graph subspace of an admissible solution of* (2.55) *and, conversely, if X is an admissible solution of* (2.55), *then the graph subspace of X is n-dimensional, $T^{\times}$-invariant, and iS-neutral.*

Theorem 2.38 naturally raises questions about the structure of all n-dimensional $T^{\times}$-invariant iS-neutral subspaces $\mathcal{M}^{\times}$. Using the iS-unitarity of $T^{\times}$, one can show (under the hypotheses of Theorem 2.38) that the intersection of such an $\mathcal{M}^{\times}$ with the root subspace of $T^{\times}$ corresponding to any unimodular eigenvalue is unique. Furthermore, the two root subspaces of $T^{\times}$ corresponding to non-unimodular eigenvalues λ and $\bar{\lambda}^{-1}$ are iS-orthogonal to each other. Combining these two observations, it follows that (under the hypotheses of Theorem 2.38 and assuming existence of admissible solutions) there is one-to-one correspondence between the admissible solutions of (2.55) and $T^{\times}$-invariant subspaces $\mathcal{N}$ such that $|\lambda| < 1$ for every $\lambda \in \sigma(T^{\times}|_{\mathcal{N}})$.

2.17 Existence Theorems for the Recursive Method

The theory evolving from the application of the Newton-Raphson method to (2.54) is a close analogue of that mentioned in Section 2.7 for the solution of (2.9). Indeed, the two developments are nicely juxta-posed in reference [30]. The presentation here is based on [63] and [30], although, again, these results have gone through several stages of refinement (see [64], [57], [6], for example).

Define the Riccati matrix function,

$$\mathcal{R}(X) := -X + A^*XA + Q - (C + B^*XA)^*(R + B^*XB)^{-1}(C + B^*XA), \tag{2.60}$$

and consider the Riccati inequality $\mathcal{R}(X) \geq 0$. As for equation (2.54), admissible solutions of the inequality are required to be hermitian matrices X for which $R + B^*XB$ is invertible.

Let $\hat{M}$, M be the sets of all admissible solutions of $\mathcal{R}(X) = 0$ and $\mathcal{R}(X) \geq 0$, respectively, so that $\hat{M} \subset M$. In contrast to Sections 2.15 and 2.16, the controllability of (A, B) is relaxed to stabilizability, but the stronger condition $R > 0$ is required.

Theorem 2.39. *Assume that $R > 0$, (A, B) is stabilizable, and M is not empty. Then there is an $X_+ \in \hat{M}$ such that $X_+ \geq X$ for all $X \in M$. Also, the spectrum of*

$$A - B(R + B^*X_+B)^{-1}(C + B^*X_+A)$$

is in the closed unit disc.

An interesting comparison theorem is obtained if we consider another Riccati function $\tilde{\mathcal{R}}(X)$ defined as in equation (2.60), but with coefficients $\tilde{A}, \tilde{B}, \tilde{Q}, \tilde{R}, \tilde{C}$. With hypotheses on $\tilde{\mathcal{R}}$ like those required for $\mathcal{R}$ in Theorem 2.39, we can assert the existence of a matrix $\tilde{X}_+$ in the class $\hat{N}$ of admissible solutions of $\tilde{\mathcal{R}}(X) = 0$ which is maximal in the class of N of admissible solutions of $\tilde{\mathcal{R}}(X) \geq 0$. Let

$$T = \begin{bmatrix} Q & C^* \\ C & R \end{bmatrix}, \quad \tilde{T} = \begin{bmatrix} \tilde{Q} & \tilde{C}^* \\ \tilde{C} & \tilde{R} \end{bmatrix}.$$

(These are the matrices of quadratic functionals to be minimized in the linear-quadratic optimal control problems.)

Theorem 2.40. [30] *Assume that $R > 0$, $\tilde{R} > 0$, (A, B) and $(\tilde{A}, \tilde{B})$ are stabilizable. Then*

(1) *If $N \neq \emptyset$ and $N \subset M$, then X_+ and $\tilde{X}_+$ exist and $X_+ \geq \tilde{X}_+$.*
(2) *If $\tilde{A} = A$, $\tilde{B} = B$, $T \geq \tilde{T}$, and $N \neq \emptyset$, then X_+ and $\tilde{X}_+$ exist and $X_+ \geq \tilde{X}_+$.*
(3) *If $T \geq 0$, then X_+ exists and $X_+ \geq 0$.*

2.18 The Real Discrete Riccati Equation

In this section, we discuss briefly the discrete Riccati equation (2.54) with real coefficients A, B, C, Q, and R (and with Q and R symmetric). Again, it will be assumed that $C = 0$ and $0 < |\lambda| < 1$ for all $\lambda \in \sigma(A)$.

It is easy to see that now the function $\Psi(z)$ given by (2.56) enjoys an additional symmetry, namely, that $\Psi(\overline{z}) = \overline{\Psi(z)}$. Equivalently, $\Psi(z)$ is a real matrix for every real z not a pole of $\Psi(z)$. Rational matrix functions with this property will be called *real*. In this context, it is natural to consider factorizations (2.57) with real $\Phi(z)$. The next theorem characterizes the admissible solutions corresponding to the real $\Phi(z)$.

Theorem 2.41. *Assume that the discrete Riccati equation* (2.55) *has real coefficients, with controllable* (A, B) *and with* $\sigma(A) \subset \mathcal{D} := \{\lambda : 0 < |\lambda| < 1\}$. *Then there is a one-to-one correspondence between the admissible solutions* X *of* (2.55) *such that the matrix* XB *is real and the set of all real rational matrix functions* $\Phi(z)$ *with all their poles in* $\mathcal{D}$ *and* $\Phi(\infty) = I$ *for which* (2.57) *is a minimal factorization for some (necessarily real) matrix* D. *This correspondence is defined on the real admissible solutions* X *of* (2.55) *by the formula* (2.58).

Proof. Use Theorem 2.36. If XB is real, then so is B^*X, and by formula (2.58), $\Phi(z)$ is real as well. Now suppose $\Phi(z)$ is real. Then $D = R + B^*XB$ is also real (because of factorization (2.57)), and consequently the function $B^*XA(zI-A)^{-1}B$ is real. Using controllability of (A, B), it is not difficult to see that B^*X is a real matrix. □

The case when $R + B^*XB > 0$ for some (and hence, for every) admissible solution X is of special interest. In this case, if the coefficients of (2.55) are real, then the maximal admissible solution X_+ is real as well (an inspection of the iterative proof given in [30] reveals this fact).

Finally we pass to the geometric description (as in Section 2.16) of real admissible solutions.

Theorem 2.42. *Let the coefficients of* (2.55) *be real, and assume the hypotheses of Theorem* 2.38. *If there is an admissible solution of* (2.55), *then there is a real admissible solution. Furthermore, there is one-to-one correspondence between the set of real admissible solutions of* (2.55) *and the set of real* $T^\times$*-invariant subspaces* $\mathcal{N}$ *such that* $|\lambda| < 1$ *for every* $\lambda \in \sigma(T^\times|_\mathcal{N})$. *Here*

$$T^\times = \begin{bmatrix} A + BR^{-1}B^*A^{*-1}Q & -BR^{-1}B^*A^{*-1} \\ -A^{*-1}Q & A^{*-1} \end{bmatrix}.$$

The proof can be obtained by using Theorem 2.38 and the subsequent remarks.

Acknowledgement. The first author was supported in part by the Natural Sciences and Engineering Research Council of Canada; the second one by NSF Grand DMS-8802836.

References

1. SINGER, M.A. and HAMMARLING, S.J., *The algebraic Riccati equation: a summary review of some available results.* Nat. Phys. Lab. Report DITC 23/83, 1983.
2. SHAYMAN, M., *Geometry of the algebraic Riccati equation, Parts I and II*, SIAM J. Control Optim. **21**(1983), 375-394 and 395-409.
3. GOHBERG, I., LANCASTER, P., and RODMAN, L., Matrices and Indefinite Scalar Products, Birkhäuser Verlag, Basel, 1983.
4. RAN, A.C.M. and RODMAN, L., *The algebraic matrix Riccati equation*, Operator Theory: Advances & Applications, Vol. 12 (pp. 351-381). Birkhauser, Basel, 1984.
5. ANDO, T., Matrix Quadratic Equations, Hokkaido University, Sapporo, Japan, 1988.
6. DORATO, P., *Theoretical developments in discrete-time control*, Automatica **19**(1983), 395-400.
7. WILLEMS, J.C., *Least squares stationary optimal control and the algebraic Riccati equation.* IEEE Trans. Automat. Control AC-**16**(1971), 621-634.
8. COPPEL, W.A., *Matrix quadratic equations.* Bull. Austral. Math. Soc. **10**(1974), 377-401.
9. POTTER, J.E., *Matrix quadratic solutions.* SIAM J. Appl. Math. **14**(1966), 496-501.
10. KUČERA, V., *A contribution to matrix quadratic equations*, IEEE Trans. Automat. Control **17**(1972), 344-347.
11. LANCASTER, P. and RODMAN, L., *Existence and uniqueness theorems for algebraic Riccati equations*, Internat. J. Control **32**(1980), 285-309.
12. CURILOV, A.N., *The frequency theorem and the $Lur'e$ equation*, Sibirsk. Mat. Z. **20**(1979), 600-611 (in Russian).
13. GOHBERG, I. and RUBINSTEIN, S., *Proper contractions and their unitary minimal completions*, Operator Theory: Advances and Applications **33**(1988), 233-247.
14. ANDERSON, B.D.O. and VONGPANITLERD, S., Network Analysis and Synthesis, Prentice-Hall, Englewood Cliffs, NJ, 1973.
15. BENDER, D.J. and LAUB, A.J., *The linear-quadratic optimal regulator problem for descriptor systems*, IEEE Trans. Automat. Control, AC-**32**(1987), 672-688.
16. BENDER, D.J. and LAUB, A.J., *The linear-quadratic optimal regulator problem for descriptor systems: Discrete-time case*, Automatica **23**(1987), 71-85.
17. MEHRMANN, V., The Linear-Quadratic Control Problem, Habilitationsschrift, University of Bielefeld, 1988.
18. MEHRMANN, V., *A symplectic orthogonal method for single input or single output discrete time optimal quadratic control problems*, SIAM J. Matrix Anal. Appl. **9**(1988), 221-247.
19. LANCASTER, P. and TISMENETSKY, M., The Theory of Matrices (second edition) with Applications, Academic Press, Orlando, 1985.
20. GOHBERG, I., LANCASTER, P., and RODMAN, L., Invariant Subspaces of Matrices with Applications, J. Wiley, 1986.
21. HAUTUS, M.L.J., *Controllability and observability conditions of linear autonomous systems*, Ned. Akad. Wet. Proc., Ser. A **12**(1969), 443-448.
22. WONHAM, W.M., Linear Multivariable Control, Springer-Verlag, Berlin, 1970 and 1979.
23. RAN, A.C.M. and RODMAN, L., *Stability of invariant maximal semidefinite subspaces II, Applications: selfadjoint rational matrix functions, algebraic Riccati equation*, Linear Algebra and Appl. **63**(1984), 133-173.
24. PAIGE, C. and VAN LOAN, C., *A Schur decomposition for Hamiltonian matrices*, Linear Algebra and Appl. **41**(1981), 11-32.
25. FAIBUSOVICH, L.E., *Algebraic Riccati equation and symplectic algebra*, Internat. J. Control **43**(1986), 781-792.
26. RAN, A.C.M. and RODMAN, L., *Stability of invariant maximal semidefinite subspaces I*, Linear Algebra and Appl. **62**(1984), 51-86.
27. GOHBERG, I., LANCASTER, P., and RODMAN, L., *On hermitian solutions of the symmetric algebraic Riccati equations*, SIAM J. Control Optim. **24**(1986), 1323-1334.
28. KLEINMAN, D.L., *On an iterative technique for Riccati equation computation*, IEEE Trans. Automat. Control **13**(1968), 114-115.

29. WIMMER, H.K., *Monotonicity of maximal solutions of algebraic Riccati equations*, Systems and Control Letters **5**(1985), 317-319.
30. RAN, A.C.M. and VREUGDENHIL, R., *Existence and comparison theorems for algebraic Riccati equations for continuous- and discrete-time systems*, Linear Algebra and Appl. **99**(1988), 63-83.
31. DELCHAMPS, D.F., *A note on the analyticity of the Riccati metric*, in: Lectures in Applied Mathematics, Vol. 18 (eds. C.I. Byrnes and C.F. Martin), pp. 37-41, AMS, Providence, RI, 1980.
32. DELCHAMPS, D.F., *Analytic stabilization and the algebraic Riccati equation*, Proc. of the 22nd IEEE Conference on Decision and Control 1983, pp. 1396-1401.
33. DELCHAMPS, D.F., *Analytic feedback control and the algebraic Riccati equation*, IEEE Trans. Automat. Control **29**(1984), 1032-1033.
34. KAMEN, E.W. and KHARGONEKAR, P.P., *On the control of linear systems whose coefficients are functions of parameters*, IEEE Trans. Automat. Control **29**(1984), 25-33.
35. RAN, A.C.M. and RODMAN, L., *On parameter dependence of solutions of algebraic Riccati equations*, Math. of Control, Signals and Systems **1**(1988), 269-284.
36. RODMAN, L., *On extremal solutions of the algebraic Riccati equation*, in: Lectures in Applied Mathematics, Vol. 18 (eds. C.I. Byrnes and C.F. Martin), pp. 311-327, AMS, Providence, RI, 1980.
37. RAN, A.C.M. and RODMAN, L., *Stability of invariant lagrangian subspaces II*, Operator Theory: Advances and Applications **40**(1989) (eds. H. Dym, S. Goldberg, M.A. Kaashoek, P. Lancaster), pp. 391-425.
38. POLDERMAN, J.W., *A note on the structure of two subsets of the parameter space in adaptive control problems*, Systems and Control Letters **7**(1986), 25-34.
39. BAUMGARTEL, H., *Analytic perturbation theory for matrices and operators* (Operator Theory: Advances and Applications, Vol. 15), Birkhäuser, Basel, 1985.
40. BART, H., GOHBERG, I., and KAASHOEK, M.A., *Minimal factorizations of matrix and operator functions*, Birkhäuser Verlag, Basel, 1979.
41. BELEVITCH, V., Classical Network Theory, San Francisco-Cambridge-Amsterdam, Holden Day, 1968.
42. KAILATH, T., Linear Systems, Prentice Hall, Englewood Cliffs, NJ, 1980.
43. RAN, A.C.M., *Minimal factorization of self-adjoint rational matrix functions*, Integral Equations and Operator Theory **5**(1982), 850-869.
44. REDHEFFER, R.M., *On a certain linear fractional transformation*, J. Mathematical Phys. **39**(1960), 260-286.
45. DEWILDE, P., *Input-output descriptions of roomy systems*, SIAM J. Control and Optim. **14**(4)(1976), 712-736.
46. BALL, J.A. and COHEN, N., *Sensitivity minimization in an H^∞ norm: Parametrization of all suboptimal solutions*, Intern. J. of Control **46**(1987), 785-816.
47. LIMEBEER, D.J.N. and HALIKIAS, G.D., *A controller degree bound for H^∞-optimal control problems of the second kind*, SIAM J. Control Optim. **26**(1988), 646-677.
48. GLOVER, K., GREEN, M., LIMEBEER, D., and DOYLE, J., *A J-spectral factorization approach of H_∞ control*, preprint.
49. DOYLE, J.C., GLOVER, K., KHARGONEKAR, P.P. and FRANCIS, F., *State-space solutions to standard H_2 and H_∞ control problems*, IEEE Trans. Automat. Control **34**(1989), 831-347.
50. DJOKOVIC, D.Z., PATERA, J., WINTERNITZ, P., and ZASSENHAUS, H., *Normal forms of elements of classical real and complex Lie and Jordan algebras*, J. Math. Phys. **24**(1983), 1363-1374.
51. THOMPSON, R.C., *Pencils of complex and real symmetric and skew matrices*, unpublished notes.
52. RAN, A.C.M. and RODMAN, L., *Stable invariant lagrangian subspaces: Factorization of symmetric rational matrix functions and other applications*, to appear in Linear Algebra and Appl.
53. RAN, A.C.M. and RODMAN, L., *Stability of invariant lagrangian subspaces I*, Operator Theory: Advances and Applications **32**(1988), 181-218.

54. ANDERSON, B.D.O. and MOORE, J.B., Optimal Filtering, Prentice-Hall, Englewood Cliffs, NJ, 1979.
55. VAUGHAN, D.R., *A nonrecursive algebraic solution for the discrete Riccati equation*, IEEE Trans. Automat. Control **15**(1970), 597-599.
56. KUČERA, V., *The discrete Riccati eqution of optimal control*, Kybernetika **8**(1972), 430-447.
57. DORATO, P. and LEVIS, A.H., *Optimal linear regulators: the discrete-time case*. IEEE Trans. Automat. Control **16**(1971), 613-620.
58. MOLINARI, B.P., *The stabilizing solution of the algebraic Riccati equation*, SIAM J. Control **11**(1973), 262-271.
59. MOLINARI, B.P., *The stabilizing solution of the discrete algebraic Riccati equation*, IEEE Trans. Automat. Control **20**(1975), 396-399.
60. LANCASTER, P., RAN, A.C.M., and RODMAN, L., *Hermitian solutions of the discrete algebraic Riccati equation*, Internat. J. Control **44**(1986), 777-802.
61. HEWER, G.A., *An iterative technique for the computation of steady-state gains for the discrete optimal regulator*, IEEE Trans. Automat. Control **16**(1971), 382-383.
62. PAPPAS, T., LAUB, A.J., and SANDELL, N.R., *On the numerical solution of the discrete-time algebraic Riccati equations*, IEEE Trans. Automat. Control **25**(1980), 631-641.
63. LANCASTER, P., RAN, A.C.M., and RODMAN, L., *An existence and monotonicity theorem for the discrete algebraic Riccati equation*, Linear and Multilinear Algebra **20**(1987), 353-361.
64. WONHAM, W.M., *On a matrix Riccati equation of stochastic control*, SIAM J. Control **6**(1968), 681-697.

3 Algebraic Riccati Equation: Hermitian and Definite Solutions

Vladimír Kučera

3.1 Introduction

Let A, B, and C be constant $n \times n$ matrices with entries in $\boldsymbol{C}$, the field of complex numbers. Let B and C be *hermitian*, i.e., $B = B^*$ and $C = C^*$, where an asterisk is used to denote the conjugate transpose of a matrix. The quadratic equation

$$XA + A^*X - XBX + C = 0 \tag{3.1}$$

for the $n \times n$ complex matrix X is called the *algebraic Riccati equation.*

3.1.1 Motivation

The motivation for the study of this equation comes from the analysis of the linear hamiltonian matrix system of differential equations

$$\dot{U}(t) - AU(t) + BV(t) = 0$$

$$\dot{V}(t) + CU(t) + A^*V(t) = 0. \tag{3.2}$$

This system is intimately connected with the Riccati matrix differential equation

$$\dot{X}(t) + X(t)A + A^*X(t) - X(t)BX(t) + C = 0 \tag{3.3}$$

by the following proposition [1].

If U, V is a solution pair of (3.2) with U non-singular on a t-interval T, then $X = VU^{-1}$ is a solution of (3.3) on T. Converselly, if X is a solution of (3.3) on T and U is a fundamental solution of

$$\dot{U}(t) = [A - BX(t)]U(t)$$

then U, $V = XU$ is a solution pair of (3.2) on T.

The constant solutions of (3.3) are just the solutions of the algebraic Riccati equation (3.1). Largely as a result of the connection between (3.2) and (3.3), the matrix equation (3.1) has found applications in many fields including system theory, signal processing and optimal control, to be discussed later.

Bittanti, Laub, Willems (Eds.)
The Riccati Equation

3.1.2 Notations and Preliminaries

For an $n \times n$ complex matrix A, we use χ_A to designate the characteristic polynomial of A and $\mathcal{L}_A^l, \mathcal{L}_A^0, \mathcal{L}_A^r$ the sum of the generalized eigenspaces of A which correspond to its left half-plane, imaginary axis, right half-plane eigenvalues (having respectively negative, zero, positive real part). We say that A is *stable* if all of its eigenvalues have negative real part and *anti-stable* if $-A$ is stable. The *partial multiplicities* of an eigenvalue λ of A are just the sizes of the Jordan blocks corresponding to λ in the Jordan form of A, or, what is the same, the degrees of the elementary divisors of $sI - A$ of type $(s-\lambda)^x$. The *multiplicity* of λ is the sum of the partial multiplicities of λ. We say that A is *cyclic* if each eigenvalue of A has a single partial multiplicity.

A subspace $\mathcal{V}$ of $\boldsymbol{C}^n$ is said to be *invariant* under A, or A-invariant, if $A\mathcal{V} \subset \mathcal{V}$. For an A-invariant subspace $\mathcal{V}$, we let $A|\mathcal{V}$ denote the restriction of A to $\mathcal{V}$. Further we let $\operatorname{Sp} A$ denote the column span of A and $\operatorname{Ker} A$ the kernel of A. All the subspaces $\mathcal{L}_A^l, \mathcal{L}_A^0, \mathcal{L}_A^r$ and $\operatorname{Sp} A, \operatorname{Ker} A$ are invariant under A. The direct sum of two subspaces $\mathcal{V}_1$ and $\mathcal{V}_2$ is denoted by $\mathcal{V}_1 \dot{+} \mathcal{V}_2$.

An asterisk is used to denote the conjugate transpose of a matrix or vector and the complex conjugate of a number. For a complex polynomial

$$q = \alpha_0 + \alpha_1 s + \ldots + \alpha_n s^n$$

we let q^* denote the conjugate polynomial defined by

$$q^* = \alpha_0^* + \alpha_1^*(-s) + \ldots + \alpha_n^*(-s)^n.$$

The inequality $X_1 \geq X_2$ between two hermitian matrices X_1, X_2 means that $X_1 - X_2$ is *non-negative definite*, and $X_1 > X_2$ means that $X_1 - X_2$ is *positive definite*.

If A and B are two complex $n \times n$ matrices, we put

$$\mathcal{R}_{A,B} = \operatorname{Sp} B + \operatorname{Sp} AB + \ldots + \operatorname{Sp} A^{n-1} B$$

and

$$\mathcal{S}_{A,B} = \mathcal{R}_{A,B} + \mathcal{L}_A^l.$$

If A and C are two complex $n \times n$ matrices, we put

$$\mathcal{O}_{A,C} = \operatorname{Ker} C \cap \operatorname{Ker} CA \cap \ldots \cap \operatorname{Ker} CA^{n-1}$$

and

$$\mathcal{D}_{A,C} = \mathcal{O}_{A,C} \cap (\mathcal{L}_A^0 + \mathcal{L}_A^r).$$

All these subspaces of $\boldsymbol{C}^n$ are invariant under A. The pair (A, B) is said to be *reachable* if $\mathcal{R}_{A,B} = \boldsymbol{C}^n$ and *stabilizable* if $\mathcal{S}_{A,B} = \boldsymbol{C}^n$. The pair (A, C) is said to be *observable* if $\mathcal{O}_{A,C} = 0$ and *detectable* if $\mathcal{D}_{A,C} = 0$.

A complex number λ (an eigenvalue of A) is called B-reachable if

$$v^* A = \lambda v^*, \quad v^* B = 0$$

implies $v = 0$, and it is called C-observable if

$$Aw = \lambda w, \quad Cw = 0$$

implies $w = 0$. Following [2] we shall say that (A, B) is *sign-reachable* if either λ or $-\lambda^*$ is B-reachable for every eigenvalue λ of A.

For later reference, we recall two propositions. Consider three $n \times n$ complex matrices A, B and C. Then the *Sylvester* equation

$$XA + BX + C = 0$$

has a unique solution X if and only if χ_A and χ_{-B} are relatively prime [3]. Now take $B = A^*, C = C^*$ and consider the *Lyapunov* equation

$$XA + A^*X + C = 0.$$

If A is stable and $C \geq 0$, a unique solution X exists, and $X \geq 0$. On the other hand, if a solution $X \geq 0$ exists and $C > 0$, then A is stable [4].

3.1.3 Classification of Solutions

The algebraic Riccati equation (3.1) can have solutions which are hermitian or non-hermitan, sign-definite or indefinite, and the set of solutions can be either finite or infinite.

In view of the connection between (3.2) and (3.3), it is not surprising to find that the solution set $\boldsymbol{G}$ of (3.1) corresponds to a certain class of n-dimensional invariant subspaces of the associated $2n \times 2n$ matrix

$$H = \begin{bmatrix} A & -B \\ -C & -A^* \end{bmatrix}. \tag{3.4}$$

This matrix has the *hamiltonian* property

$$JH = -H^*J \tag{3.5}$$

where

$$J = \begin{bmatrix} 0 & I \\ -I & 0 \end{bmatrix}. \tag{3.6}$$

It follows that H is similar to $-H^*$, and therefore the spectrum of H is symmetrical with respect to the imaginary axis.

The development of these results is due to many people including MacFarlane, Potter, Mårtensson, Kučera, Van Swieten and Shayman [5-10].

Theorem 3.1.1 ([10]). There is a one-to-one correspondence between the solution set $\boldsymbol{G}$ of (3.1) and the set of n-dimensional H-invariant subspaces of $\boldsymbol{C}^{2n}$ which are complementary to the n-dimensional subspace $\text{Sp}\begin{bmatrix} 0 \\ I \end{bmatrix}$. This correspondence assigns the invariant subspace $\mathcal{S}_X = \mathrm{Sp}\begin{bmatrix} I \\ X \end{bmatrix}$ to the solution X. The matrix of $H|\mathcal{S}_X$, the restriction of H to $\mathcal{S}_X$, with respect to the basis given by the columns of $\begin{bmatrix} I \\ X \end{bmatrix}$ is $A - BX$.

Proof. If $\mathcal{S}$ is an n-dimensional subspace of $\boldsymbol{C}^{2n}$ which is complementary to $\mathrm{Sp}\begin{bmatrix} 0 \\ I \end{bmatrix}$, then there exists an $n \times n$ matrix X such that $\mathcal{S} = \mathrm{Sp}\begin{bmatrix} I \\ X \end{bmatrix}$. If $\mathcal{S}$ is also H-invariant, then there exists an $n \times n$ matrix L such that

$$\begin{bmatrix} A & -B \\ -C & -A^* \end{bmatrix}\begin{bmatrix} I \\ X \end{bmatrix} = \begin{bmatrix} I \\ X \end{bmatrix} L.$$

The top equation implies that $L = A - BX$. Substituting for L in the second equation then gives $-C - A^*X = X(A - BX)$, which shows that X is a solution of (3.1).

Conversely, if X satisfies (3.1),

$$\begin{bmatrix} A & -B \\ -C & -A^* \end{bmatrix}\begin{bmatrix} I \\ X \end{bmatrix} = \begin{bmatrix} I \\ X \end{bmatrix}(A - BX)$$

which shows that $\mathcal{S}_X$ is H-invariant and that $A - BX$ is the matrix of $H|\mathcal{S}_X$ with respect to the basis given by the columns of $\begin{bmatrix} I \\ X \end{bmatrix}$. □

Example 3.1.1 The scalar quadratic equation

$$\alpha x^2 + \beta x + \gamma = 0$$

with α, β, γ real and $\alpha \neq 0$ gives rise to the hamiltonian matrix

$$H = \begin{bmatrix} \frac{\beta}{2} & \alpha \\ -\gamma & -\frac{\beta}{2} \end{bmatrix}.$$

Let $\delta = \beta^2 - 4\alpha\gamma$. If $\delta \neq 0$ there are two 1-dimensional H-invariant subspaces of $\boldsymbol{C}^2$, namely

$$\mathcal{S}_1 = \mathrm{Sp}\begin{bmatrix} 1 \\ \frac{-\beta+\sqrt{\delta}}{2\alpha} \end{bmatrix}, \quad \mathcal{S}_2 = \mathrm{Sp}\begin{bmatrix} 1 \\ \frac{-\beta-\sqrt{\delta}}{2\alpha} \end{bmatrix},$$

which correspond to the solutions

$$x_1 = \frac{-\beta + \sqrt{\delta}}{2\alpha}, \quad x_2 = \frac{-\beta - \sqrt{\delta}}{2\alpha}.$$

If $\delta = 0$ then there exists only one such subspace,

$$\mathcal{S} = \mathrm{Sp}\begin{bmatrix} 1 \\ \frac{-\beta}{2\alpha} \end{bmatrix},$$

which corresponds to the unique solution

$$x = \frac{-\beta}{2\alpha}. \qquad \square$$

Having in mind the applications in optimal control and signal processing, we shall be concerned with the solutions of (3.1) that are hermitian and either non-negative or non-positive definite.

3.2 Hermitian Solutions

We shall study the set of hermitian solutions of (3.1), which we denote by $\boldsymbol{H}$, *under the condition that* $B \geq 0$.

We recall the matrix J defined in (3.6). A subspace $\mathcal{S}$ of $\boldsymbol{C}^{2n}$ such that $u^* J v = 0$ for all $u, v \in \mathcal{S}$ will be called J*- neutral.* This property can be used to characterize the subset $\boldsymbol{H}$ of $\boldsymbol{G}$.

Theorem 3.2.1 ([10]). *An element X of $\boldsymbol{G}$ belongs to $\boldsymbol{H}$ if and only if the associated H-invariant subspace $\mathcal{S}_X = \mathrm{Sp}\begin{bmatrix} I \\ X \end{bmatrix}$ of $\boldsymbol{C}^{2n}$ is J-neutral.*

Proof. We invoke Theorem 3.1.1 and note that

$$[I \quad X^*] \begin{bmatrix} 0 & I \\ -I & 0 \end{bmatrix} \begin{bmatrix} I \\ X \end{bmatrix} = 0$$

if and only if $X = X^*$. $\square$

3.2.1 Existence of Unique Solutions

Each element X of $\boldsymbol{H}$ gives rise to a factorization of the characteristic polynomial χ_H of H as

$$\chi_H = (-1)^n q q^* \tag{3.7}$$

where $q = \chi_{A-BX}$. This follows from

$$\begin{bmatrix} I & 0 \\ X & I \end{bmatrix}^{-1} \begin{bmatrix} A & -B \\ -C & -A^* \end{bmatrix} \begin{bmatrix} I & 0 \\ X & I \end{bmatrix} = \begin{bmatrix} A - BX & -B \\ 0 & -(A - BX)^* \end{bmatrix} \tag{3.8}$$

and was already noted by Roth [11].

We shall now study existence of the hermitian solutions X of (3.1) that are *unique* in the sense that no other hermitian solution of (3.1) gives rise to the same polynomial q. We shall proceed through a series of auxiliary results.

Lemma 3.2.1 ([12], [10]). *Let (A, B) be reachable. Then every n-dimensional H-invariant J-neutral subspace of $\mathbf{C}^{2n}$ is complementary to* Sp $\begin{bmatrix} 0 \\ I \end{bmatrix}$.

Proof. Let $\mathcal{S}$ be an n-dimensional H-invariant J-neutral subspace of $\mathbf{C}^{2n}$ and $\begin{bmatrix} U \\ V \end{bmatrix}$ a basis for $\mathcal{S}$. We shall first show that $\mathcal{S} \cap \text{Sp} \begin{bmatrix} 0 \\ I \end{bmatrix}$ is H-invariant. Since $\mathcal{S}$ is J-neutral,

$$[U^* \quad V^*] \; J \begin{bmatrix} U \\ V \end{bmatrix} = 0,$$

which implies that $V^*U = U^*V$. Since $\mathcal{S}$ is H-invariant, there exists an $n \times n$ matrix L such that

$$\begin{bmatrix} A & -B \\ -C & -A^* \end{bmatrix} \begin{bmatrix} U \\ V \end{bmatrix} = \begin{bmatrix} U \\ V \end{bmatrix} L.$$

In particular, this implies that $AU - BV = UL$. Let $z \in \text{Ker}\, U$. Then $0 = V^*Uz = U^*Vz$. Now,

$$z^*V^*AUz - z^*V^*BVz = z^*V^*ULz,$$

which implies that $BVz = 0$. Then $0 = AUz - BVz = ULz$, so $Lz \in \text{Ker}\, U$. Thus $\text{Ker}\, U$ is L-invariant. Now,

$$\mathcal{S} \cap \text{Sp} \begin{bmatrix} 0 \\ I \end{bmatrix} = \left\{ \begin{bmatrix} U \\ V \end{bmatrix} z : z \in \text{Ker}\, U \right\}.$$

Since $H \begin{bmatrix} U \\ V \end{bmatrix} = \begin{bmatrix} U \\ V \end{bmatrix} L$, the L-invariance of $\text{Ker}\, U$ implies the H-invariance of $\mathcal{S} \cap \text{Sp} \begin{bmatrix} 0 \\ I \end{bmatrix}$.

Now we shall show that $\text{Sp} \begin{bmatrix} 0 \\ I \end{bmatrix}$ contains no non-trivial H-invariant subspace provided (A, B) is reachable. Let $\mathcal{W}$ be any H-invariant subspace which is contained in $\text{Sp} \begin{bmatrix} 0 \\ I \end{bmatrix}$ and let $\begin{bmatrix} 0 \\ W \end{bmatrix}$ be a basis for $\mathcal{W}$. The H-invariance of $\mathcal{W}$ implies that there exists a matrix N such that

$$H \begin{bmatrix} 0 \\ W \end{bmatrix} = \begin{bmatrix} -BW \\ -A^*W \end{bmatrix} = \begin{bmatrix} 0 \\ W \end{bmatrix} N.$$

This shows that $BW = 0$ and $A^*W = -WN$, which implies that

$$W^*B = 0, \quad W^*AB = 0, \quad W^*A^2B = 0, \ldots$$

Hence $W = 0$ whenever (A, B) is reachable.

To conclude, the reachability of (A, B) implies that $\mathcal{S}$ is complementary to $\text{Sp} \begin{bmatrix} 0 \\ I \end{bmatrix}$. □

Lemma 3.2.2 ([13], [14]). *Let F and G be complex $n \times n$ matrices and let G be hermitian, $G \geq 0$ and (F, G) reachable. Then the partial multiplicities of the pure imaginary eigenvalues of*

$$M = \begin{bmatrix} F & -G \\ 0 & -F^* \end{bmatrix}$$

(if any) are all even.

Proof. Let $i\omega$ be a pure imaginary eigenvalue of F. Let $U(s), V(s)$ be two invertible complex rational $2n \times 2n$ matrices such that $U(s), U^{-1}(s)$ and $V(s), V^{-1}(s)$ have no pole at $s = i\omega$. We write

$$sI - M \sim D(s) \tag{3.9}$$

if

$$U(s)(sI - M)V(s) = D(s).$$

We can choose $U(s)$ and $V(s)$ so as to display the elementary divisors of F corresponding to $i\omega$,

$$sI - F \sim \operatorname{diag}[1, \ldots, 1, (s - i\omega)^{r_1}, \ldots, (s - i\omega)^{r_\gamma}], \quad 0 < r_1 \leq \ldots \leq r_\gamma.$$

Then

$$sI - M \sim \begin{bmatrix} I & 0 & 0 \\ 0 & F_2(s) & G_2(s) \\ 0 & 0 & -F_2^*(s) \end{bmatrix}$$

where

$$F_2(s) \sim \operatorname{diag}[(s - i\omega)^{r_1}, \ldots, (s - i\omega)^{r_\gamma}]$$

and $G_2(s)$ is a rational matrix such that $G_2(i\omega) \geq 0$. In fact $G_2(i\omega) > 0$, for (F, G) reachable implies

$$\operatorname{rank}[F_2(i\omega) \;\; G_2(i\omega)] = \gamma$$

while $F_2(i\omega) = 0$.

We further obtain

$$sI - M \sim \begin{bmatrix} I & 0 \\ 0 & W(s) \end{bmatrix}$$

where

$$W(s) = -F_2^*(s) \;\; G_2^{-1}(s) \;\; F_2(s).$$

It follows from $G_2^{-1}(i\omega) > 0$ that the principal minors of $W(s)$ are non-zero. Since $F_2(s)$ is diagonal, the greatest common divisor of all $k \times k$ minors of $W(s)$

is $(s - i\omega)^{d_k}$, where

$$d_k = \sum_{j=1}^{k} 2r_j, \quad k = 1, 2, \ldots, \gamma.$$

Therefore

$$sI - M \sim \operatorname{diag}[1, \ldots, 1, (s - i\omega)^{2r_1}, \ldots, (s - i\omega)^{2r_\gamma}].$$

Since (3.9) preserves the elementary divisors corresponding to $i\omega$, this completes the proof. □

Lemma 3.2.3 ([15], [13]). *Suppose (A, B) is reachable. Let χ_H be factorized as in (3.7) where the polynomials q and q^* have at most pure imaginary roots in common. Then equation (3.1) has a unique hermitian solution X such that $\chi_{A-BX} = q$ if and only if the partial multiplicities of the pure imaginary eigenvalues of H (if any) are all even.*

Proof. Let X be a unique hermitian solution of (3.1) such that $\chi_{A-BX} = q$. Then (3.8) holds,

$$\begin{bmatrix} I & 0 \\ X & I \end{bmatrix}^{-1} H \begin{bmatrix} I & 0 \\ X & I \end{bmatrix} = \begin{bmatrix} A - BX & -B \\ 0 & -(A - BX)^* \end{bmatrix}.$$

Since (A, B) is reachable, $(A - BX, B)$ is also. Then Lemma 3.2.2 can be applied to $F = A - BX$ and $G = B$, showing that the partial multiplicities of the pure imaginary eigenvalues of (3.8), and hence of H, are all even.

Conversely, suppose that H satisfies the condition of the lemma. We shall first prove the existence of an n-dimensional H-invariant J-neutral subspace $\mathcal{S}$ of $\boldsymbol{C}^{2n}$ which is complementary to $\operatorname{Sp}\begin{bmatrix} 0 \\ I \end{bmatrix}$. Since H is hamiltonian, its spectrum is symmetrical with respect to the imaginary axis. Let $i\omega_1, \ldots, i\omega_\alpha, \lambda_1, -\lambda_1^*, \ldots, \lambda_\beta, -\lambda_\beta^*$ be the distinct eigenvalues of H (ω_j real and λ_k non-imaginary). Then

$$\boldsymbol{C}^{2n} = \mathcal{E}_{i\omega_1} \dot{+} \ldots \dot{+} \mathcal{E}_{i\omega_\alpha} \dot{+} (\mathcal{E}_{\lambda_1} \dot{+} \mathcal{E}_{-\lambda_1^*}) \dot{+} \ldots \dot{+} (\mathcal{E}_{\lambda_\beta} \dot{+} \mathcal{E}_{-\lambda_\beta^*})$$

where $\mathcal{E}_\lambda$ is the generalized eigenspace of dimension m_λ corresponding to the eigenvalue λ.

We shall define subspaces $\mathcal{N}_{i\omega_j} \subset \mathcal{E}_{i\omega_j}, j = 1, 2, \ldots, \alpha$ as follows. Write

$$\mathcal{E}_{i\omega_j} = \mathcal{P}_1 \dot{+} \mathcal{P}_2 \dot{+} \ldots \dot{+} \mathcal{P}_\gamma$$

where $\mathcal{P}_1, \mathcal{P}_2, \ldots, \mathcal{P}_\gamma$ are cyclic subspaces whose dimensions are equal to the partial multiplicities of the eigenvalue $i\omega_j$ of H. By assumption, they are all even, say $2r_1, 2r_2, \ldots, 2r_\gamma$. Let e_k be a generator of $\mathcal{P}_k$, i.e., the vectors $e_k, (sI - H)e_k, \ldots, (sI - H)^{2r_k - 1} e_k$ form a basis of $\mathcal{P}_k$ and $(sI - H)^{2r_k} e_k = 0$. We let $\mathcal{Q}_k$ denote the r_k-dimensional space spanned by $(sI - H)^\mu e_k, k = 1, 2, \ldots, \gamma$ and $\mu = r_k + 1, \ldots, 2r_k$, and put

$$\mathcal{N}_{i\omega_j} = \mathcal{Q}_1 \dot{+} \mathcal{Q}_2 \dot{+} \ldots \dot{+} \mathcal{Q}_\gamma.$$

Now we let $\mathcal{S}$ denote the subspace of $\boldsymbol{C}^{2n}$ defined by

$$\mathcal{S} = \mathcal{N}_{i\omega_1} \dot{+} \ldots \dot{+} \mathcal{N}_{i\omega_\alpha} \dot{+} \mathcal{E}_{\lambda_1} \dot{+} \ldots \dot{+} \mathcal{E}_{\lambda_\beta}. \tag{3.10}$$

Obviously, $\mathcal{S}$ is n-dimensional and H-invariant. We shall show that it is J-neutral. For $\mathcal{N}_{i\omega_j}$ it suffices to show that

$$e_k^*(sI - H)^{*\mu} J(sI - H)^\nu e_l = 0.$$

For definiteness we may assume that $r_k \le r_l$. In view of (3.5),

$$e_k^*(sI - H)^{*\mu} J(sI - H)^\nu e_l = (-1)^\mu e_k^* J(sI - H)^{\mu+\nu} e_l. \tag{3.11}$$

Since $\mu + \nu \ge 2r_k$, the right-hand side must vanish. Hence $\mathcal{N}_{i\omega_j}, j = 1, 2, \ldots, \alpha$ are J-neutral. Now, let E be a basis for $\mathcal{E}_{\lambda_j}$. Then there exists a matrix L having the eigenvalue λ_j such that $HE = EL$. In view of (3.5),

$$0 = E^*(JH + H^*J)E = (E^*JE)L + L^*(E^*JE).$$

Since $\lambda_j \ne -\lambda_j^*$, we have $E^*JE = 0$, showing that $\mathcal{E}_{\lambda_j}, j = 1, 2, \ldots, \beta$ are J-neutral. Finally, the eigenspaces corresponding to distinct eigenvalues are clearly J-neutral; hence so must be $\mathcal{S}$.

Since (A, B) is reachable, it follows from Lemma 3.2.1 that $\mathcal{S}$ is complementary to Sp $\begin{bmatrix} 0 \\ I \end{bmatrix}$. Then, using Theorem 3.2.1, one can associate with $\mathcal{S}$ an hermitian solution X of (3.1) such that $\mathcal{S} = \text{Sp} \begin{bmatrix} 0 \\ I \end{bmatrix}$.

Let $q = \chi_{A-BX}$. Then (3.7) holds. By the construction of $\mathcal{S}$, the polynomial q has roots $i\omega_1, \ldots, i\omega_\alpha, \lambda_1, \ldots, \lambda_\beta$ and hence only pure imaginary ones in common with q^*. It follows from (3.11) that $\mathcal{N}_{i\omega_j}$ is the *unique* H-invariant J-neutral subspace of $\mathcal{E}_{i\omega_j}$ whose dimension is half that of $\mathcal{E}_{i\omega_j}$. Therefore the space $\mathcal{S}$ corresponding to q is uniquely determined, showing that X is a unique solution of (3.1) such that $\chi_{A-BX} = q$. □

Lemma 3.2.3 solves the existence of unique hermitian solutions of (3.1) under the assumption that (A, B) is reachable. The general result will be obtained by reduction to the reachable case. Let T be a complex non-singular $n \times n$ matrix, and put

$$\bar{A} = TAT^{-1}, \quad \bar{B} = TBT^*, \quad \bar{C} = (T^*)^{-1}CT^{-1} \tag{3.12}$$

and

$$\bar{X} = (T^*)^{-1}XT^{-1}.$$

Then X is a solution of (3.1) if and and only if $\bar{X}$ is a solution of

$$\bar{X}\bar{A} + \bar{A}^*\bar{X} - \bar{X}\bar{B}\bar{X} + \bar{C} = 0. \tag{3.13}$$

The hamiltonian matrix

$$\bar{H} = \begin{bmatrix} \bar{A} & -\bar{B} \\ -\bar{C} & -\bar{A}^* \end{bmatrix}$$

of (3.13) is related to H by $\bar{H} = V^{-1}HV$, where

$$V = \begin{bmatrix} T^{-1} & 0 \\ 0 & T^* \end{bmatrix}. \tag{3.14}$$

Since $\mathcal{R}_{A,B}$ is A-invariant, the matrix T in (3.12) can be chosen such that

$$\bar{A} = \begin{bmatrix} A_1 & A_{12} \\ 0 & A_2 \end{bmatrix}, \quad \bar{B} = \begin{bmatrix} B_1 & 0 \\ 0 & 0 \end{bmatrix}$$

where $B_1 \geq 0$ and (A_1, B_1) is *reachable*. Put

$$h = \chi_{A_2} \quad (= \chi_A \ / \ \chi_{A|\mathcal{R}_{A,B}}).$$

The roots of the polynomial h are precisely those eigenvalues of A which are not B-reachable.

Let $\bar{C}$ and $\bar{X}$ be partitioned compatibly with (3.14) as

$$\bar{C} = \begin{bmatrix} C_1 & C_{12} \\ C_{12}^* & C_2 \end{bmatrix}, \quad \bar{X} = \begin{bmatrix} X_1 & X_{12} \\ X_{12}^* & X_2 \end{bmatrix}.$$

Then (3.13) decomposes into one quadratic and two linear equations,

$$X_1A_1 + A_1^*X_1 - X_1B_1X_1 + C_1 = 0 \tag{3.15}$$

$$X_{12}A_2 + (A_1 - B_1X_1)^*X_{12} + (X_1A_{12} + C_{12}) = 0 \tag{3.16}$$

$$X_2A_2 + A_2^*X_2 + (X_{12}^*A_{12} + A_{12}^*X_{12} - X_{12}^*B_1X_{12} + C_2) = 0. \tag{3.17}$$

The blocks X_{12} and X_2 have no effect on

$$\chi_{A-BX} = \chi_{A_1-B_1X_1}\ \chi_{A_2}.$$

Finally, we let n_1, n_2 denote the size of A_1, A_2 and put

$$H_1 = \begin{bmatrix} A_1 & -B_1 \\ -C_1 & -A_1^* \end{bmatrix}.$$

Then

$$\chi_H = (-1)^{n_2} h \chi_{H_1} h^*. \tag{3.18}$$

Lemma 3.2.4 ([16]). *Let q be a complex polynomial of degree n such that q has at most pure imaginary roots in common with q^*. Then equation (3.1) has a unique hermitian solution X such that $\chi_{A-BX} = q$ if and only if (3.15) has a unique hermitian solution X_1 such that $\chi_{A_1-B_1X_1} = q_1 = q/h$, and $h = \chi_{A_2}$ is relatively prime to h^*.*

Proof. Let X be a unique hermitian solution of (3.1) such that $\chi_{A-BX} = q$. Then $\bar{X}$ is a unique hermitian solution of (3.13) with $\chi_{\bar{A}-\bar{B}\bar{X}} = q$. In particular,

X_1 is a unique hermitian solution of (3.15) such that $\chi_{A_1-B_1X_1} = q_1 = q/h$. Furthermore, X_2 is a unique hermitian solution of the Lyapunov equation (3.17). Hence the matrices A_2 and $-A_2^*$ have no common eigenvalue, i.e., $h = \chi_{A_2}$ and $h^* = (-1)^{n_2}\chi_{A_2^*}$ are relatively prime.

To prove the converse, let X_1 be a unique hermitian solution of the Riccati algebraic equation (3.15) such that $\chi_{A_1-B_1X_1} = q_1$, where q_1 is a complex polynomial having at most pure imaginary roots in common with q_1^* and satisfying

$$\chi_{H_1} = (-1)^{n_1} q_1 q_1^*.$$

We can assume that q_1^* is relatively prime to h. For if it were not, we could interchange the roles of q_1 and q_1^* and, by virtue of the reachability of (A_1, B_1), apply Lemma 3.2.3 to conclude that another unique hermitian solution X_1 of (3.15) exists having already the requisite property. Therefore, the Sylvester equation (3.16) possesses a unique solution X_{12}. The assumption that h is relatively prime to h^* implies that the Lyapunov equation (3.17) has a unique hermitian solution X_2. Hence equation (3.13) has a unique hermitian solution $\bar{X}$ such that $\chi_{\bar{A}-\bar{B}\bar{X}} = hq_1$. We conclude that (3.1) has a unique hermitian solution X such that $\chi_{A-BX} = hq_1$ and hq_1 has at most pure imaginary roots in common with $h^*q_1^*$. □

The main result concerning the existence of unique hermitian solutions to (3.1) is as follows.

Theorem 3.2.2 ([15], [16]). *Let q be any polynomial that satisfies (3.7) and has at most pure imaginary roots in common with q^*. Then equation (3.1) has a unique hermitian solution X such that $\chi_{A-BX} = q$ if and only if the following conditions are satisfied.*

(i) (A, B) *is sign-reachable.*
(ii) *The partial multiplicities of the pure imaginary eigenvalues of H (if any) are all even.*

Proof. We observe that (A, B) is sign-reachable if and only if $h = \chi_{A_2}$ is relatively prime to h^*. Then, the assertion of the theorem is an immediate consequence of Lemmas 3.2.3 and 3.2.4. □

The genesis of Theorem 3.2.2 has had a long history. Existence and uniqueness of two special hermitian solutions of (3.1), namely those corresponding to a stable and to an anti-stable polynomial q, seems to have been first established by Reid [17] under the assumptions that (A, B) is reachable and H has no pure imaginary eigenvalue. Existence of unique hermitian solutions that correspond to *any* q was addressed much later. Assuming that H has no pure imaginary eigenvalue, Coppel [18] related the existence of such a solution to the absence of common factors in h and q^*. On the other hand, Churilov [15] assumed that (A, B) is sign-reachable and related the existence of such a solution to the absence of odd partial multiplicities of the pure imaginary eigenvalues of H. This is already the sufficiency part of

Theorem 3.2.2. The necessity part was then proved by Wimmer [16], who also established a link between the existing results.

3.2.2 Unmixed Solutions

An element X of $\boldsymbol{H}$ will be called *unmixed* if

(i) χ_{A-BX} has at most pure imaginary roots in common with χ^*_{A-BX}, and
(ii) the partial multiplicities of these pure imaginary roots as eigenvalues of H are all even.

The concept of unmixed hermitian solutions is due to Shayman [10]. The following examples illustrate the two requirements in their definition.

Example 3.2.1 ([19]). Consider the equation (3.1) where

$$A = \begin{bmatrix} 0 & 1 \\ -1 & 0 \end{bmatrix}, \quad B = \begin{bmatrix} 1 & 1 \\ 1 & 1 \end{bmatrix}, \quad C = \begin{bmatrix} 1 & 1 \\ 1 & 1 \end{bmatrix}.$$

We have

$$H = \begin{bmatrix} 0 & 1 & -1 & -1 \\ -1 & 0 & -1 & -1 \\ -1 & -1 & 0 & 1 \\ -1 & -1 & -1 & 0 \end{bmatrix}$$

and

$$\chi_H = (s-1)^2(s+1)^2$$

The generalized eigenspace of H corresponding to the eigenvalue 1 is

$$\mathcal{E}_1 = \mathrm{Sp} \begin{bmatrix} 1 & 0 \\ 0 & 1 \\ -1 & 0 \\ 0 & -1 \end{bmatrix}$$

and that corresponding to the eigenvalue -1 is

$$\mathcal{E}_{-1} = \mathrm{Sp} \begin{bmatrix} 1 & 0 \\ 0 & 1 \\ 1 & 0 \\ 0 & 1 \end{bmatrix}.$$

There are precisely three 2-dimensional H-invariant subspaces of $\boldsymbol{C}^4$, namely $\mathcal{S}_1 = \mathcal{E}_1$, $\mathcal{S}_2 = \mathcal{E}_{-1}$ and

$$\mathcal{S}_3 = \mathrm{Sp} \begin{bmatrix} 1 & 0 \\ 0 & 1 \\ -1 & 0 \\ 0 & 1 \end{bmatrix},$$

which are all J-neutral and complementary to

$$\mathrm{Sp}\begin{bmatrix}0&0\\0&0\\1&0\\0&1\end{bmatrix}.$$

By Theorem 3.2.1, the solution set $\boldsymbol{H}$ of (3.1) consists of three solutions,

$$X_1=\begin{bmatrix}-1&0\\0&-1\end{bmatrix},\quad X_2=\begin{bmatrix}1&0\\0&1\end{bmatrix},\quad X_3=\begin{bmatrix}-1&0\\0&1\end{bmatrix}.$$

We note that X_1 and X_2 are unmixed while X_3 is not, for $\chi_{A-BX_3}=s^2-1$ fails to satisfy the requirement (i). □

Example 3.2.2 Consider the equation (3.1) where

$$A=\begin{bmatrix}0&0\\0&0\end{bmatrix},\quad B=\begin{bmatrix}1&1\\1&1\end{bmatrix},\quad C=\begin{bmatrix}1&0\\0&0\end{bmatrix}.$$

The matrix H now has the characteristic polynomial

$$\chi_H=s^2(s^2-1).$$

The eigenspace of H corresponding to the eigenvalue 0 is

$$\mathcal{E}_0=\mathrm{Sp}\begin{bmatrix}0&0\\1&0\\0&1\\0&-1\end{bmatrix}$$

and those corresponding to the eigenvalues 1 and -1 are

$$\mathcal{E}_1=\mathrm{Sp}\begin{bmatrix}1\\1\\-1\\0\end{bmatrix},\quad \mathcal{E}_{-1}=\mathrm{Sp}\begin{bmatrix}1\\1\\1\\0\end{bmatrix}.$$

There are now two families of 2-dimensional H-invariant subspaces of $\boldsymbol{C}^4$ that are J-neutral and complementary to

$$\mathrm{Sp}\begin{bmatrix}0&0\\0&0\\1&0\\0&1\end{bmatrix},$$

namely

$$\mathcal{S}_1(\alpha)=\mathrm{Sp}\begin{bmatrix}1&0\\1&1\\-1&\alpha\\0&-\alpha\end{bmatrix},\quad \mathcal{S}_2(\alpha)=\mathrm{Sp}\begin{bmatrix}1&0\\1&1\\1&\alpha\\0&-\alpha\end{bmatrix},$$

where $\begin{bmatrix} 0 \\ 1 \\ \alpha \\ -\alpha \end{bmatrix}$ is a vector from $\mathcal{E}_0$ for any real α. By Theorem 3.2.1, the solution set $\boldsymbol{H}$ of (3.1) consists of two families of solutions,

$$X_1(\alpha) = \begin{bmatrix} -1-\alpha & \alpha \\ \alpha & -\alpha \end{bmatrix}, \quad X_2(\alpha) = \begin{bmatrix} 1-\alpha & \alpha \\ \alpha & -\alpha \end{bmatrix}.$$

Though $\chi_{A-BX_1(\alpha)} = s(s-1)$ and $\chi_{A-BX_2(\alpha)} = s(s+1)$ for any α, $\boldsymbol{H}$ contains no unmixed solution. In fact, requirement (ii) is violated since the partial multiplicities of the eigenvalue 0 of H are 1,1. □

The unmixed solutions enjoy several interesting properties. For one thing, they are *unique*: if X is an unmixed solution of (3.1) then $\chi_{A-BX} = \chi_{A-BX'}$ implies $X = X'$ for any element X' of $\boldsymbol{H}$. This is a consequence of Theorem 3.2.2. Futhermore, they occur in pairs whenever (A, B) is reachable. For let X_1 be an unmixed element of $\boldsymbol{H}$ and let $\chi_{A-BX_1} = q$. By virtue of reachability, Lemma 3.2.3 implies the existence of a unique element X_2 of $\boldsymbol{H}$ such that χ_{A-BX_2} is any complex polynomial of degree n which has at most pure imaginary roots in common with $\chi^*_{A-BX_2}$. In particular we take $\chi_{A-BX_2} = q^*$. Then X_2 is an unmixed element of $\boldsymbol{H}$. We say that X_1 and X_2 are a *pair of opposite unmixed solutions* of (3.1).

There is one special pair of opposite unmixed solutions that is of particular importance. It corresponds to a factorization (3.7) in which q has all its roots with non-positive real part and q^* with non-negative real part. We shall designate these solutions X_+ and X_- and deduce the following results from Theorem 3.2.2.

Corollary 3.2.1 *There exists a unique hermitian solution X_+ of (3.1) such that all eigenvalues of $A - BX_+$ have non-positive real part if and only if (A, B) is stabilizable and the partial multiplicities of the pure imaginary eigenvalues of H (if any) are all even.*

Proof. A special case of Theorem 3.2.2 where all roots of h have negative real part, i.e., where (A, B) is stabilizable. □

Corollary 3.2.2 *There exists a unique hermitian solution X_- of (3.1) such that all eigenvalues of $A - BX_-$ have non-negative real part if and only if $(-A, B)$ is stabilizable and the partial multiplicities of the pure imaginary eigenvalues of H (if any) are all even.*

Proof. A special case of Theorem 3.2.2 where all roots of h have positive real part, i.e., where $(-A, B)$ is stabilizable. □

Corollary 3.2.3 *The class $\boldsymbol{H}$ of hermitian solutions contains both elements X_+ and X_- if and only if (A, B) is reachable and the partial multiplicities of the pure imaginary eigenvalues of H (if any) are all even.*

Proof. Both (A, B) and $(-A, B)$ are stabilizable if and only if (A, B) is reachable. □

Simple examples are most illustrative.

Example 3.2.3 ([19]). Consider the equation (3.1) where

$$A = \begin{bmatrix} 0 & 0 \\ 0 & 1 \end{bmatrix}, \quad B = \begin{bmatrix} 1 & 0 \\ 0 & 1 \end{bmatrix}, \quad C = \begin{bmatrix} 1 & 0 \\ 0 & 3 \end{bmatrix}$$

and determine its unmixed solutions.

We have

$$\chi_H = s^4 - 5s^2 + 4$$

and the following eigenspaces of H,

$$\mathcal{E}_1 = \text{Sp} \begin{bmatrix} 1 \\ 0 \\ -1 \\ 0 \end{bmatrix}, \quad \mathcal{E}_{-1} = \text{Sp} \begin{bmatrix} 1 \\ 0 \\ 1 \\ 0 \end{bmatrix}$$

$$\mathcal{E}_2 = \text{Sp} \begin{bmatrix} 0 \\ 1 \\ 0 \\ -1 \end{bmatrix}, \quad \mathcal{E}_{-2} = \text{Sp} \begin{bmatrix} 0 \\ 1 \\ 0 \\ 3 \end{bmatrix}$$

are associated with the eigenvalues $1, -1$ and $2, -2$, respectively. Applying Theorem 3.2.1 we obtain the following two pairs of unmixed solutions:

$$X_+ = \begin{bmatrix} 1 & 0 \\ 0 & 3 \end{bmatrix}, \quad X_- = \begin{bmatrix} -1 & 0 \\ 0 & -1 \end{bmatrix},$$

which correspond to the factorization

$$\chi_H = (s^3 - 3s + 2)(s^2 + 3s + 2),$$

and

$$X_{-1,2} = \begin{bmatrix} 1 & 0 \\ 0 & -1 \end{bmatrix}, \quad X_{-2,1} = \begin{bmatrix} -1 & 0 \\ 0 & 3 \end{bmatrix},$$

which correspond to the factorization

$$\chi_H = (s^2 - s - 2)(s^2 + s - 2).$$

There exist no further solutions of (3.1). □

3.2.3 Classification Theorems

Theorems 3.1.1 and 3.2.1 describe the set $\boldsymbol{H}$ of hermitian solutions of (3.1) by relating it to the set of n-dimensional H-invariant J-neutral subspaces of $\boldsymbol{C}^{2n}$

which are complementary to the n-dimensional subspace Sp $\begin{bmatrix} 0 \\ I \end{bmatrix}$. They do not explain, however, how the requirements of neutrality and complementarity cut out a subset from the set of all n-dimensional H-invariant subspaces.

We shall present another, simpler classification for the solution set $\boldsymbol{H}$ by relating it to a certain subset of the set of invariant subspaces of the matrix $A - BX_+$ or the matrix $A - BX_-$, *under the assumption that* X_+ *and* X_- *both exist.*

Lemma 3.2.5 ([18]). *Suppose that* $\boldsymbol{H}$ *is non-empty and* (A, B) *is reachable. Let* X_1, X_2 *be any two elements of* $\boldsymbol{H}$ *and put* $D = X_2 - X_1$. *Let* π, ν, δ *be the number of eigenvalues (counted according to multiplicities) of* D *with respectively positive, negative, zero real part. Then,*

(i) Ker D *is invariant under* $A - BX_1$
(ii) $A - BX_1$ *has* π *eigenvalues with positive real part and* ν *eigenvalues with negative real part which are not associated with* Ker D.

Proof. The difference D is a solution of the quadratic equation

$$D(A - BX_1) + (A - BX_1)^*D - DBD = 0. \tag{3.19}$$

If $u \in \operatorname{Ker} D$ then, on postmultiplying (3.19) by u, we obtain $D(A - BX_1)u = 0$. Thus Ker D is invariant under $A - BX_1$, proving (i).

By replacing $A-BX_1, B, D$ by $T(A-BX_1)T^{-1}, TBT^*, (T^*)^{-1}DT^{-1}$, which does not alter the hypotheses, we can asume that

$$D = \begin{bmatrix} D_1 & 0 \\ 0 & 0 \end{bmatrix}$$

where D_1 is non-singular. Ker D now consists of all vectors whose first $\pi + \nu$ coordinates vanish. Correspondingly, let

$$A - BX_1 = \begin{bmatrix} A_1 & 0 \\ A_{21} & A_2 \end{bmatrix}, \quad B = \begin{bmatrix} B_1 & B_{21}^* \\ B_{21} & B_2 \end{bmatrix}.$$

Then we have from (3.19)

$$DA_1 + A_1^*D - B_1 = 0. \tag{3.20}$$

We are going to show now that A_1 has no pure imaginary eigenvalue. Suppose on the contrary that $A_1^*v_1 = i\omega v_1$, where ω is real and $v_1 \neq 0$. Then from (3.20) we obtain

$$v_1^*B_1v_1 = 0.$$

Since $B_1 \geq 0$ this implies $B_1v_1 = 0$, and also $B_{21}v_1 = 0$. Thus if we put

$$v = \begin{bmatrix} v_1 \\ 0 \end{bmatrix}$$

then

$$v^*(A - BX_1) = -i\omega v^*, \quad v^*B = 0.$$

Hence

$$v^*B = 0, \quad v^*AB = 0, \quad v^*A^2B = 0, \ldots$$

Since $v \neq 0$, this contradicts the hypothesis of reachability.

Put

$$A_1(\varepsilon) = A_1 + \varepsilon D_1.$$

Then, by (3.20), $A_1(\varepsilon)$ is a solution of the equation

$$D_1 A_1(\varepsilon) + A_1^*(\varepsilon) D_1 - D_1(B_1 + 2\varepsilon I) D_1 = 0.$$

Since $D_1(B_1+2\varepsilon I)D_1 > 0$ for $\varepsilon > 0$, an inertia theorem [20] implies that $A_1(\varepsilon)$ has no pure imaginary eigenvalues and further that it has π eigenvalues with positive real part, ν eigenvalues with negative real part. Since A_1 has no pure imaginary eigenvalue, the same holds by continuity for it. This proves (ii). □

Theorem 3.2.3 ([21]). *Suppose that* $\boldsymbol{H}$ *contains* X_+ *and* X_-. *If* X *is any element of* $\boldsymbol{H}$, *then*

$$X_+ \geq X \geq X_-.$$

Proof. Take $X_1 = X_+, \quad X_2 = X$ in Lemma 3.2.5. Since $A - BX_+$ has no eigenvalue with positive real part, so has $X - X_+$. Hence $X - X_+ \leq 0$. Similarly, taking $X_1 = X_-, \quad X_2 = X$ one can show that $X - X_- \geq 0$. □

That is why X_+ and X_- are called the *extreme solutions* of (3.1); X_+ is the maximal element of $\boldsymbol{H}$ while X_- is the minimal element of $\boldsymbol{H}$. We put

$$\Delta = X_+ - X_-$$

and call it the *gap* of (3.1).

Lemma 3.2.6 ([18]). *Suppose that* $\boldsymbol{H}$ *contains* X_+ *and* X_-. *If* X *is any element of* $\boldsymbol{H}$, *then*

$$\operatorname{Ker} \Delta = \mathcal{L}^0_{A-BX} \tag{3.21}$$

and

$$A - BX_+ | \operatorname{Ker} \Delta = A - BX_- | \operatorname{Ker} \Delta. \tag{3.22}$$

Proof. If $v \in \operatorname{Ker} \Delta$ then $X_+ v = Xv = X_- v$, since

$$0 \leq v^*(X_+ - X)v \leq v^*(X_+ - X_-)v = 0.$$

By Lemma 3.2.5 (i), $\operatorname{Ker} \Delta$ is invariant under $A - BX_+$. Hence it is equally invariant under $A - BX$. The restriction of $A - BX_+$ to $\operatorname{Ker} \Delta$ has only pure imaginary eigenvalues. By Lemma 3.2.5 (ii), $A - BX_+$ has no other pure imaginary eigenvalue. If $A - BX$ had an additional pure imaginary eigenvalue then, by Lemma 3.2.5 (ii) again, it would be associated with the kernel of $X_+ - X$ and therefore be an eigenvalue of $A - BX_+$. This proves (3.21).

Since

$$(A - BX_+) = (A - BX_-) - B\Delta$$

it is obvious that $A - BX_+ = A - BX_-$ on Ker Δ, which is (3.22). □

We shall find it useful to denote

$$A_+ = A - BX_+, \quad A_- = A - BX_-$$

and

$$\mathcal{U}_+ = \mathcal{L}^l_{A_+}, \quad \mathcal{U}_- = \mathcal{L}^r_{A_-}$$

and

$$\mathcal{U}_0 = \mathcal{L}^0_{A_+} = \mathcal{L}^0_{A_-} (= \operatorname{Ker} \Delta).$$

Note that

$$\mathcal{L}^l_{A_-} = \mathcal{L}^r_{A_+} = 0.$$

The next theorem describes a classification of $\boldsymbol{H}$.

Theorem 3.2.4 ([21], [18]). *Suppose that $\boldsymbol{H}$ contains X_+ and X_-. Let $\mathcal{V}$ be any A_+ - invariant subspace contained in $\mathcal{U}_+$ and $\tilde{\mathcal{V}}$ the Δ-orthogonal complement of $\mathcal{V}$ in $\boldsymbol{C}^n$. Then*

$$\boldsymbol{C}^n = \mathcal{V} \dot{+} \tilde{\mathcal{V}}$$

and let P denote the projection onto $\mathcal{V}$ along $\tilde{\mathcal{V}}$. There is a one-to-one correspondence between the solution set $\boldsymbol{H}$ of (3.1) and the set of invariant subspaces of $A_+|\mathcal{U}_+$. This correspondence assigns the solution

$$X = X_+P + X_-(I - P) \tag{3.23}$$

to the invariant subspace $\mathcal{V}$. Moreover,

$$\mathcal{L}^l_{A-BX} = \mathcal{V}, \; \mathcal{L}^0_{A-BX} = \mathcal{U}_0, \; \mathcal{L}^r_{A-BX} = \tilde{\mathcal{V}} \cap \mathcal{U}_-.$$

Proof. Evidently $\mathcal{U}_0 \subset \tilde{\mathcal{V}}$. If $u \in \mathcal{V} \cap \tilde{\mathcal{V}}$ then $u^*\Delta u = 0$. Since $\Delta \geq 0$ it follows that $\Delta u = 0$. Since $\mathcal{U}_0 \cap \mathcal{V} = 0$ this implies $u = 0$. Thus $\mathcal{V} \cap \tilde{\mathcal{V}} = 0$. We have also

$$\dim \tilde{\mathcal{V}} = n - \dim \Delta\mathcal{V}$$

and, by $\mathcal{U}_0 \cap \mathcal{V} = 0$, $\dim \Delta\mathcal{V} = \dim\mathcal{V}$. It follows that

$$\mathcal{V} \dot{+} \tilde{\mathcal{V}} = \boldsymbol{C}^n.$$

The difference Δ satisfies the equation

$$\Delta A_- + A_+^* \Delta = 0 \tag{3.24}$$

which is derived from (3.1). If $v \in \mathcal{V}$ and $\tilde{v} \in \tilde{\mathcal{V}}$ then, by (3.24),

$$v^*\Delta A_-\tilde{v} = -v^*A_+^*\Delta\tilde{v} = 0,$$

since $\mathcal{V}$ is invariant under A_+. Thus $\tilde{\mathcal{V}}$ is invariant under A_-. Hence the projection P satisfies

$$PA_+P = A_+P \tag{3.25}$$

and $(I-P)A_-(I-P) = A_-(I-P)$; that is,

$$PA_-P = PA_-. \tag{3.26}$$

Since $\tilde{\mathcal{V}}$ is Δ-orthogonal to $\mathcal{V}$ we have also $P^*\Delta(I-P) = 0$; that is $P^*\Delta = P^*\Delta P$. Since the right side is hermitian this gives

$$P^*\Delta = \Delta P. \tag{3.27}$$

By (3.25) and (3.26),

$$(I-P)A_+ = (I-P)A_+(I-P) =$$

$$= (I-P)(A_- - B\Delta)(I-P) = A_-(I-P) - (I-P)B\Delta(I-P).$$

Therefore, by (3.24),

$$\Delta(I-P)A_+ + A_+^*\Delta(I-P) + \Delta(I-P)B\Delta(I-P) = 0.$$

If we define X by (3.23) then

$$D = X_+ - X = \Delta(I-P).$$

By (3.27), D is hermitian, and hence X is also. Moreover X is a solution of (3.1) if D is a solution of the equation

$$DA_+ + A_+^*D + DBD = 0.$$

But we have just shown that this is the case.

Furthermore,

$$Xu = X_+u, \quad u \in \mathcal{V}$$

$$Xu = X_-u, \quad u \in \tilde{\mathcal{V}}.$$

Therefore $\mathcal{V}$ and $\tilde{\mathcal{V}}$ are invariant under $A-BX$ and all eigenvalues of the restriction of $A-BX$ to $\mathcal{V}, \tilde{\mathcal{V}}$ have respectively negative, non-negative real parts. This proves that $\mathcal{V}$ is uniquely determined by X and that $\mathcal{L}^l_{A-BX} = \mathcal{V}, \mathcal{L}^r_{A-BX} = \tilde{\mathcal{V}} \cap \mathcal{U}_-$ and $\mathcal{L}^0_{A-BX} = \tilde{\mathcal{V}} \cap \mathcal{U}_0 = \mathcal{U}_0$.

Conversely, let X be any element of $\boldsymbol{H}$. Let $\mathcal{V} = \mathcal{L}^l_{A-BX}$ and let $\tilde{\mathcal{V}}$ be the Δ-orthogonal complement of $\mathcal{V}$ in $\boldsymbol{C}^n$. It follows that

$$Xu = X_+u, \quad u \in \mathcal{V}$$

$$Xu = X_-u, \quad u \in \tilde{\mathcal{V}}.$$

Thus if P is the projection onto $\mathcal{V}$ along $\tilde{\mathcal{V}}$ then X satisfies (3.23). Moreover $\mathcal{V}$ is invariant under A_+ and $\mathcal{U}_0 \cap \mathcal{V} = 0$. Therefore $\mathcal{V} \subset \mathcal{U}_+$. Since $X_+ - X = \Delta(I-P)$ is hermitian, (3.27) holds. Therefore, since P is a projection, $P^*\Delta = P^*\Delta P$; that is,

$$P^* \Delta(I-P) = 0.$$

Thus $\tilde{\mathcal{V}}$ is Δ-orthogonal to $\mathcal{V}$. Since

$$\boldsymbol{C}^* = \mathcal{V} \dot{+} \tilde{\mathcal{V}}$$

it follows from the first part of the proof that $\tilde{\mathcal{V}}$ is the subspace of *all* u that are Δ-orthogonal to $\mathcal{V}$. This completes the proof. □

A dual result exists in which X_+ is replaced by its opposite X_-.

Theorem 3.2.5 ([22]). *Suppose that* $\boldsymbol{H}$ *contains* X_+ *and* X_-. *Let* $\mathcal{W}$ *be any* A_-*-invariant subspace contained in* $\mathcal{U}_-$ *and* $\tilde{\mathcal{W}}$ *the* Δ*-orthogonal complement of* $\mathcal{W}$ *in* $\boldsymbol{C}^n$. *Then*

$$\boldsymbol{C}^n = \mathcal{W} \dot{+} \tilde{\mathcal{W}}$$

and let P' *denote the projection onto* $\mathcal{W}$ *along* $\tilde{\mathcal{W}}$. *There is a one-to-one correspondence between the solution set* $\boldsymbol{H}$ *of (3.1) and the set of invariant subspaces of* $A_-|\mathcal{U}_-$. *This correspondence assigns the solution*

$$X = X_- P' + X_+(I - P') \tag{3.28}$$

to the invariant subspace $\mathcal{W}$. *Moreover*

$$\mathcal{L}^r_{A-BX} = \mathcal{W}, \quad \mathcal{L}^0_{A-BX} = \mathcal{U}_0, \quad \mathcal{L}^l_{A-BX} = \tilde{\mathcal{W}} \cap \mathcal{U}_+.$$ □

The classification for $\boldsymbol{H}$ (Theorem 3.2.4) was first obtained by Willems [21] under the assumption that H has no pure imaginary eigenvalue. Then Ker $\Delta = 0$ and $\mathcal{U}_+ = \boldsymbol{C}^n$, $\mathcal{U}_- = \boldsymbol{C}^n$. This restriction was relaxed by Coppel [18]. The dual classification (Theorem 3.2.5) was used by Callier and Willems [22] to describe the subset of non-negative definite elements of $\boldsymbol{H}$. Still other classifications are possible, however. Shayman [10] noticed that the extreme solutions X_+ and X_- can be replaced by an arbitrary pair of opposite unmixed solutions and each such pair gives a classification for $\boldsymbol{H}$ analogous to Theorems 3.2.4 and 3.2.5.

The following result is a consequence of the correspondences (3.23) and (3.28).

Theorem 3.2.6 ([18]). *Suppose that* $\boldsymbol{H}$ *contains* X_+ *and* X_-. *Let* X_1, X_2 *be elements of* $\boldsymbol{H}$ *corresponding to the invariant subspaces* $\mathcal{V}_1, \mathcal{V}_2$ *of* $A_+|\mathcal{U}_+$ *(or* $\mathcal{W}_1, \mathcal{W}_2$ *of* $A_-|\mathcal{U}_-$*). Then* $X_1 \geq X_2$ *if and only if* $\mathcal{V}_1 \supset \mathcal{V}_2$ *(or if and only if* $\mathcal{W}_1 \subset \mathcal{W}_2$*).*

Proof. Suppose first that $\mathcal{V}_1 \supset \mathcal{V}_2$. Then, by Theorem 3.2.4, $\tilde{\mathcal{V}}_1 \subset \tilde{\mathcal{V}}_2$. On $\mathcal{V}_2$ we have $X_1 u = X_+ u = X_2 u$. On $\tilde{\mathcal{V}}_1$ we have $X_1 u = X_- u = X_2 u$. Therefore $\mathcal{V}_2 \dot{+} \tilde{\mathcal{V}}_1$ is contained in Ker $(X_1 - X_2)$ and is invariant under $A - BX_1$ and $A - BX_2$. The remaining eigenvalues of $A - BX_1$ have negative real part, whereas the remaining eigenvalues of $A - BX_2$ have non-negative real part. Hence

$$\text{Ker } (X_1 - X_2) = \mathcal{V}_2 \dot{+} \tilde{\mathcal{V}}_1.$$

It follows from Lemma 3.2.5 (ii) that $X_1 - X_2 \geq 0$.

Conversely, suppose $X_1 \geq X_2$. The kernel of $X_1 - X_2$ is invariant under $A - BX_2$, and hence has a unique decomposition

$$\text{Ker } (X_1 - X_2) = \mathcal{K}_- \dot{+} \mathcal{K}_+$$

where $\mathcal{K}_-, \mathcal{K}_+$ are invariant under $A - BX_2$ and all eigenvalues of the restriction of $A - BX_2$ to $\mathcal{K}_-, \mathcal{K}_+$ have respectively negative, non-negative real part. Hence we have $\mathcal{K}_- \subset \mathcal{V}_1, \mathcal{K}_- \subset \mathcal{V}_2$. By Lemma 3.2.5 (ii) the eigenvalues of $A - BX_2$ not belonging to Ker $(X_1 - X_2)$ have positive real part. Therefore $\dim \mathcal{K}_- = \dim \mathcal{V}_2$. Thus $\mathcal{K}_- = \mathcal{V}_2$ and $\mathcal{V}_2 \subset \mathcal{V}_1$. □

It follows at once that the solution set $\boldsymbol{H}$ is a complete *lattice* with respect to the usual ordering of hermitian matrices. This is rather remarkable, since the set of all hermitian $n \times n$ matrices is not a lattice if $n > 1$.

The maximal solution X_+ is associated with the A_+-invariant subspace $\mathcal{V} = \mathcal{U}_+$ or with $\mathcal{W} = 0$, whereas the minimal solution X_- is associated with $\mathcal{V} = 0$ or with the A_--invariant subsapce $\mathcal{W} = \mathcal{U}_-$.

3.2.4 Cardinality of Solutions

We shall now pay attention to cardinality of the solution set $\boldsymbol{H}$, *assuming that* $\boldsymbol{H}$ *contains both* X_+ *and* X_-. Let β be the number of distinct eigenvalues of $A - BX_+|\mathcal{U}_+$ and let $m_1, m_2, \ldots, m_\beta$ be the multiplicities of these eigenvalues. Owing to the symmetries in H, the matrix $A - BX_-|\mathcal{U}_-$ exhibits the same structure of eigenvalues.

Theorem 3.2.7 ([21], [10]). *Suppose that* **H** *contains* X_+ *and* X_-. *Then* **H** *has finite cardinality if and only if* $A_+|\mathcal{U}_+$ *or* $A_-|\mathcal{U}_-)$ *is cyclic. In this case,* **H** *contains exactly* $(m_1 + 1) \ldots (m_\beta + 1)$ *elements.*

Proof. Since $\boldsymbol{H}$ is in one-to-one correspondence with the set of invariant subspaces (of all possible dimensions) of $A_+|\mathcal{U}_+$, it has finite cardinality if and only if there is a finite number of eigenvectors associated with each eigenvalue of $A_+|\mathcal{U}_+$. This is the case if and only if $A_+|\mathcal{U}_+$ is cyclic.

Now suppose that $A_+|\mathcal{U}_+$ is cyclic. If λ_j is an eigenvalue of $A_+|\mathcal{U}_+$ with multiplicity $m_j, j = 1, 2, \ldots, \beta$ then the generalized eigensubspace associated with λ_j has dimension m_j. Hence it contains exactly $m_j + 1$ subspaces that are invariant under A_+, having the dimensions $0, 1, \ldots, m_j$. This proves the theorem for $A_+|\mathcal{U}_+$. Using the dual classification of $\boldsymbol{H}$, $A_+|\mathcal{U}_+$ can be replaced by $A_-|\mathcal{U}_-$ yielding the same result. □

Corollary 3.2.4 ([10]). *Suppose that* **H** *contains* X_+ *and* X_-. *Then, whether or not* $A_+|\mathcal{U}_+$ *is cyclic,* **H** *contains exactly* 2^β *unmixed elements.*

Proof. In forming an unmixed solution, it is either the zero or the whole generalized eigenspace associated with each non-imaginary eigenvalue of $A_+|\mathcal{U}_+$ that may be included in $\mathcal{V}$. Hence 2 choices instead of $m_j + 1$ are possible for each eigenvalue λ_j of $A_+|\mathcal{U}_+$, the cyclicity being irrelevant. □

Corollary 3.2.5 ([13]). *Suppose that* $\boldsymbol{H}$ *is non-empty and* (A, B) *is reachable. Then* $\boldsymbol{H}$ *contains a single element* $X_+ = X_-$ *if and only if all the eigenvalues of* H *are pure imaginary.*

Proof. We have $X_+ = X_-$ if and only if

$$\mathcal{L}^l_{A_+} = 0 = \mathcal{L}^r_{A_-} .$$

□

If $A_+|\mathcal{U}_+$ is not cyclic, at least one of its eigenvalues is associated with an infinite family of invariant subspaces, so that $\boldsymbol{H}$ is infinite. We note, however, that $\boldsymbol{H}$ is never countably infinite. It is either finite or contains at least one continuous family of solutions. The topological properties of $\boldsymbol{H}$ are studied by Shayman [10].

Example 3.2.4 ([21]). Consider the equation (3.1) where

$$A = \begin{bmatrix} 0 & 0 \\ 0 & 0 \end{bmatrix}, \quad B = \begin{bmatrix} 1 & 0 \\ 0 & 1 \end{bmatrix}, \quad C = \begin{bmatrix} 1 & 0 \\ 0 & 1 \end{bmatrix}$$

and classify the solution set $\boldsymbol{H}$.

We have

$$\chi_H = (s-1)^2(s+1)^2$$

and the eigenspaces corresponding to the eigenvalues 1, -1 of $\boldsymbol{H}$ read

$$\mathcal{E}_1 = \mathrm{Sp}\begin{bmatrix} 1 & 0 \\ 0 & 1 \\ -1 & 0 \\ 0 & -1 \end{bmatrix}, \quad \mathcal{E}_{-1} = \mathrm{Sp}\begin{bmatrix} 1 & 0 \\ 0 & 1 \\ 1 & 0 \\ 0 & 1 \end{bmatrix}.$$

Hence

$$X_+ = \begin{bmatrix} 1 & 0 \\ 0 & 1 \end{bmatrix}, \quad X_- = \begin{bmatrix} -1 & 0 \\ 0 & -1 \end{bmatrix}$$

are the extreme solutions.

We calculate

$$A_+ = \begin{bmatrix} -1 & 0 \\ 0 & -1 \end{bmatrix}, \quad A_- = \begin{bmatrix} 1 & 0 \\ 0 & 1 \end{bmatrix}$$

and observe that $\mathcal{U}_+ = \boldsymbol{C}^2$, $\mathcal{U}_- = \boldsymbol{C}^2$. The set of invariant subspaces of A_+ (other than 0 or $\boldsymbol{C}^2$, which correspond to X_+ or X_-) is the family of 1-dimensional subspaces $\mathcal{V}_\vartheta$ of $\boldsymbol{C}^2$ parametrized by their azimuth angle ϑ. Since

$$\Delta = \begin{bmatrix} 2 & 0 \\ 0 & 2 \end{bmatrix}$$

we conclude that P is the othogonal projection onto $\mathcal{V}_\vartheta$, namely

$$P = \begin{bmatrix} \cos^2 \vartheta & \sin \vartheta \cos \vartheta \\ \cos \vartheta \sin \vartheta & \sin^2 \vartheta \end{bmatrix}.$$

Then, using (3.23), the solution X_ϑ that corresponds to $\mathcal{V}_\vartheta$ is

$$X_\vartheta = \begin{bmatrix} \cos 2\vartheta & \sin 2\vartheta \\ \sin 2\vartheta & -\cos 2\vartheta \end{bmatrix}.$$

Therefore $\boldsymbol{H}$ consists of X_+, X_- and the continuous family of solutions X_ϑ. By Theorem 3.2.6, $\boldsymbol{H}$ is a complete lattice and $X_+ \geq X_\vartheta \geq X_-$ for every ϑ. □

3.3 Definite Solutions

We shall now be concerned with the set of hermitian solutions of (3.1) which are definite. We use $\boldsymbol{P}$ to designate the set of non-negative definite solutions and $\boldsymbol{N}$ to designate the set of non-positive definite solutions. These two sets will be studied *under the condition that* $B \geq 0$ *and* $C \geq 0$.

3.3.1 Existence of Extreme Solutions

The starting point of our analysis are Corollaries 3.2.1 and 3.2.2. The additional assumption that $C \geq 0$, when combined with stabilizability of (A, B) or $(-A, B)$, will be seen to prevent the matrix H from having a pure imaginary eigenvalue with an odd partial multiplicity. One has the following results.

Theorem 3.3.1 ([23], [16], [24]). *The solution set* $\boldsymbol{P}$ *contains a unique element* X_+ *such that every eigenvalue of* $A - BX_+$ *has non-positive real part if and only if* (A, B) *is stabilizable.*

Proof. Suppose (A, B) is stabilizable. Then there exists a (not necesarily hermitian) matrix X_0 such that $A - BX_0$ is stable. Let us construct by induction a sequence $X_1, X_2, \ldots$ of hermitian non-negative definite matrices which satisfy the following conditions:

(a)$_k$ $A - BX_k$ is stable,
(b)$_k$ $X_{k+1}(A - BX_k) + (A - BX_k)^* X_{k+1} = -C - X_k B X_k - \frac{1}{k+1} I$,
(c)$_k$ $X_{k+1} \leq X_k$.

Since $A - BX_0$ is stable, there is a unique non-negative definite matrix X_1 that satisfies the Lyapunov equation

$$X_1(A - BX_0) + (A - BX_0)^* X_1 = -C - X_0 B X_0 - I,$$

which can be labeled $(b)_0$. The requirement $(a)_k$ then uniquely determines a non-negative definite X_{k+1} satisfying $(b)_k$.

We shall first prove that $(a)_{k+1}$ is satisfied. In fact, by $(b)_k$,

$$X_{k+1}(A - BX_{k+1}) + (A - BX_{k+1})^* X_{k+1}$$

$$\leq -X_{k+1}BX_{k+1} - (X_k - X_{k+1})B(X_k - X_{k+1}) - \frac{1}{k+1}I < 0.$$

Since $X_{k+1} \geq 0$, we conclude that $A - BX_{k+1}$ is stable. Next let us prove $(c)_{k+1}$. We have, by $(b)_k$,

$$(X_k - X_{k+1})(A - BX_k) + (A - BX_k)^*(X_k - X_{k+1})$$

$$\leq -(X_k - X_{k+1})B(X_k - X_{k+1}) - (\frac{1}{k} - \frac{1}{k+1})I < 0.$$

Since $A - BX_k$ is stable, it follows that $X_k - X_{k+1} \geq 0$. This completes the inductive construction.

Since $X_1, X_2, \ldots$ is a non-increasing sequence of hermitian non-negative definite matrices, it converges to an hermitian non-negative definite matrix X_+. It follows from (a)'s and (b)'s that X_+ is a solution of (3.1) and that $A - BX_+$ has all its eigenvalues with non-negative real part.

Conversely, let X_+ exist having the requisite properties. Then (A, B) is stabilizable by Corollary 3.2.1. □

Theorem 3.3.2 *The solution set* $\boldsymbol{N}$ *contains a unique element* X_- *such that every eigenvalue of* $A - BX_-$ *has non-negative real part if and only if* $(-A, B)$ *is stabilizable.*

Proof. Apply Theorem 3.3.1 to determine the solution $-X_+$ of the equation

$$-XA - A^*X - XBX + C = 0.$$ □

The role of stabilizability in the analysis of hermitian non-negative definite solutions of (3.1) was discovered by Wonham [25]. A restricted version of Theorem 3.3.1, under the assumption that H have no pure imaginary eigenvalue, can be found in Kučera [8]. The sufficiency part of Theorem 3.3.1, as it is stated here, was obtained by Molinari [23]; our proof follows that given later by Ando [24]. The necessity part of Theorem 3.3.1 is a consequence of Corollary 3.2.1 due to Wimmer [16].

By uniqueness, the solutions X_+ and X_- obtained in Theorems 3.3.1 and 3.3.2 are actually the *extreme* elements of $\boldsymbol{H}$ studied in the preceding section. The condition $C \geq 0$ secures that X_+ belongs to $\boldsymbol{P}$ and X_- belongs to $\boldsymbol{N}$. The existence of X_+, however, excludes the existence of X_- and vice versa, unless (A, B) is reachable.

We also note the following results.

Theorem 3.3.3 ([23]). *If $\boldsymbol{P}$ contains X_+, then*

$$\operatorname{Ker} X_+ = \mathcal{D}_{-A,C}. \tag{3.29}$$

If $\boldsymbol{N}$ contains X_-, then

$$\operatorname{Ker} X_- = \mathcal{D}_{A,C}. \tag{3.30}$$

Proof. We shall prove (3.30). If $X_- u = 0$, then from (3.1) we see first that $Cu = 0$ and then that $X_- Au = 0$. Hence we have $\operatorname{Ker} X_- \subset \mathcal{O}_{A,C}$. Further, $\operatorname{Ker} X_- \subset \mathcal{L}_A^r \dot{+} \mathcal{L}_A^0$, since the eigenvalues of A restricted to $\operatorname{Ker} X_-$ must also be eigenvalues of A_-. Together,

$$\operatorname{Ker} X_- \subset \mathcal{D}_{A,C}.$$

To show the converse we consider a basis where

$$X_- = \begin{bmatrix} X_1 & 0 \\ 0 & 0 \end{bmatrix}$$

and $X_1 < 0$. When A, B and C are partitioned compatibly as

$$A = \begin{bmatrix} A_1 & A_{12} \\ A_{21} & A_2 \end{bmatrix}, \quad B = \begin{bmatrix} B_1 & B_{12} \\ B_{21}^* & B_2 \end{bmatrix}, \quad C = \begin{bmatrix} C_1 & C_{12} \\ C_{21}^* & C_2 \end{bmatrix}$$

then equation (3.1) splits as follows,

$$X_1 A_1 + A_1^* X_1 - X_1 B_1 X_1 + C_1 = 0$$
$$X_1 A_{12} + C_{12} = 0$$
$$C_2 = 0.$$

Since $C \geq 0$ we see first that $C_{12} = 0$ and then that $A_{12} = 0$. Hence

$$\mathcal{D}_{A,C} \subset \operatorname{Ker} X_-$$

and the proof of (3.30) is complete.

The proof of (3.29) is analogous. □

We already know that X_+ and X_- both exist if and only if (A, B) is reachable. Since both (A, C) and $(-A, C)$ are detectable if and only if (A, C) is observable, we complete the picture by noting that X_+ and X_- are both non-singular (that is, $X_+ > 0$ and $X_- < 0$) if and only if (A, C) is observable.

3.3.2 Classification Theorems

We shall now prove classification results for the solution sets $\boldsymbol{P}$ and $\boldsymbol{N}$ which are similar to those for $\boldsymbol{H}$. We note that $X_+ \leq 0$ only on Ker X_+ and $X_- \geq 0$ only on Ker X_-. Thus, if $\boldsymbol{H}$ contains both X_+ and X_-, the set $\boldsymbol{P}$ can be studied by restricting $\boldsymbol{H}$ to KerX_- and the set $\boldsymbol{N}$ can be studied by restricting $\boldsymbol{H}$ to Ker X_+. We have $A - BX_+ = A$ on $\operatorname{Ker} X_+$ and $A - BX_- = A$ on $\operatorname{Ker} X_-$. Therefore $\mathcal{V}$ is an A_+-invariant subspace contained in $\mathcal{U}_+ \cap \operatorname{Ker} X_+$ if and only if $\mathcal{V}$ is an

A-invariant subspace contained in $\mathcal{O}_{A,C} \cap \mathcal{L}_A^l$ and, similarly, $\mathcal{W}$ is an A_--invariant subspace contained in $\mathcal{U}_- \cap \operatorname{Ker} X_-$ if and only if $\mathcal{W}$ is an A-invariant subspace contained in $\mathcal{O}_{A,C} \cap \mathcal{L}_A^r$. We also have

$$\Delta\mathcal{V} = X_-\mathcal{V}, \quad \Delta\mathcal{W} = X_+\mathcal{W}$$

and the classification theorems for $\boldsymbol{P}$ and $\boldsymbol{N}$ can be easily deduced from Theorems 3.2.5 and 3.2.4, respectively. What we are looking for, however, is the classification of $\boldsymbol{P}$ and $\boldsymbol{N}$ when X_+ and X_- exist only individually.

Theorem 3.3.4 ([26], [24]). *Suppose that $\boldsymbol{P}$ contains X_+. Let $\mathcal{W}$ be any A-invariant subspace contained in $\mathcal{O}_{A,C} \cap \mathcal{L}_A^r$ and $\tilde{\mathcal{W}}$ the X_+-orthogonal complement of $\mathcal{W}$ in $\boldsymbol{C}^n$. Then*

$$\boldsymbol{C}^n = \mathcal{W} \dot{+} \tilde{\mathcal{W}}$$

and let Q denote the projection onto $\mathcal{W}$ along $\tilde{\mathcal{W}}$. There is a one-to-one correspondence between the solution set $\boldsymbol{P}$ of (3.1) and the set of invariant subspaces of $A|\mathcal{O}_{A,C} \cap \mathcal{L}_A^r$. This correspondence assigns the solution

$$X = X_+(I - Q) \tag{3.31}$$

to the invariant subspace $\mathcal{W}$. Moreover,

$$\mathcal{L}_{A-BX}^r = \mathcal{W}, \ \mathcal{L}_{A-BX}^0 = \mathcal{U}_0, \ \mathcal{L}_{A-BX}^l = \tilde{\mathcal{W}} \cap \mathcal{U}_+.$$

Proof. We shall first show that each element X of $\boldsymbol{P}$ can be associated with an A-invariant subspace $\mathcal{W}$ contained in $\mathcal{O}_{A,C} \cap \mathcal{L}_A^r$. In fact, we take

$$\mathcal{W} = \mathcal{L}_{A-BX}^r, \quad \tilde{\mathcal{W}} = \mathcal{L}_{A-BX}^l \dot{+} \mathcal{L}_{A-BX}^0$$

and let Q denote the projection onto $\mathcal{W}$ along $\tilde{\mathcal{W}}$. From (3.1),

$$X(A - BX) + (A - BX)^*X = -XBX - C.$$

Since $\mathcal{W}$ is invariant under $A - BX$,

$$Q(A - BX)Q = (A - BX)Q.$$

Then

$$QXQ[(A - BX)Q] + [(A - BX)Q]^*QXQ = -Q(XBX + C)Q$$

and all the eigenvalues of $(A - BX)Q$ have positive real part. Since $-Q(XBX + C)Q \leq 0$, it follows from Lyapunov theory that $QXQ \leq 0$. Then $XQ = 0$ and hence $CQ = 0$. Therefore $\mathcal{W} \subset \operatorname{Ker} C$ and $A - BX = A$ on $\mathcal{W}$. Thus $\mathcal{W}$ is A-invariant, contained in $\mathcal{O}_{A,C} \cap \mathcal{L}_A^r$, and uniquely determined by X.

Suppose conversely that $\mathcal{W}$ is an A-invariant subspace in $\mathcal{O}_{A,C} \cap \mathcal{L}_A^r$. Then

$$\mathcal{M} = \left\{ \begin{bmatrix} w \\ 0 \end{bmatrix} : w \in \mathcal{W} \right\}$$

is H-invariant. We let $\tilde{\mathcal{M}}$ denote the J-orthogonal complement of $\mathcal{M}$ in $\boldsymbol{C}^{2n}$. By (3.5), $J\mathcal{M}$ is H^*-invariant so that $\tilde{\mathcal{M}}$ is H-invariant. Therefore the subspace

$$\mathcal{N} = \tilde{\mathcal{M}} \cap \mathrm{Sp}\begin{bmatrix} I \\ X_+ \end{bmatrix}$$

is also H-invariant.

Now we put $\mathcal{S} = \mathcal{M} + \mathcal{N}$ and show that $\mathcal{S}$ is an n-dimensional H-invariant J-neutral subspace of $\boldsymbol{C}^{2n}$ which is complementary to $\mathrm{Sp}\begin{bmatrix} 0 \\ I \end{bmatrix}$. Clearly $\mathcal{S}$ is H-invariant. Since $\mathcal{M}$ is J-orthogonal to $\mathcal{N}$ and since both $\mathcal{M}$ and $\mathcal{N}$ are J-neutral, $\mathcal{S}$ is J-neutral. Since $\mathcal{M}$ and $\mathcal{N}$ are linearly independent,

$$\begin{aligned}\dim \mathcal{S} &= \dim \mathcal{M} + \dim \mathcal{N} \\ &= \dim \mathcal{M} + \dim \mathrm{Sp}\begin{bmatrix} I \\ X_+ \end{bmatrix} - \dim J\mathcal{M} \\ &= \dim \mathrm{Sp}\begin{bmatrix} 0 \\ I \end{bmatrix}\end{aligned}$$

so that $\mathcal{S}$ is n-dimensional. Finally, suppose that

$$\begin{bmatrix} w \\ 0 \end{bmatrix} \in \mathcal{M}, \quad \begin{bmatrix} v \\ X_+ v \end{bmatrix} \in \mathcal{N} \tag{3.32}$$

and $v + w = 0$. Since $\mathcal{M}$ is J-orthogonal to $\mathcal{N}$ we have $w^* X_+ v = 0$, hence $v^* X_+ v = 0$. Then $X_+ \geq 0$ implies $X_+ v = 0$, and hence

$$\begin{bmatrix} w \\ 0 \end{bmatrix} + \begin{bmatrix} v \\ X_+ v \end{bmatrix} = 0$$

showing that $\mathcal{S}$ is complementary to $\mathrm{Sp}\begin{bmatrix} 0 \\ I \end{bmatrix}$.

By Theorem 3.2.1, the subspace $\mathcal{S}$ gives rise to an hermitian solution X of (3.1) given by

$$\mathcal{S} = \mathrm{Sp}\begin{bmatrix} I \\ X \end{bmatrix}.$$

We put

$$\tilde{\mathcal{W}} = \left\{ v \in \boldsymbol{C}^n : \begin{bmatrix} v \\ X_+ v \end{bmatrix} \in \mathcal{N} \right\}.$$

Then $\tilde{\mathcal{W}}$ is the X_+-orthogonal complement of $\mathcal{W}$ in $\boldsymbol{C}^n$ and $\boldsymbol{C}^n = \mathcal{W} \dot{+} \tilde{\mathcal{W}}$. In view of (3.32) any vector

$$\begin{bmatrix} u \\ Xu \end{bmatrix} \in \mathcal{S}$$

can be written as

$$\begin{bmatrix} u \\ Xu \end{bmatrix} = \begin{bmatrix} w \\ 0 \end{bmatrix} + \begin{bmatrix} v \\ X_+ v \end{bmatrix}$$

where $w \in \mathcal{W}$ and $v \in \tilde{\mathcal{W}}$. Since

$$Qw = w, \quad Qv = 0$$

we obtain

$$Xu = X_+(I - Q)v = X_+(I - Q)u.$$

This proves (3.31) and $X_+ \geq 0$ implies $X \geq 0$.

Furthermore

$$Xu = 0, \quad u \in \mathcal{W}$$

$$Xu = X_+u, \quad u \in \tilde{\mathcal{W}}.$$

Therefore $\mathcal{W}$ and $\tilde{\mathcal{W}}$ are invariant under $A - BX$ and all eigenvalues of the restriction of $A - BX$ to $\mathcal{W}$, $\tilde{\mathcal{W}}$ have respectively positive, non-positive real parts. This proves that

$$\mathcal{L}^r_{A-BX} = \mathcal{W},\ \mathcal{L}^0_{A-BX} = \mathcal{U}_0,\ \mathcal{L}^l_{A-BX} = \tilde{\mathcal{W}} \cap \mathcal{U}_+. \qquad \square$$

The corresponding result for $\boldsymbol{N}$ is as follows.

Theorem 3.3.5 *Suppose that* $\boldsymbol{N}$ *contains* X_-. *Let* $\mathcal{V}$ *be any* A*-invariant subspace contained in* $\mathcal{O}_{A,C} \cap \mathcal{L}^l_A$ *nad* $\tilde{\mathcal{V}}$ *the* X_-*-orthogonal complement of* $\mathcal{V}$ *in* $\boldsymbol{C}^n$. *Then*

$$\boldsymbol{C}^n = \mathcal{V} \dot{+} \tilde{\mathcal{V}}$$

and let Q' *denote the projection onto* $\mathcal{V}$ *along* $\tilde{\mathcal{V}}$. *There is a one-to-one correspondence between the solution set* $\boldsymbol{N}$ *of (3.1) and the set of invariant subspaces of* $A|\mathcal{O}_{A,C} \cap \mathcal{L}^l_A$. *This correspondence assigns the solution*

$$X = X_-(I - Q') \tag{3.33}$$

to the invariant subspace $\mathcal{V}$. *Moreover,*

$$\mathcal{L}^l_{A-BX} = \mathcal{V},\ \mathcal{L}^0_{A-BX} = \mathcal{U}_0,\ \mathcal{L}^r_{A-BX} = \tilde{\mathcal{V}} \cap \mathcal{U}_-. \qquad \square$$

The classification for $\boldsymbol{P}$ (Theorem 3.3.4) was first obtained by Kučera [26] using the geometry of a (generalized) eigenvector basis of the hamiltonian matrix H. A basis-free formulation in terms of projections, inspired by the results of Willems [21], was given by Callier and Willems [22]. The proof presented here is adapted from [26] and [24]. The classification of $\boldsymbol{N}$ (Theorem 3.3.5) is a dual result; for a direct proof see [24].

The results which follow are conseqences of the correspondences (3.31) and (3.33).

Theorem 3.3.6 ([26], [18]). *Suppose that* $\boldsymbol{P}$ *contains* X_+. *Let* X_1, X_2 *be elements of* $\boldsymbol{P}$ *corresponding to the invriant subspaces* $\mathcal{W}_1, \mathcal{W}_2$ *of* $A|\mathcal{O}_{A,C} \cap \mathcal{L}^r_A$. *Then* $X_1 \geq X_2$ *if and only if* $\mathcal{W}_1 \subset \mathcal{W}_2$.

Proof. Suppose first that $\mathcal{W}_1 \subset \mathcal{W}_2$. Then, by Theorem 3.3.4, $\tilde{\mathcal{W}}_1 \supset \tilde{\mathcal{W}}_2$. On $\mathcal{W}_1$ we have $X_1u = X_2u = 0$. On $\tilde{\mathcal{W}}_2$ we have $X_1u = X_2u = X_+u$. Therefore

$\mathcal{W}_1 \dot{+} \tilde{\mathcal{W}}_2$ is contained in Ker$(X_1 - X_2)$ and is invariant under $A - BX_1$ and $A - BX_2$. The remaining eigenvalues of $A - BX_1$ have non-positive real part, whereas the remaining eigenvalues of $A - BX_2$ have positive real part. Hence Ker $(X_1 - X_2) = \mathcal{W}_1 \dot{+} \tilde{\mathcal{W}}_2$.

The difference $X_1 - X_2$ satisfies the quadratic equation

$$(X_1 - X_2)(A - BX_2) + (A - BX_2)^*(X_1 - X_2) \\ - (X_1 - X_2)B(X_1 - X_2) = 0. \tag{3.34}$$

Consider a basis in which

$$X_1 - X_2 = \begin{bmatrix} D & 0 \\ 0 & 0 \end{bmatrix}$$

and D is non-singular. Correspondingly, write

$$B = \begin{bmatrix} B_1 & B_{21}^* \\ B_{21} & B_2 \end{bmatrix}, \quad A - BX_2 = \begin{bmatrix} A_1 & 0 \\ A_{21} & A_2 \end{bmatrix}$$

Then (3.34) reduces to

$$DA_1 + A_1^* D - B_1 = 0. \tag{3.35}$$

Since $B_1 \geq 0$ and all the eigenvalues of A_1 have positive real part, it follows from Lyapunov theory that $D \geq 0$. Hence $X_1 - X_2 \geq 0$.

Conversely, suppose $X_1 \geq X_2$. The kernel of $X_1 - X_2$ is invariant under $A - BX_2$, and hence has a unique decomposition

$$\text{Ker}\,(X_1 - X_2) = \mathcal{K}_+ \dot{+} \mathcal{K}_-$$

where $\mathcal{K}_+, \mathcal{K}_-$ are invariant under $A - BX_2$ and all eigenvalues of the restriction of $A - BX_2$ to $\mathcal{K}_+, \mathcal{K}_-$ have respectively positive, non-positive real part. Hence we have $\mathcal{K}_+ \subset \mathcal{W}_1$, $\mathcal{K}_+ \subset \mathcal{W}_2$. Since $X_1 - X_2 \geq 0$, the eigenvalues of $A - BX_2$ not belonging to Ker $(X_1 - X_2)$ have negative real part by (3.35). Therefore $\dim \mathcal{K}_+ = \dim \mathcal{W}_1$. Thus $K_+ = \mathcal{W}_1$ and $\mathcal{W}_1 \subset \mathcal{W}_2$. □

Theorem 3.3.7 *Suppose that* $\boldsymbol{N}$ *contains* X_-. *Let* X_1, X_2 *be elements of* $\boldsymbol{N}$ *corresponding to the invariant subspaces* $\mathcal{V}_1$, $\mathcal{V}_2$ *of* $A|\mathcal{O}_{A,C} \cap \mathcal{L}_A^l$. *Then* $X_1 \leq X_2$ *if and only if* $\mathcal{V}_1 \subset \mathcal{V}_2$. □

It follows at once that $\boldsymbol{P}$ and $\boldsymbol{N}$ are complete *lattices* with respect to the ordering of hermitian matrices. The maximal solution X_+ of $\boldsymbol{P}$ is associated with $\mathcal{W} = 0$ and the minimal solution X_- of $\boldsymbol{N}$ with $\mathcal{V} = 0$. The *minimal element* of $\boldsymbol{P}$, which corresponds to the A-invariant subspace $\mathcal{W} = A|\mathcal{O}_{A,C} \cap \mathcal{L}_A^r$, will be denoted by $X_\square$ while the *maximal element* of $\boldsymbol{N}$, which corresponds to the A-invariant subspace $\mathcal{V} = A|\mathcal{O}_{A,C} \cap \mathcal{L}_A^l$, will be denoted by $X_\diamond$.

We shall now pay attention to the cardinality of the solution sets $\boldsymbol{P}$ and $\boldsymbol{N}$. Let ρ be the number of distinct eigenvalues of $A|\mathcal{O}_{A,C} \cap \mathcal{L}_A^r$ and let $p_1, p_2, \ldots, p_\rho$ be the multiplicities of these eigenvalues. Similarly, let σ be the number of distinct eigenvalues of $A|\mathcal{O}_{A,C} \cap \mathcal{L}_A^l$ and let $q_1, q_2, \ldots, q_\sigma$ be the multiplicities of these eigenvalues.

Theorem 3.3.8 ([26]). *Suppose that* $\boldsymbol{P}$ *contains* X_+. *Then* $\boldsymbol{P}$ *has finite cardinality if and only if* $A|\mathcal{O}_{A,C} \cap \mathcal{L}_A^r$ *is cyclic. In this case,* $\boldsymbol{P}$ *contains exactly* $(p_1+1)\ldots(p_\rho+1)$ *elements.*

Proof. Analogous to Theorem 3.2.7. □

Theorem 3.3.9 *Suppose that* $\boldsymbol{N}$ *contains* X_-. *Then* $\boldsymbol{N}$ *has finite cardinality if and only if* $A|\mathcal{O}_{A,C} \cap \mathcal{L}_A^l$ *is cyclic. In this case,* $\boldsymbol{N}$ *contains exactly* $(q_1+1)\ldots(q_\sigma+1)$ *elements.*

Proof. Analogous to Theorem 3.2.7. □

If $A|\mathcal{O}_{A,C} \cap \mathcal{L}_A^r$ is not cyclic, at least one of its eigenvalues is associated with an infinite family of invariant subspaces, so that $\boldsymbol{P}$ is infinite. Similarly, if $A|\mathcal{O}_{A,C} \cap \mathcal{L}_A^l$ is not cyclic then $\boldsymbol{N}$ contains at least one continuous family of solutions. The geometric structure of $\boldsymbol{P}$ and $\boldsymbol{N}$ is inherited from $\boldsymbol{H}$, see [10].

Example 3.3.1 Consider the equation (3.1) where

$$A = \begin{bmatrix} 1 & 1 \\ 0 & 1 \end{bmatrix}, \quad B = \begin{bmatrix} 0 & 0 \\ 0 & 1 \end{bmatrix}, \quad C = \begin{bmatrix} 0 & 0 \\ 0 & 0 \end{bmatrix}$$

and classify the solution sets $\boldsymbol{P}$ and $\boldsymbol{N}$.

We have

$$X_+ = \begin{bmatrix} 8 & 4 \\ 4 & 4 \end{bmatrix}, \quad X_- = \begin{bmatrix} 0 & 0 \\ 0 & 0 \end{bmatrix}.$$

As to the set $\boldsymbol{P}$, note that

$$\mathcal{O}_{A,C} \cap \mathcal{L}_A^r = \boldsymbol{C}^2.$$

There are three A-invariant subspaces

$$\mathcal{W}_+ = 0, \quad \mathcal{W}_1 = \text{Sp}\begin{bmatrix} 1 \\ 0 \end{bmatrix}, \quad \mathcal{W}_\Box = \boldsymbol{C}^2$$

whose X_+- orthogonal complements

$$\tilde{\mathcal{W}}_+ = \boldsymbol{C}^2, \quad \tilde{\mathcal{W}}_1 = \text{Sp}\begin{bmatrix} -1 \\ 2 \end{bmatrix}, \quad \tilde{\mathcal{W}}_\Box = 0$$

yield the projections

$$Q_+ = 0, \quad Q_1 = \begin{bmatrix} 1 & 1/2 \\ 0 & 0 \end{bmatrix}, \quad Q_\Box = I.$$

Using (3.31) we find the elements of $\boldsymbol{P}$ to be

$$X_+ = \begin{bmatrix} 8 & 4 \\ 4 & 4 \end{bmatrix}, \quad X_1 = \begin{bmatrix} 0 & 0 \\ 0 & 2 \end{bmatrix}, \quad X_\square = \begin{bmatrix} 0 & 0 \\ 0 & 0 \end{bmatrix}.$$

By Theorem 3.3.6, $\boldsymbol{P}$ is a complete lattice and

$$X_+ \geq X_1 \geq X_\square.$$

As to the set $\boldsymbol{N}$, we note that

$$\mathcal{O}_{A,C} \cap \mathcal{L}_A^l = 0$$

and hence X_- is the only non-positive definite solution of (3.1). □

Example 3.3.2 Consider the equation (3.1) where

$$A = \begin{bmatrix} 0 & 0 \\ 0 & 0 \end{bmatrix}, \quad B = \begin{bmatrix} 1 & 0 \\ 0 & 0 \end{bmatrix}, \quad C = \begin{bmatrix} 0 & 0 \\ 0 & 0 \end{bmatrix}.$$

It is seen that neither (A, B) nor $(-A, B)$ is stabilizable, hence neither X_+ nor X_- exists, and our preceding analysis does not apply.

The set $\boldsymbol{H}$ consists of one continuous family of solutions,

$$X_\alpha = \begin{bmatrix} 0 & 0 \\ 0 & \alpha \end{bmatrix}$$

for any real α. Therefore both $\boldsymbol{P}$ and $\boldsymbol{N}$ are infinite sets; $\boldsymbol{P}$ is unbounded from above while $\boldsymbol{N}$ is unbounded from below. □

3.3.3 Stabilizing Solutions

An hermitian solution X of (3.1) is said to be *stabilizing* if $A - BX$ is stable, and *anti-stabilizing* if $-(A - BX)$ is stable. It is clear that the stabilizing solution, if it exists, is the maximal hermitian solution of (3.1), and if $C \geq 0$ it is non-negative definite. Similarly the anti-stabilizing solution, if it exists, is the minimal hermitian solution of (3.1) and if $C \geq 0$ it is non-positive definite.

The existence of a stabilizing, anti-stabilizing solution therefore depends on $A - BX_+$, $A - BX_-$ having only eigenvalues with negative, positive real part respectively.

Corollary 3.3.1 ([8]). *The following statements are equivalent:*

(i) $\boldsymbol{P}$ *contains a stabilizing solution.*
(ii) (A, B) *is stabilizable and* H *has no pure imaginary eigenvalue.*
(iii) (A, B) *is stabilizable and the pure imaginary eigenvalues of* A*, if any, are all* C*-observable.*

Proof. The set $\boldsymbol{P}$ contains a stabilizing solution, namely X_+, if and only if all eigenvalues of $A - BX_+$ have negative real part. By Corollary 3.2.1, this is equivalent to (ii).

We now show the equivalence of (ii) and (iii). If $Au = i\omega u,\ \ Cu = 0$ for some real ω and $u \neq 0$, then

$$H \begin{bmatrix} u \\ 0 \end{bmatrix} = i\omega \begin{bmatrix} u \\ 0 \end{bmatrix}.$$

Hence H has a pure imaginary eigenvalue. Conversely, suppose H has a pure imaginary eigenvalue, say $i\omega$. Since (A, B) is stabilizable, X_+ exists and by (3.8), $A_+ v = i\omega v$ for some $v \neq 0$. Now it follows from

$$X_+ A_+ + A_+^* X_+ + (C + X_+ B X_+) = 0$$

that

$$v*(C + X_+ B X_+)v = 0.$$

Since $B \geq 0$ and $C \geq 0$ this implies $Cv = BX_+ v = 0$, hence $Cv = 0$ and $Av = i\omega v$. □

Corollary 3.3.2 *The following statements are equivalent:*

(i) *$\boldsymbol{N}$ contains an anti-stabilizing solution.*
(ii) *$(-A, B)$ is stabilizable and H has no pure imaginary eigenvalue.*
(iii) *$(-A, B)$ is stabilizable and the pure imaginary eigenvalues of A, if any, are all C-observable.* □

These two results now appear to be easy corollaries of the general theorems. Historically, however, they were the first results dealing with existence of solutions to (3.1) in the absence of reachability of (A, B) and observability of (A, C).

Corollary 3.3.3 ([8]). *The set $\boldsymbol{P}$ contains a single element $X_+ = X_\square$ and this element is stabilizing if and only if (A, B) is stabilizable and (A, C) is detectable.*

Proof. By Theorem 3.3.6, one has $X_+ = X_\square$ if and only if

$$\mathcal{O}_{A,C} \cap \mathcal{L}_A^r = 0.$$

Combining this with Corollary 3.3.1 shows that

$$\mathcal{O}_{A,C} \cap \mathcal{L}_A^r = \mathcal{D}_{A,C}$$

and proves the claim. □

Corollary 3.3.4 *The set $\boldsymbol{N}$ contains a single element $X_- = X_\diamond$ and this element is anti-stabilizing if and only if $(-A, B)$ is stabilizable and $(-A, C)$ is detectable.* □

3.4 Applications to Optimal Control

As an example of the occurrence of Riccati matrix differential and algebraic equations in the variational context of optimal control, we shall consider a *linear-qudratic optimal control problem.* It consists in finding a control function $u(t)$ with values in $\mathbf{C}^n$ which minimizes quadratic cost functional

$$\eta(x, u|t_0, t_1) = \int_{t_0}^{t_1} [x^*(t)Cx(t) + u^*(t)u(t)]dt + x^*(t_1)Dx(t_1) \tag{3.36}$$

where $x(t)$ is the state, with values in $\mathbf{C}^n$, of the linear system

$$\dot{x}(t) = Fx(t) + Gu(t), \ x(t_0) = a \tag{3.37}$$

and C, D and F, G are constant matrices with entries in $\mathbf{C}$ such that $C = C^*$, $C \geq 0$ and $D = D^*$, $D \geq 0$.

It has been known for some time [27] that a unique optimal control

$$u(t) = -G^*X(t, t_1, D)x(t) \tag{3.38}$$

exists for all finite $t_1 - t_0 > 0$ and $X(t, t_1, D)$ is the solution of the Riccati matrix differential equation (3.3),

$$\dot{X}(t) + X(t)A + A^*X(t) - X(t)BX(t) + C = 0, \tag{3.39}$$

subject to the terminal condition

$$X(t_1) = D \tag{3.40}$$

where

$$A = F, \ \ B = GG^*.$$

The optimal control (3.38) is a linear state feedback, which gives rise to the closed-loop system

$$\dot{x}(t) = [A - BX(t, t_1, D)]x(t) \tag{3.41}$$

and affords the minimum cost

$$\bar{\eta}(t_0, t_1) = a^*X(t_0, t_1, D)a.$$

Let us now consider the effect of letting the interval $t_1 - t_0$ tend to infinity in (3.36); this is called the *infinite-horizon* linear-quadratic optimal control problem. If (A, B) is stabilizable and (A, C) is detectable, Kwakernaak and Sivan [28] showed that the infinite-horizon optimal control is given by

$$u(t) = -G^*X_+x(t)$$

where X_+ satisfies the algebraic Riccati equation (3.1),

$$XA + A^*X - XBX + C = 0.$$

The matrix X_+ is the only hermitian non-negative definite solution of (3.1) and $A - BX_+$ is stable (Corollary 3.3.3). Hence the infinite-horizon optimal system

$$\dot{x}(t) = (A - BX_+)x(t)$$

is stable. Moreover it is well known [12] that for any D, the solution $X(t_0, t_1, D)$ of (3.39) and (3.40) tends to X_+ as the interval $t_1 - t_0$ tends to infinity.

The situation becomes much more complicated if (A, C) is not detectable. In this case the algebraic Riccati equation (3.1) has more than one hermitian non-negative definite solution. In fact, the solution set is a complete lattice (Theorem 3.3.6) with a largest element X_+ and a smallest element $X_\square$. Each element of the set is associated with an A-invariant subspace contained in $\mathcal{O}_{A,C} \cap \mathcal{L}_A^r$ (Theorem 3.3.4), with X_+ corresponding to 0 and $X_\square$ to the whole subspace $\mathcal{O}_{A,C} \cap \mathcal{L}_A^r$.

Obviously, for a suitable choice of the terminal conditon D, each hermitian non-negative definite solution of (3.1) is a limiting solution of (3.39) when $t_1 - t_0$ tends to infinity. On the other hand, $X(t_0, t_1, D)$ may fail to converge for some D. If it does converge to, say, X_D then Callier and Willems [22] proved that X_D is the hermitian non-negative definite soluiton of (3.1) which is associated with the largest A-invariant subspace contained in the intersection of Ker D and $\mathcal{O}_{A,C} \cap \mathcal{L}_A^r$. In particular, $X_D = X_+$ if and only if

$$\text{Ker } D \cap (\mathcal{O}_{A,C} \cap \mathcal{L}_A^r) = 0$$

and $X_D = X_\square$ if and only if

$$(\mathcal{O}_{A,C} \cap \mathcal{L}_A^r) \subset \text{Ker } D.$$

For example, $D = 0$ always leads to $X_\square$ whereas $D > 0$ to X_+.

The infinite-horizon optimal control

$$u(t) = -G^* X_D x(t)$$

gives rise to the closed-loop system

$$\dot{x}(t) = (A - BX_D)x(t) \tag{3.42}$$

and affords the minimum cost

$$\bar{\eta} = a^* X_D a. \tag{3.43}$$

The optimal system (3.42) will be stable if and only if $X_D = X_+$ and X_+ is stable. This is the case if and only if

$$\text{Ker } D \cap \mathcal{D}_{A,C} = 0,$$

see [29]. Thus the unstable eigenvalues left in (3.42) are precisely those of A which are neither C-observable nor D-observable.

For practical reasons we often want a, the initial state of (3.42), to belong to $\mathcal{L}^l_{A-BX_D}$. This can be achieved by a judicious choice of C and D. Each time an extra eigenvalue of $A - BX_D$ is stabilized, however, the minimum cost (3.43) increases. This is a consequence of the lattice structure of the set of hermitian non-negative definite solutions and illustrates the trade-off between stability and optimality in the control system (3.42).

References

1. Reid, W.T. (1972). *Riccati Differential Equations*. Academic Press, New York.
2. Faibusovich, L.E. (1986). Algebraic Riccati equation and symplectic algebra. *Int. J. Control*, 43, 781-792.
3. Gantmacher, F.R. (1959). *The Theory of Matrices*. Chelsea, New York.
4. Barnett, S. (1983). *Polynomials and Linear Control Systems*. Dekker, New York.
5. MacFarlane, A.G.J. (1963). An eigenvector solution of the optimal linear regulator problem. *J. Electron. Contr.*, 14, 496-501.
6. Potter, J.E. (1966). Matrix quadratic solutions. *SIAM J. Appl. Math.*, 14, 496-501.
7. Mårtensson, K. (1971). On the matrix Riccati equation. *Inform. Sci.*, 3, 17-49.
8. Kučera, V. (1972). A contribution to matrix quadratic equaitons. *IEEE Trans. Automat. Contr.*, AC-17, 344-347.
9. Van Swieten, A.C.M. (1977).*Qualitative behaviour of dynamical games with feedback strategies*. Ph.D. Thesis, University of Groningen, The Netherlands.
10. Shayman, M.A. (1983). Geometry of the algebraic Riccati equation, Parts I and II. *SIAM J. Contr. Optimiz.*, 21, 375-394 and 395-409.
11. Roth, W.E. (1950). On the matrix equatin $X^2 + AX + XB + C = 0$. *Proc. Amer. Math. Soc.*, 1, 586-589.
12. Kučera, V. (1973). A review of the matrix Riccati equation. *Kybernetika*, 9, 42-61.
13. Lancaster, P. and L. Rodman (1980). Existence and uniqueness theorems for the algebraic Riccati equation. *Int. J. Control*, 32, 285-309.
14. Wimmer, H.K. (1982). The algebraic Riccati equation without complete controllability. *SIAM J. Alg. Disc. Meth.*, 3, 1-12.
15. Churilov, A.N. (1978). On the solutions of quadratic matrix equations (in Russian). *Nonlinear Vibrations and Control Theory* (Udmurt State University, Izhevsk), 2, 24-33.
16. Wimmer, H.K. (1984). The algebraic Riccati equation: Conditions for the existence and uniqueness of solutions. *Linear Alg. Appl.*, 58, 441-452.
17. Reid, W.T. (1963). Riccati matrix differential equations and non-oscillation criteria for associated linear diferential systems. *Pacific J. Math.*, 13, 665-685.
18. Coppel, W.A. (1974). Matrix quadratic equations. *Bull. Austral. Math. Soc.*, 10, 377-401.
19. Rodriguez-Canabal, J. (1973). The geometry of the Riccati equation. *Stochastics*, 1, 129-149.
20. Lancaster, P. (1970). Explicit solutions of linear matrix equations. *SIAM Review*, 12, 544-566.
21. Willems, J.C. (1971). Least squares stationary optimal control and the algebraic Riccati equation. *IEEE Trans. Automat. Contr.*, AC-16, 621-634.
22. Callier, F.M. and J.L. Willems (1981). Criterion for the convergence of the solution of the Riccati differential equation. *IEEE Trans. Automat. Contr.*, AC-26, 1232-1242.
23. Molinari, B. P. (1977). The time-invariant linear-quadratic optimal control. *Automatica*, 13, 347-357.
24. Ando, T. (1988). *Matrix quadratic equations*. Lecture Notes, Hokkaido University, Sapporo, Japan.
25. Wonham, W.M. (1968). On a matrix Riccati equation of stochastic control. *SIAM J. Control*, 6, 681-697.

26. Kučera, V. (1972). On nonnegative definite solutions to matrix quadratic equations. *Automatica*, 8, 413-423.
27. Kalman, R.E. (1961). Contributions to the theory of optimal control. *Bol. Soc. Mat. Mex.*, 2, 102-119.
28. Kwakernaak, H. and R. Sivan (1972). *Linear Optimal Control Systems*. Wiley, New York.
29. Willems, J.L. and F.M. Callier (1983). Large finite horizon and infinite horizon LQ-optimal control problems. *Opt. Contr. Appl. Math.*, 4, 31-45.

4 A Geometric View of the Matrix Riccati Equation

Mark A. Shayman

4.1 Introduction

This is a tutorial paper which describes a "geometric approach" to the description of the phase portrait of the Riccati differential equation. As such, no new results are presented. Our intention is to show how the geometric viewpoint gives insight into many of the properties of the Riccati differential equation. It is not our purpose to present a comprehensive exposition of all that is presently known on the subject. Instead, we will willingly make (mostly) generic assumptions and focus on the properties of the differential equation under these simplifying assumptions. For more details, some generalizations and additional references, the reader is referred to the papers [4.14] (time-invariant coefficients) and [4.13] (periodic coefficients).

The matrix Riccati differential equation (RDE) refers to the quadratic differential equation

$$\dot{K} = B_{21} + B_{22}K - KB_{11} - KB_{12}K \tag{4.1}$$

defined on the vector space $\mathbb{R}^{m \times n}$ of real $m \times n$ matrices. Here, B_{ij} is a (possibly time-varying) matrix of the appropriate dimensions. Of particular interest is the Riccati equation which arises in optimal control and filtering problems which has the form

$$\dot{K} = -Q - A'K - KA + KLK \tag{4.2}$$

defined on the space $\mathbb{R}^{n \times n}$ of real $n \times n$ matrices. A, L, Q are (possibly time-varying) real $n \times n$ matrices with L and Q symmetric and L usually nonnegative definite. The vector space $S(n)$ of real symmetric $n \times n$ matrices in an invariant manifold for this differential equation, and it is the restriction to $S(n)$ which is important in the applications. Thus, we will regard this differential equation as defined on $S(n)$, and call it the *symplectic Riccati differential equation* (SRDE) for reasons which will become apparent.

It has been known at least since the time of Poincaré that the behavior of certain differential equations in the plane could be clarified by extending the domain to the projective plane. It was observed by C. Schneider [4.11] that the natural compactification of the domain $\mathbb{R}^{m \times n}$ for the RDE is the *Grassmann manifold* $G^n(\mathbb{R}^{n+m})$ of n-dimensional subspaces of $\mathbb{R}^{n+m}$. We refer to the Riccati equation on $G^n(\mathbb{R}^{n+m})$

Bittanti, Laub, Willems (Eds.)
The Riccati Equation

as the *extended Riccati differential equation* (ERDE). The natural compactification of the domain $S(n)$ for the SRDE has been described by R. Hermann and C. Martin [4.3] [4.8]. It is the *Lagrange-Grassmann manifold* $\mathcal{L}(n)$ which consists of those n-dimensional subspaces of $\mathbb{R}^{2n}$ on which a certain nondegenerate skew-symmetric bilinear form vanishes. We refer to the symplectic Riccati equation on $\mathcal{L}(n)$ as the *extended symplectic Riccati differential equation* (ESRDE).

The "geometric viewpoint" of extending the Riccati equation to the Grassmann (or Lagrange-Grassmann) manifold is extremely useful for studying many properties of the differential equation. A principal advantage of this approach is that the flow on the Grassmann manifold is extremely simple. It is given by a one-parameter subgroup of the general linear group (or of the symplectic group) acting in the natural way on the Grassmann manifold. This permits the (extended) Riccati equation to be viewed as a *linear flow together with a projection.* The simplicity of this flow permits a complete description of the phase portrait (qualitative behavior of all solutions) under generic assumptions. The key features of the phase portrait are invariant tori of various dimensions to which every solution converges. The stable and unstable manifolds of the tori are unions of Schubert cells for the Grassmann (respectively, Lagrange-Grassmann) manifolds.

It must be emphasized that the phase portraits of the ERDE and ESRDE are not the same as the phase portraits of the RDE and SRDE, although they are closely related. The extended differential equations are constructed by adding some points to compactify the state-space, and then extending the flow through these "points at infinity." Since the extended state-space is compact, the extended vector field is necessarily complete. Consequently, the ERDE and ESRDE do not exhibit finite escape times. Of course, it is well-known that finite escape times are typical of certain solutions of the RDE and SRDE.

A second complication is that the extension of the differential equation can result in additional equilibrium points (or more generally, invariant tori) among the points at infinity. These equilibrium points do not correspond to equilibrium points for the original RDE or SRDE. Consequently, the task of recovering the phase portrait of the original RDE or SRDE from the phase portrait of the extended equation is not trivial. However, in the case of the SRDE in the presence of some system-theoretic assumptions (nonnegativity of the quadratic term and controllability), this is indeed possible, resulting in a complete description of the phase portrait on the space $S(n)$ of symmetric matrices.

The geometric viewpoint is also useful in reinterpreting and extending certain "classical" results in the theory of the SRDE. The classical results which will be discussed are the following: (1) Radon's formula $K(t) = Y(t)X(t)^{-1}$ for the solution to the RDE [4.10]; (2) the eigenvector method for computing the solutions of the algebraic Riccati equation (sometimes referred to as Potter's method [4.9]); (3) the classification of the solutions of the algebraic Riccati equation via projections due to J.C. Willems [4.16]; (4) the role of controllability in the existence of solutions for the algebraic Riccati equation; (5) the role of observability in the nonsingularity of solutions of the algebraic Riccati equation; (6) the domains of attraction for the extremal equilibriums points of the SRDE; (7) the inertia theorem for the solutions of the algebraic Riccati equation due to Wimmer [4.17].

4.2 Extension to the Grassmann Manifold

One of the questions which led Count Riccati to become interested in quadratic differential equations was the following: Suppose that a trajectory in the plane satisfies a linear differential equation. What can be said about the time-evolution of the slope of the line through the origin determined by the trajectory?

Consider the linear (possibly time-varying) differential equation in the plane described by

$$\begin{pmatrix} \dot{x} \\ \dot{y} \end{pmatrix} = \begin{pmatrix} b_{11} & b_{12} \\ b_{21} & b_{22} \end{pmatrix} \begin{pmatrix} x \\ y \end{pmatrix} \tag{4.3}$$

Let $k = y/x$ denote the slope of the line through the origin which is determined by the point (x, y). It is trivial to verify that k satisfies the differential equation

$$\dot{k} = b_{21} + (b_{22} - b_{11})k - b_{12}k^2 \tag{4.4}$$

Thus, the slope evolves as the solution of a Riccati differential equation.

Suppose that we are interested in the line itself rather than in its slope. Let $[x, y]$ denote the line through the origin which is determined by the point (x, y). In contrast to the slope k, the line $[x, y]$ is not an element of a Euclidean space R^n. Rather, $[x, y]$ is an element of the differentiable manifold $\mathsf{RP}(1)$, 1-dimensional real projective space – i.e., the manifold consisting of all the lines through the origin in R^2. Since $\mathsf{RP}(1)$ is not a Euclidean space, we would need to make a choice of local coordinates before we could write down a differential equation which describes the evolution of $[x, y]$. Furthermore, the resulting differential equation would be valid only on some open subset of $\mathsf{RP}(1)$ which is homeomorphic to an open subset of R^1.

However, there is another way that we can describe the evolution of $[x, y]$. Instead of giving a differential equation, we can display the flow explicitly. Let B denote the (possibly time-varying) coefficient matrix for the linear differential equation, and let $\Phi(t, t_0)$ denote the corresponding transition matrix. Let $\mathrm{Sp}\begin{pmatrix} x \\ y \end{pmatrix}$ denote the span of the vector (x, y) – i.e., the line $[x, y]$. Then the line evolves according to the flow $\mathrm{Sp}\begin{pmatrix} x(t) \\ y(t) \end{pmatrix}$ where

$$\begin{pmatrix} x(t) \\ y(t) \end{pmatrix} := \Phi(t, t_0) \begin{pmatrix} x(t_0) \\ y(t_0) \end{pmatrix}. \tag{4.5}$$

There are several features of the flow $\mathrm{Sp}\begin{pmatrix} x(t) \\ y(t) \end{pmatrix}$ which are worth noting. First of all, this flow can be viewed as the composition of a projection map Π : $\mathsf{R}^2 \to \mathsf{RP}(1)$, $\Pi((x,y)) := [x,y]$ with the linear flow $\Phi(t,t_0)\begin{pmatrix} x(t_0) \\ y(t_0) \end{pmatrix}$ on R^2. Secondly, the trajectory of the slope $k(t)$ is easily recovered from the flow $[x(t),y(t)]$ on $\mathsf{RP}(1)$ by the formula $k(t) = y(t)x(t)^{-1}$ which is valid as long as $x(t)$ is nonzero – i.e., as long as the line $[x(t),y(t)]$ is not vertical. Thirdly, note that the flow on $\mathsf{RP}(1)$ is defined for all time whereas the solution $k(t)$ can escape to infinity in finite time. This happens when $x(t)$ first becomes zero or, equivalently, when the line $[x(t),y(t)]$ first becomes vertical.

Some comments about the structure of $\mathsf{RP}(1)$ are useful. Each point in $\mathsf{RP}(1)$ is a line through the origin in R^2, and each such line can be indexed by the angle it makes with the x-axis. This angle varies from 0 to π. However, the angles 0 and π represent the same line, namely the x-axis. Thus, topologically $\mathsf{RP}(1)$ can be regarded as the closed interval $[0,\pi]$ with the two endpoints identified – i.e., the circle S^1. R^1 can be imbedded in $\mathsf{RP}(1)$ by the mapping ψ : $\mathsf{R}^1 \to \mathsf{RP}(1)$ with $\psi(k) := [1,k]$. The image of this mapping is the subset of $\mathsf{RP}(1)$ which consists of all lines which are not vertical. If $[x,y]$ represents such a line – i.e., $x \neq 0$, then $\psi^{-1}([x,y]) = \psi^{-1}([1,yx^{-1}]) = yx^{-1}$. By this imbedding, we may identify R^1 with the open and dense subset of $\mathsf{RP}(1)$ which consists of all the nonvertical lines. The flow $[x(t),y(t)]$ on $\mathsf{RP}(1)$ corresponds to the flow $k(t)$ on R^1 via $k(t) = y(t)x(t)^{-1}$ as long as the trajectory in $\mathsf{RP}(1)$ remains in this subset.

Let us consider some examples.

Example 4.1. Let $B = \begin{pmatrix} 1 & 0 \\ 0 & 2 \end{pmatrix}$. The equilibrium points of the flow on $\mathsf{RP}(1)$ are the lines spanned by the eigenvectors of B, namely $S_1 = [1,0]$ and $S_2 = [0,1]$. To investigate the stability properties of these equilibria, let the initial point (i.e., initial line) S_0 be any point other than S_1. Since S_0 is not horizontal, we can express it as $[r,1]$ for some real number r. The resulting trajectory

$$S(t) = \mathrm{Sp}\, e^{Bt}\begin{pmatrix} r \\ 1 \end{pmatrix} = \mathrm{Sp}\begin{pmatrix} e^t r \\ e^{2t} \end{pmatrix} = \mathrm{Sp}\begin{pmatrix} e^{-t} r \\ 1 \end{pmatrix} \to \mathrm{Sp}\begin{pmatrix} 0 \\ 1 \end{pmatrix}$$

as $t \to \infty$. Thus, S_1 is unstable while S_2 is asymptotically stable with domain of attraction consisting of every point other than S_1.

The corresponding Riccati differential equation is $\dot{k} = k$. This has a unique equilibrium point at the origin, and this equilibrium point is unstable. Every nonconstant solution goes to infinity asymptotically.

The difference between the two phase portraits is easily explained. $\mathsf{RP}(1)$ contains an extra point which does not correspond to any point of R^1 – namely the point $[0,1]$ which is not in the image of the imbedding ψ. This point is the extra equilibrium point of the flow on $\mathsf{RP}(1)$. The asymptotic stability of this equilibrium point is reflected in the phase portrait of the Riccati equation on R^1 in the fact that all the nonconstant solutions asymptotically approach infinity. In light of this, it is natural to regard the extra point $[0, 1]$ as a "point at infinity" which has been added to the real line. □

Example 4.2. Redefine $B = \begin{pmatrix} 2 & 0 \\ 0 & 1 \end{pmatrix}$. The equilibrium points S_1 and S_2 are unchanged, but their stability properties are reversed. The corresponding Riccati differential equation is $\dot{k} = -k$ which has the origin as a globally asymptotically stable equilibrium point. Since the origin corresponds to S_1 via the imbedding ψ, we again see that corresponding equilibrium points have the same stability properties. □

Example 4.3. Redefine $B = \begin{pmatrix} 1 & 1 \\ 0 & 2 \end{pmatrix}$. The equilibrium points of the flow on $\mathsf{RP}(1)$ are the lines spanned by the eigenvectors of B, namely $S_1 = [1, 0]$ and $S_2 = [1, 1]$. A straightforward calculation shows that every initial condition other than S_1 results in convergence to S_2. The corresponding Riccati equation is $\dot{k} = k - k^2$ which has an unstable equilibrium point $k_1 = 0$ and an asymptotically stable equilibrium point $k_2 = 1$. Since k_1 and k_2 correspond to S_1 and S_2 via the imbedding ψ, the corresponding equilibrium points have the same stability properties.

Even though the equiliibrium points and their stability properties correspond via ψ, there is an important difference between the two phase portraits. It is not hard to check that the domain of attraction of k_2 is given by the condition $k_0 > 0$. If $k_0 < 0$, the resulting trajectory has a finite escape time. Thus, the imbedding ψ can map trajectories in R^1 which escape in finite time to trajectories in $\mathsf{RP}(1)$ which converge. The explanation for this lack of correspondence is that the corresponding trajectories cross the point $[0, 1]$ in the process of converging to S_2. Since $[0, 1]$ is outside the image of ψ, the correspondence between the trajectories $k(t)$ and $S(t)$ is lost – $k(t)$ passing to infinity at this time while $S(t)$ continues on its asymptotic approach to S_2. □

It is worth noting that as far as the phase portraits on $\mathsf{RP}(1)$ are concerned, Examples 4.1–4.3 are essentially the same. In each case there are two equilibrium points, one of which having a region of attraction which includes every nonequilibrium point. The equivalence of the three phase portraits is explained as follows: Let B be a constant 2×2 matrix, let P be a nonsingular 2×2 matrix, and let $\bar{B} := PBP^{-1}$. Then it is trivial to check that if $S(t)$ is a trajectory of the flow corresponding to the coefficient matrix B, then $PS(t)$ is a trajectory of the flow corresponding to the coefficient matrix $\bar{B}$. In other words, the nonsingular linear transformation P on R^2 induces a diffeomorphism of the manifold $\mathsf{RP}(1)$ (where $P[x, y]$ is the image of the line $[x, y]$ under the linear transformation P) which maps trajectories to trajectories. Since the matrices B in Examples 4.1–4.3 have the same eigenvalues, the three phase portraits are related by such transformations. This brings up the question as to why the three corresponding phase portraits on the real line are so different. The answer is that the diffeomorphism of $\mathsf{RP}(1)$ induced by P will generally not leave invariant the subset which is the image of the imbedding ψ. Consequently, the diffeomorphism cannot be restricted to this subset to obtain a diffeomorphism of the real line. To say this another way, the diffeomorphism induced by P can map the point at infinity $[0,1]$ to a point of the form $[1,k]$ in the image of ψ. If $[0,1]$ is an equilibrium point for B, then $[1,k]$ is an equilibrium point for $\bar{B}$. The Riccati equation corresponding to $\bar{B}$ has an equilibrium point at k whereas the Riccati equation corresponding to B has no equilibrium point corresponding to $[0,1]$ since it is outside the image of ψ. In going from Example 4.1 to Example 4.2, the transformation P interchanges which equilibrium point is "at infinity" and hence changes the stability property of the single equilibrium point for the Riccati equation. In going from Example 4.1 to Example 4.3, P moves the asymptotically stable equilibrium point from the point $[0,1]$ at infinity to a point $[1,k_2]$ in the image of ψ. Thus, the Riccati equation in Example 4.3 has an asymptotically stable equilibrium point while the one in Example 4.1 does not.

Example 4.4. Redefine $B = \begin{pmatrix} a & -b \\ b & a \end{pmatrix}$, where $b \neq 0$. Letting $S_0 = [1,0]$ yields

$$S(t) = \text{Sp}\, e^{Bt} \begin{pmatrix} 1 \\ 0 \end{pmatrix} = \text{Sp} \begin{pmatrix} e^{at}\cos bt \\ e^{at}\sin bt \end{pmatrix} = \text{Sp} \begin{pmatrix} \cos bt \\ \sin bt \end{pmatrix}.$$

Thus, $S(t)$ is periodic with period π/b. The corresponding Riccati equation is

$$\dot{k} = b + bk^2$$

The initial condition corresponding to S_0 is $k_0 = 0$ which yields the solution $k(t) = \tan bt$. This solution is not periodic. Rather, the solution has finite escape time $\pi/(2b)$ which is when the solution $S(t)$ crosses the point at infinity $[0,1]$.

It is useful to interpret the periodic solution $S(t)$ from the point of view that the flow on $\mathsf{RP}(1)$ is a linear flow on R^2 composed with the projection Π. The trajectory of the linear flow corresponding to the initial condition $(1,0) \in \mathsf{R}^2$ is

$$\begin{pmatrix} e^{at}\cos bt \\ e^{at}\sin bt \end{pmatrix}$$

which is not periodic if a is nonzero. $S(t)$ is obtained from this trajectory by applying the projection Π which identifies as the same all points in R^2 which are on the same line through the origin. Thus, the radial distance from the origin becomes irrelevant and a periodic solution results which is independent of the value of a. □

In Example 4.4, the periodic solution on $\mathsf{RP}(1)$ does not correspond to a periodic solution of the Riccati equation on R^1 but rather to a solution which escapes in finite time. This is necessarily the case for a (time-invariant) scalar Riccati equation since a time-invariant differential equation on the real line cannot have a periodic solution. Equivalently, this can be explained by noting that $\mathsf{RP}(1)$ is a circle, so every periodic solution on $\mathsf{RP}(1)$ must traverse every point of $\mathsf{RP}(1)$ including the point $[0,1]$ at infinity. In higher dimensions, this topological obstruction to periodic solutions is not present. Under typical conditions, the periodic solutions on the higher-dimensional analogues of $\mathsf{RP}(1)$ will indeed correspond (via an imbedding

ψ) to periodic solutions of the Riccati equation. In addition, in higher dimensions, the periodic motion exhibited in Example 4.4 can occur in several independent coordinates simultaneously. This results in invariant sets for the Riccati equation which are products of circles – i.e., invariant tori. Under reasonable assumptions, the phase portrait of the Riccati equation is characterized by the presence of a finite number of invariant tori of various dimensions. The 0-dimensional tori are the equilibrium points, the 1-dimensional tori are the isolated periodic orbits, while the higher-dimensional tori contain trajectories which are either periodic or almost periodic. Furthermore, every other solution either escapes in finite time or converges to one of the invariant tori.

In the scalar case, the Riccati equation was extended from R^1 to $\mathsf{RP}(1)$ via the imbedding ψ which identifies the real number k with the line $[1, k]$ – i.e., with the 1-dimensional subspace in R^2 given by $\mathrm{Sp}\begin{pmatrix} 1 \\ k \end{pmatrix}$. In higher dimensions, K is an $m \times n$ matrix. The corresponding imbedding is obtained by identifying the matrix K with the n-dimensional subspace $\mathrm{Sp}\begin{pmatrix} I \\ K \end{pmatrix}$ – i.e., the column space of the matrix $\begin{pmatrix} I \\ K \end{pmatrix}$. Thus, K is identified with an n-dimensional subspace of R^{n+m}.

Let $G^n(\mathsf{R}^{n+m})$ denote the *Grassmann manifold* of all n-dimensional subspaces of R^{n+m}. $G^n(\mathsf{R}^{n+m})$ is a compact real-analytic manifold of dimension mn. Define an imbedding

$$\psi : \mathsf{R}^{m\times n} \to G^n(\mathsf{R}^{n+m}), \qquad \psi(K) := \mathrm{Sp}\begin{pmatrix} I_n \\ K \end{pmatrix}. \tag{4.6}$$

The image $G_0^n(\mathsf{R}^{n+m})$ of ψ consists of those n-dimensional subspaces which are complementary to the m-dimensional subspace $\mathrm{Sp}\begin{pmatrix} 0 \\ I_m \end{pmatrix}$. It is easy to show that $G_0^n(\mathsf{R}^{n+m})$ is an open and dense subset of $G^n(\mathsf{R}^{n+m})$. It is natural to identify $\mathsf{R}^{m\times n}$ with $G_0^n(\mathsf{R}^{n+m})$ and to regard $G^n(\mathsf{R}^{n+m}) - G_0^n(\mathsf{R}^{n+m})$ as a hypersurface of points "at infinity" which have been added to $\mathsf{R}^{m\times n}$ to compactify the space.

Define a flow on $G^n(\mathsf{R}^{n+m})$ given by

$$S(t, S_0, t_0) := \Phi(t, t_0)(S_0) \tag{4.7}$$

where $\Phi(t, t_0)$ is the transition matrix associated with

$$B(t) = \begin{pmatrix} B_{11}(t) & B_{12}(t) \\ B_{21}(t) & B_{22}(t) \end{pmatrix}$$

and $\Phi(t, t_0)(S_0)$ denotes the image of the n-dimensional subspace S_0 under the nonsingular linear transformation $\Phi(t, t_0)$. Let $K(t, K_0, t_0)$ denote the flow of the Riccati equation (4.1). It is easy to derive the relationship between $S(t, S_0, t_0)$ and $K(t, K_0, t_0)$. Let $\begin{pmatrix} X(t) \\ Y(t) \end{pmatrix} := \Phi(t, t_0)\begin{pmatrix} I \\ K_0 \end{pmatrix}$. It has been known at least since the work of Radon [4.10] (and is easily verified by differentiation) that

$$K(t, K_0, t_0) = Y(t)X(t)^{-1} \tag{4.8}$$

which is valid as long as the indicated inverse continues to exist. Using this formula, we have

$$\psi(K(t, K_0, t_0)) = \mathrm{Sp}\begin{pmatrix} I \\ Y(t)X(t)^{-1} \end{pmatrix} = \Phi(t, t_0)\mathrm{Sp}\begin{pmatrix} I \\ K_0 \end{pmatrix}.$$

Thus,

$$\psi(K(t, K_0, t_0)) = S(t, \psi(K_0), t_0). \tag{4.9}$$

Equation (4.9) is valid as long as $X(t)^{-1}$ continues to exist – i.e., as long as the solution $K(t, K_0, t_0)$ continues to exist. This is equivalent to the trajectory $S(t, S_0, t_0)$ remaining in the subset $G_0^n(\mathbb{R}^{n+m})$ of $G^n(\mathbb{R}^{n+m})$.

Equation (4.9) can be viewed as a "geometric version" of Radon's formula. The content of (4.9) is that the imbedding ψ (which identifies $\mathbb{R}^{m\times n}$ with $G_0^n(\mathbb{R}^{n+m})$) maps the trajectories of the Riccati equation (4.1) onto the trajectories of the restriction of the flow $S(t, S_0, t_0)$ to $G_0^n(\mathbb{R}^{n+m})$. Thus, modulo the identification ψ, the Riccati equation can be regarded as the flow $S(t, S_0, t_0)$ restricted to the open and dense subset $G_0^n(\mathbb{R}^{n+m})$. Consequently, the flow of the Riccati equation can be regarded loosely as a linear flow composed with the projection Π. The reason that we say "loosely" is that the identification between $K(t, K_0, t_0)$ and $S(t, \psi(K_0), t_0)$ is valid only as long as $S(t, \psi(K_0), t_0)$ remains in the subset $G_0^n(\mathbb{R}^{n+m})$.

This suggests a two step strategy to characterizing the phase portrait of the Riccati equation: (1) Determine the phase portrait of the flow $S(t, S_0, t_0)$ on $G^n(\mathbb{R}^{n+m})$. (2) Identify which of the important dynamic phenomena (e.g., equilibrium points, periodic orbits, invariant tori) are actually contained in the subset $G_0^n(\mathbb{R}^{n+m})$ and hence represent dynamic phenomena for the Riccati equation on the space of matrices. We will see that in the case of constant coefficients, the first step can be carried out for a general Riccati equation (4.1) under generic assumptions. However, to obtain a satisfactory result in the second step, it appears to be necessary to specialize to the symplectic Riccati equation (4.2) defined on the space $S(n)$ of symmetric matrices. In addition, system-theoretic assumptions (especially controllability) play an important role in the second step whereas they play no role whatsoever in step one.

Let us now consider the additional structure present in the case of the symplectic Riccati equation (4.2). In this case, K is $n \times n$, so the extended space is $G^n(\mathbb{R}^{2n})$. The matrix $B(t)$ is the $2n \times 2n$ matrix

$$\begin{pmatrix} A(t) & -L(t) \\ -Q(t) & -A(t)' \end{pmatrix}.$$

For reasons of tradition, we will denote this matrix by $H(t)$ rather than $B(t)$. $H(t)$ is an infinitesimally symplectic (i.e., Hamiltonian) matrix since it satisfies the condition $JH + H'J = 0$ where J denotes the $2n \times 2n$ skew-symmetric matrix $\begin{pmatrix} 0 & I \\ -I & 0 \end{pmatrix}$. Consequently, the transition matrix $\Phi(t, t_0)$ is symplectic – i.e., satisfies the condition

$$\Phi(t, t_0)' J \Phi(t, t_0) = J$$

It follows from the special structure of the coefficients in (4.2) that the subspace $S(n)$ of symmetric matrices is an invariant manifold for the symplectic Riccati equation. One would expect there to be a corresponding invariant submanifold of $G^n(\mathsf{R}^{2n})$ for the ESRDE. This is indeed the case. Define a skew-symmetric bilinear form ω on R^{2n} by

$$\omega(x, y) := x' J y \tag{4.10}$$

The *Lagrange-Grassmann manifold* $\mathcal{L}(n)$ consists of those n-dimensional subspaces of R^{2n} on which ω vanishes identically. I.e.,

$$\mathcal{L}(n) := \{S \in G^n(\mathsf{R}^{2n}) \mid \omega(x, y) = 0, \ \forall\, x, y \in S\}.$$

It is trivial to check that the imbedding ψ restricts to an imbedding of $S(n)$ into $\mathcal{L}(n)$ which we will again denote by ψ. The image $\mathcal{L}_0(n)$ consists of those elements of $\mathcal{L}(n)$ which are complementary to the n-dimensional subspace $\mathrm{Sp}\begin{pmatrix} 0 \\ I \end{pmatrix}$. We identify $S(n)$ with this image and view $\mathcal{L}(n) - \mathcal{L}_0(n)$ as a hypersurface of points which have been added to $S(n)$ at infinity to compactify the space. Since $\Phi(t, t_0)$ is symplectic, $\mathcal{L}(n)$ is an invariant manifold for the flow $S(t, S_0, t_0)$ on $G^n(\mathsf{R}^{2n})$. By virtue of (4.9) and the identification of $S(n)$ with $\mathcal{L}_0(n)$, the restriction of $S(t, S_0, t_0)$ to $\mathcal{L}(n)$ can be viewed as an extension of the Riccati flow $K(t, K_0, t_0)$ through the points at infinity. By the *extended symplectic Riccati differential equation* (ESRDE), we mean either the flow on $G^n(\mathsf{R}^{2n})$ or its restriction to $\mathcal{L}(n)$, depending on the context.

We follow a three step procedure to obtain the phase portrait of the SRDE: (1) Characterize the phase portrait of the ERDE on $G^n(\mathsf{R}^{n+m})$ and hence the phase portrait of the ESRDE on $G^n(\mathsf{R}^{2n})$ as a special case. (2) Obtain the phase portrait of the ESRDE on $\mathcal{L}(n)$ from that on $G^n(\mathsf{R}^{2n})$ by restricting to the invariant manifold $\mathcal{L}(n)$. (3) Recover the phase portrait of the SRDE on $S(n)$ from that of the ESRDE on $\mathcal{L}(n)$ by sorting out which dynamic phenomena for the ESRDE occur in the subset $\mathcal{L}_0(n)$ and hence correspond to dynamic phenomena for the SRDE.

4.3 Phase Portrait of the Constant-Coefficient ERDE

In this section, we describe the principal features of the phase portrait of the constant-coefficient extended Riccati differential equation on the Grassmann manifold $G^n(\mathsf{R}^{n+m})$. Since the differential equation is time-invariant, we set $t_0 = 0$ and use $S(t, S_0)$ to denote $S(t, S_0, 0)$. We make the following assumptions:

(1) The $n+m$ eigenvalues of B are distinct.
(2) If λ_i and λ_j are a pair of eigenvalues with the same real part, then $\lambda_i = \bar{\lambda}_j$ where overbar denotes complex conjugation.

Each of these assumptions holds generically.

We fix some notation. Let p denote the number of real eigenvalues of B, and let q denote the number of conjugate pairs of nonreal eigenvalues of B. (So $p+2q = n+m$.) Let $r = p+q$, and let $E_1, \cdots, E_r$ denote the primary components of B ordered according to increasing real part of the corresponding eigenvalue(s). Thus, each E_j is either 1-dimensional or 2-dimensional. Also, if $i < j$ and if λ_i and λ_j are eigenvalues which correspond to E_i and E_j respectively, then Re $\lambda_i <$Re λ_j.

One of the principal features of the phase portrait of the ERDE is the presence of invariant tori of various dimensions. The existence of invariant tori of dimension at least two was described by Hermann and Martin [4.4]. It will prove convenient to view equilibrium points and isolated periodic orbits as 0-dimensional and 1-dimensional invariant tori, respectively.

Let $l = (l_1, \cdots, l_r)$ be such that $l_1 + \ldots + l_r = n$ and $0 \leq l_j \leq \dim E_j$, $j = 1, \cdots, r$. Let $G^{l_j}(E_j)$ denote the Grassmann manifold of all l_j-dimensional subspaces of E_j, and let

$$T(l) = \{S_1 \oplus \cdots \oplus S_r \mid S_j \in G^{l_j}(E_j),\ j = 1, \cdots, r\}$$

Then $T(l)$ is isomorphic to the product $G^{l_1}(E_1) \times \cdots \times G^{l_r}(E_r)$. Since E_j has dimension equal to one or two, $G^{l_j}(E_j)$ consists of a single point unless dim $E_j = 2$ and $l_j = 1$ in which case it is isomorphic to $\mathrm{RP}(1)$ – i.e., a circle. Thus, $T(l)$ is a torus, and the dimension of $T(l)$ is equal to

$$\sum_{j=1}^{r} l_j(\dim E_j - l_j) \tag{4.11}$$

Let $S_0 \in T(l)$. Then $S_0 = S_1 \oplus \cdots \oplus S_r$ with S_j an l_j-dimensional subspace of E_j. We have

$$S(t, S_0) = e^{Bt}(S_1) \oplus \cdots \oplus e^{Bt}(S_r)$$

which belongs to $T(l)$ since $S_j \subset E_j$ and E_j is B-invariant. Thus, the torus $T(l)$ is both positively and negatively invariant with respect to the flow of the ERDE.

Next we characterize the flow on the invariant torus $T(l)$. If $l_j = 0$ or $l_j = \dim E_j$, then $G^{l_j}(E_j)$ consists of a single point, so $e^{Bt}(S_j)$ must be constant. The only other case is where $l_j = 1$ and dim $E_j = 2$. The motion $e^{Bt}(S_j)$ is analogous to that in Example 4.4 – i.e., periodic with period equal to π/ω_j where the pair of complex conjugate eigenvalues associated with E_j have imaginary parts $\pm\omega_j$. It follows immediately that if dim $T(l) = 1$, then the motion on $T(l)$ is periodic, while if dim $T(l) \geq 2$, the motion is either periodic or almost periodic depending on whether or not the relevant imaginary parts are commensurable.

Theorem 4.1. *$T(l)$ is a torus of dimension given by (4.11) and is both positively and negatively invariant under the flow of the ERDE. If $k = 0$, then $T(l)$ is an equilibrium point; if $k = 1$, then $T(l)$ is an isolated periodic orbit. If $k \geq 2$, the motion on $T(l)$ is either periodic or almost periodic.*

Let $T(l)$ be an invariant torus, and let $W^s(T(l))$ and $W^u(T(l))$ be the stable and unstable manifolds respectively for $T(l)$. I.e.,

$$W^s(T(l)) = \{S_0 \in G^n(\mathsf{R}^{n+m}) \mid S(t, S_0) \to T(l) \text{ as } t \to \infty\}$$
$$W^u(T(l)) = \{S_0 \in G^n(\mathsf{R}^{n+m}) \mid S(t, S_0) \to T(l) \text{ as } t \to -\infty\}$$

Define a flag (i.e., strictly increasing sequence) of subspaces $0 = M_0 \subset M_1 \subset \cdots \subset M_r = \mathsf{R}^{n+m}$ by

$$M_j = \sum_{i=1}^{j} E_i, \quad j = 1, \cdots, r.$$

Define a second flag of subspaces $0 = N_0 \subset N_1 \subset \cdots \subset N_r = \mathsf{R}^{n+m}$ by

$$N_j = \sum_{i=1}^{j} E_{r-i+1}, \quad j = 1, \cdots, r.$$

We refer to $\{M_j\}_1^r$ as the *stable flag* associated with the ERDE, and we refer to $\{N_j\}_1^r$ as the *unstable flag* associated with the ERDE. The next theorem describes the sets $W^s(T(l))$ and $W^u(T(l))$.

Theorem 4.2.

$$W^s(T(l)) = \{S \in G^n(\mathsf{R}^{n+m}) \mid \dim S \cap M_j = \sum_{i=1}^{j} l_i, \ j = 1, \cdots, r\}$$

$$W^u(T(l)) = \{S \in G^n(\mathsf{R}^{n+m}) \mid \dim S \cap N_j = \sum_{i=1}^{j} l_{r-i+1}, \ j = 1 \cdots, r\}$$

The proof of Theorem 4.2 can be found in [4.14]. It is clear from the characterization of $W^s(T(l))$ and $W^u(T(l))$ in Theorem 4.2 that given any $S_0 \in G^n(\mathsf{R}^{n+m})$, there exist l, l' such that $S_0 \in W^s(T(l))$ and $S_0 \in W^u(T(l'))$. I.e., every trajectory converges to an invariant torus in forward time and to an invariant torus in backward time. In particular, every solution of the ERDE is asymptotically constant, periodic, or almost periodic.

Although we omit the proof of Theorem 4.2, we indicate the type of argument involved by way of an example.

Example 4.5. We consider an example where $n = 2$ and $m = 2$. Thus, the ERDE is a flow on $G^2(\mathsf{R}^4)$, a 4-dimensional manifold. We take $r = 3$ and the dimensions of E_1, E_2, E_3 to be 2,1,1. Thus, B has a pair of complex conjugate eigenvalues and two real eigenvalues. Furthermore, the complex eigenvalues have real part which is less than that of the real eigenvalues. The particular values of the eigenvalues are not especially relevant. To be concrete, we choose them to be $1 \pm i, 2, 3$. By a change of basis, we may take B to be in real canonical form:

$$B = \begin{pmatrix} 1 & 1 & 0 & 0 \\ -1 & 1 & 0 & 0 \\ 0 & 0 & 2 & 0 \\ 0 & 0 & 0 & 3 \end{pmatrix}.$$

Thus, the stable flag is given by $M_1 = \mathrm{Sp}\{e_1, e_2\}$, $M_2 = \mathrm{Sp}\{e_1, e_2, e_3\}$ where e_i is the ith standard basis vector for $\mathbb{R}^4$. We partition $G^2(\mathbb{R}^4)$ into 6 subsets (called Schubert cells). There are 6 types of column-echelon forms for 4×2 matrices of rank 2:

$$\begin{pmatrix} 1 & 0 \\ 0 & 1 \\ 0 & 0 \\ 0 & 0 \end{pmatrix} \quad \begin{pmatrix} 1 & 0 \\ 0 & x_{22} \\ 0 & 1 \\ 0 & 0 \end{pmatrix} \quad \begin{pmatrix} 1 & 0 \\ 0 & x_{22} \\ 0 & x_{32} \\ 0 & 1 \end{pmatrix}$$

$$\begin{pmatrix} x_{11} & x_{12} \\ 1 & 0 \\ 0 & 1 \\ 0 & 0 \end{pmatrix} \quad \begin{pmatrix} x_{11} & x_{12} \\ 1 & 0 \\ 0 & x_{32} \\ 0 & 1 \end{pmatrix} \quad \begin{pmatrix} x_{11} & x_{12} \\ x_{21} & x_{22} \\ 1 & 0 \\ 0 & 1 \end{pmatrix}$$

It is easy to see that a given 2-dimensional subspace of $\mathbb{R}^4$ is spanned by the columns of a unique matrix of exactly one of the 6 types. Thus, $G^2(\mathbb{R}^4)$ is partitioned into 6 Schubert cells which have dimensions 0,1,2,2,3,4 respectively.

Let $l = (1, 1, 0)$. By Theorem 4.1, $T(l)$ is a 1-dimensional invariant torus – i.e., an isolated periodic orbit. By Theorem 4.2, its stable manifold consists of those subspaces S with the property that $\dim S \cap M_1 = 1$ and $\dim S \cap M_2 = 2$. Thus, the stable manifold is the union of the second and fourth Schubert cells. To give some indication of how the proof of Theorem 4.2 goes, we will show this directly.

Suppose that S_0 belongs to the fourth Schubert cell. Then S_0 is of the form Sp X where

$$X = \begin{pmatrix} x_{11} & x_{12} \\ 1 & 0 \\ 0 & 1 \\ 0 & 0 \end{pmatrix}.$$

Then $S(t, S_0)$ is spanned by

$$e^{Bt}X = \begin{pmatrix} x_{11}e^t \cos t + e^t \sin t & x_{12}e^t \cos t \\ -x_{11}e^t \sin t + e^t \cos t & -x_{12}e^t \sin t \\ 0 & e^{2t} \\ 0 & 0 \end{pmatrix}$$

or equivalently by

$$\begin{pmatrix} x_{11}\cos t + \sin t & x_{12}e^{-t}\cos t \\ -x_{11}\sin t + \cos t & -x_{12}e^{-t}\sin t \\ 0 & 1 \\ 0 & 0 \end{pmatrix}$$

It is clear from this that as $t \to \infty$, $S(t, S_0)$ approaches the trajectory given by

$$\mathrm{Sp}\begin{pmatrix} x_{11}\cos t + \sin t & 0 \\ -x_{11}\sin t + \cos t & 0 \\ 0 & 1 \\ 0 & 0 \end{pmatrix}$$

which is a periodic solution which traverses the 1-dimensional torus $T(l)$. The case where S_0 belongs to the second Schubert cell is treated similarly.

4.4 Phase Portrait of the Constant-Coefficient ESRDE

In this section, we describe the phase portrait of the ESRDE on the Lagrange-Grassmann manifold $\mathcal{L}(n)$. We make the following assumptions:

(1) The $2n$ eigenvalues of the Hamiltonian matrix H are distinct.
(2) If λ_i and λ_j are a pair of eigenvalues with the same real part, then $\lambda_i = \bar{\lambda}_j$.
(3) H has no imaginary axis eigenvalues.

Note that we place no individual restrictions on the matrices A, L, Q other than requiring that L and Q be symmetric. In particular, L could be 0 or indefinite.

If we regard the ESRDE as a flow on $G^n(\mathsf{R}^{2n})$, then the phase portrait is obtained as a special case of the results in Section III. As mentioned earlier, $\mathcal{L}(n)$ is an invariant manifold so the phase portrait of the flow on $\mathcal{L}(n)$ can be obtained by intersecting the phase portrait on $G^n(\mathsf{R}^{2n})$ with the submanifold $\mathcal{L}(n)$. To be more specific, the recurrent trajectories are contained in the intersections of the various invariant tori $T(l)$ with $\mathcal{L}(n)$ while the stable and unstable manifolds are the intersections of the various stable and unstable manifolds $W^s(T(l))$, $W^u(T(l))$ with $\mathcal{L}(n)$. Consequently, the main task at hand is to describe the structure of the intersections $T(l) \cap \mathcal{L}(n)$, $W^s(T(l)) \cap \mathcal{L}(n)$, $W^u(T(l)) \cap \mathcal{L}(n)$.

Since H is a Hamiltonian matrix, its eigenvalues come in pairs. If λ is an eigenvalue, then so is $-\lambda$. We redefine p, q, r as follows: Let $2p$ denote the number of real eigenvalues of H, and let $4q$ denote the number of nonreal eigenvalues of H. Let $E_1, \cdots, E_r, E_{r+1}, \cdots, E_{2r}$ denote the primary components of H. (Thus, when we apply results from Section III, we must use $2r$ in place of r.) The order of the indexing is determined by increasing real part of the corresponding eigenvalues. This implies that if E_j corresponds to a real eigenvalue λ, then E_{2r-j+1} corresponds to $-\lambda$. Similarly, if E_j corresponds to a conjugate pair $\lambda, \bar{\lambda}$, then E_{2r-j+1} corresponds to $-\lambda, -\bar{\lambda}$. In particular, dim E_j =dim E_{2r-j+1}. Note also that $E_1, \cdots, E_r$

correspond to eigenvalues with negative real part, while $E_{r+1}, \cdots, E_{2r}$ correspond to eigenvalues with positive real part.

The following result describes the intersections of the various invariant tori with the Lagrange-Grassmann manifold. If U is any subspace of $\mathbb{R}^{2n}$, let $U^{\#}$ denote the subspace which is orthogonal to U relative to the nondegenerate skew-symmetric bilinear form ω. I.e.,

$$U^{\#} := [J(U)]^{\perp},$$

where $\perp$ denotes orthogonal complement with respect to the standard inner product on $\mathbb{R}^{2n}$.

Proposition 4.1. *$T(l) \cap \mathcal{L}(n)$ is nonempty if and only if*

$$l_j + l_{2r-j+1} = \dim E_j, \qquad j = 1, \cdots, r$$

in which case it is given by

$$\{S_1 \oplus \cdots \oplus S_{2r} \in T(l) \mid S_{2r-j+1} = S_j^{\#} \cap E_{2r-j+1},\ j = 1, \cdots, r\}.$$

Proof. Since H is a Hamiltonian matrix with no imaginary axis eigenvalues, there exists a symplectic matrix P such that

$$\bar{H} := P^{-1}HP = \begin{pmatrix} \bar{A} & 0 \\ 0 & -\bar{A}' \end{pmatrix}$$

with $\bar{A}$ a Hurwitz matrix. In fact, we can choose P such that $\bar{A}$ is in real canonical form with the 1×1 and 2×2 blocks along the diagonal ordered according to increasing real part of the associated eigenvalue(s).

Let $\bar{E}_1, \cdots, \bar{E}_{2r}$ denote the primary components for $\bar{H}$. From the special structure of $\bar{H}$, it is clear that a given primary component $\bar{E}_j$ is of the form $\mathrm{Sp}\{e_i\}$ if 1-dimensional, or of the form $\mathrm{Sp}\{e_i, e_{i+1}\}$ if 2-dimensional. In the first case, $E_{2r-j+1} = \mathrm{Sp}\{e_{i+n}\}$, while in the second case, $E_{2r-j+1} = \mathrm{Sp}\{e_{i+n}, e_{i+1+n}\}$.

Let $\bar{S} = \bar{S}_1 \oplus \cdots \oplus \bar{S}_{2r} \in \bar{T}(l) \cap \mathcal{L}(n)$, where we use $\bar{T}(l)$ to indicate an invariant torus for $\bar{H}$. Since $\bar{S}^{\#} = \bar{S}$, it follows that

$$\bar{S}_{2r-j+1} \subset \bar{S}_j^{\#} \cap \bar{E}_{2r-j+1} \tag{4.12}$$

From the structure of $\bar{H}$, $J(\bar{E}_j) = \bar{E}_{2r-j+1}$, so $J(\bar{S}_j) \subset \bar{E}_{2r-j+1}$, which implies that

$$\dim \bar{S}_j^{\#} \cap \bar{E}_{2r-j+1} = \dim \bar{E}_{2r-j+1} - \dim \bar{S}_j \tag{4.13}$$

It follows that

$$\bar{S} = \sum_{j=1}^{r} \bar{S}_j \oplus (\bar{S}_j^{\#} \cap \bar{E}_{2r-j+1}) := \bar{S}' \tag{4.14}$$

since $\bar{S} \subset \bar{S}'$ by (4.12) and $\dim \bar{S}' = n$ by (4.13). From (4.13) and (4.14), we conclude that $\dim \bar{S}_j + \dim \bar{S}_{2r-j+1} = \dim \bar{E}_{2r-j+1}$ – i.e., $l_j + l_{2r-j+1} =$

dim $\bar{E}_j$, $\forall j$. Finally, we note that if $\bar{S} \in \bar{T}(l)$ is of the form (4.14), then $\bar{S}^\# = \bar{S}$, so $\bar{S} \in \bar{T}(l) \cap \mathcal{L}(n)$. This completes the proof in the special case where the Hamiltonian matrix has the special form of $\bar{H}$.

If $T(l)$, $\bar{T}(l)$ denote invariant tori for H, $\bar{H}$ respectively, then $T(l) = P\bar{T}(l)$. Since $\mathcal{L}(n)$ is invariant under the symplectic transformation P, it follows immediately that

$$T(l) \cap \mathcal{L}(n) = P(\bar{T}(l) \cap \mathcal{L}(n)).$$

Thus, $T(l) \cap \mathcal{L}(n)$ is nonempty if and only if

$$l_j + l_{2r-j+1} = \dim E_j, \ \forall j$$

since this is the condition for $\bar{T}(l) \cap \mathcal{L}(n)$ to be nonempty. If this is the case, the elements of $T(l) \cap \mathcal{L}(n)$ are of the form

$$P \sum_{j=1}^{r} \bar{S}_j \oplus (\bar{S}_j^\# \cap \bar{E}_{2r-j+1}).$$

Letting $S_j := P(\bar{S}_j)$ and noting that

$$P(\bar{S}_j^\# \cap \bar{E}_{2r-j+1}) = S_j^\# \cap E_{2r-j+1}$$

completes the proof. □

Suppose that $T(l) \cap \mathcal{L}(n)$ is nonempty – i.e., $l_j + l_{2r-j+1} = \dim E_j$, $j = 1, \cdots, r$. The dimension of $T(l)$ is equal to the number of indices j, $j = 1, \cdots, 2r$ such that $l_j = 1$ and $\dim E_j = 2$. By Proposition 4.1, the dimension of $T(l) \cap \mathcal{L}(n)$ is equal to the number of indices j, $j = 1, \cdots, r$ such that $l_j = 1$ and $\dim E_j = 2$. This implies that

$$\dim T(l) \cap \mathcal{L}(n) = \frac{1}{2} \dim T(l)$$

We can determine exactly how many invariant tori $T(l) \cap \mathcal{L}(n)$ are of dimension k. Exactly q of the left half-plane primary components $E_1, \cdots, E_r$ are 2-dimensional. Thus, there are $\binom{q}{k}$ choices for the k indices j, $j = 1, \cdots, r$ such that $l_j = 1$ and $\dim E_j = 2$. For each of the remaining indices j, $j = 1, \ldots, r$, either $l_j = 0$ or $l_j = \dim E_j$. Finally, the indices $l_{r+1}, \cdots, l_{2r}$ are uniquely determined by the condition $l_j + l_{2r-j+1} = \dim E_j$. We obtain the following result:

Theorem 4.3. *The invariant tori for the ESRDE on $\mathcal{L}(n)$ are $\{T(l) \cap \mathcal{L}(n) \mid l_j + l_{2r-j+1} = \dim E_j,\ j = 1, \cdots, r\}$. There are $\binom{q}{k} 2^{r-k}$ invariant tori of dimension k.*

Corollary 4.1. *The ESRDE has 2^r equilibrium points and $q2^{r-1}$ isolated periodic orbits.*

The stable and unstable manifolds for the ESRDE on $\mathcal{L}(n)$ are obtained by intersecting those for the ESRDE on $G^n(\mathbb{R}^{2n})$ with $\mathcal{L}(n)$. Thus, it follows immediately from Theorem 4.2 that

$$W^s(T(l) \cap \mathcal{L}(n)) = \{S \in \mathcal{L}(n) \mid \dim S \cap M_j = \sum_{i=1}^{j} l_i,\ j = 1, \cdots, 2r\}$$

$$W^u(T(l) \cap \mathcal{L}(n)) = \{S \in \mathcal{L}(n) \mid \dim S \cap N_j = \sum_{i=1}^{j} l_{2r-i+1}, j = 1, \cdots, 2r\}.$$

Now, for any $S \in \mathcal{L}(n)$, there exists l' such that $S \in W^s(T(l') \cap \mathcal{L}(n))$. If l is such that $l_j + l_{2r-j+1} = \dim E_j$, $\forall j$ and $\dim S \cap M_j = \sum_{i=1}^{j} l_i$, $j = 1, \cdots, r$, then $l_j = l'_j$, $j = 1, \cdots, r$ and hence $l_j = l'_j$, $j = 1, \cdots, 2r$. Thus, $S \in W^s(T(l) \cap \mathcal{L}(n))$. This together with an analogous argument for the unstable manifolds gives

Theorem 4.4.

$$W^s(T(l) \cap \mathcal{L}(n)) = \{S \in \mathcal{L}(n) \mid \dim S \cap M_j = \sum_{i=1}^{j} l_i,\ j = 1, \cdots, r\}$$

$$W^u(T(l) \cap \mathcal{L}(n)) = \{S \in \mathcal{L}(n) \mid \dim S \cap N_j = \sum_{i=1}^{j} l_{2r-i+1}, j = 1, \cdots, r\}$$

As mentioned in Section III, $W^s(T(l))$ and $W^u(T(l))$ are unions of Schubert cells for $G^n(\mathbb{R}^{2n})$. It turns out that the intersection of each of these Schubert cells with $\mathcal{L}(n)$ is either empty or a Schubert cell for $\mathcal{L}(n)$. Thus, the stable and unstable manifolds for the invariant tori $T(l) \cap \mathcal{L}(n)$ are each unions of Schubert cells for $\mathcal{L}(n)$. Details may be found in [4.14].

4.5 Phase Portrait of the Constant-Coefficient SRDE

In this section we apply the results for the ESRDE on $\mathcal{L}(n)$ to obtain a complete description of the phase portrait for the symplectic Riccati equation on the space $S(n)$ of symmetric matrices. Since the SRDE corresponds to the restriction of the ESRDE to the subset $\mathcal{L}_0(n)$ via the imbedding ψ, the basic problem is to sort out which of the dynamic behavior for the ESRDE on $\mathcal{L}(n)$ is confined to $\mathcal{L}_0(n)$. It is at this stage that system-theoretic conditions enter into the problem. For the results in Section II on the ERDE, the only assumptions made concerned the eigenvalues of B. For the results in Section III, the infinitesimally symplectic structure of H also played an important role. However, no additional assumptions on A, L, Q were made. In particular, the results for the ESRDE apply to the case where $L = 0$ (Liapunov differential equation) and to the case where L is indefinite (Riccati equation associated with a zero-sum differential game) as well as to the Riccati equations arising from standard control and filtering problems. In contrast, the results in this section depend on assumptions such as the nonnegativity of L and controllability of the associated linear system.

The role of controllability (and weakened notions such as stabilizability and sign-controllability) for the existence of solutions to the *algebraic* Riccati equation (i.e., equilibrium points of the SRDE) is well-understood. (See e.g., [4.16], [4.5], [4.2], [4.6], [4.12].) A geometric description of the role of controllability for the equilibrium points of the SRDE is provided by the following results. They are well-known and do not require any of the assumptions on the eigenvalues of H made in the preceding section. The proof of the first result is an easy exercise.

Proposition 4.2. *The largest subspace of* $Sp\begin{pmatrix} 0 \\ I_n \end{pmatrix}$ *which is* H*-invariant is*

$$\{\begin{pmatrix} 0 \\ y \end{pmatrix} \mid L(A')^k y = 0, \quad k = 0, \cdots, n-1\}.$$

The proof of the second result can be found in [4.5].

Proposition 4.3. *Suppose that* L *is nonnegative definite and let* $S \in \mathcal{L}(n)$ *be* H*-invariant. Then* $S \cap Sp\begin{pmatrix} 0 \\ I_n \end{pmatrix}$ *is* H*-invariant.*

Proposition 4.4. *If* L *is nonnegative definite and* (A, L) *is controllable, then every element* $S \in \mathcal{L}(n)$ *which is* H*-invariant belongs to the subset* $\mathcal{L}_0(n)$.

Since the equilibrium points of the ESRDE are the H-invariant elements of $\mathcal{L}(n)$, the following result is immediate.

Theorem 4.5. *Suppose that* L *is nonnegative definite and* (A, L) *is controllable. Then the imbedding* ψ *gives a one-to-one correspondence between the equilibrium points of the SRDE on* $S(n)$ *and the equilibrium points of the ESRDE on* $\mathcal{L}(n)$.

It should be noted that Proposition 4.3 and Theorem 4.5 fail without the assumption that L is nonnegative definite. Thus, controllability does not guarantee the existence of equilibrium points for the SRDE arising from a zero-sum differential game.

The geometric content of Theorem 4.5 is that the role of controllability is to guarantee that the SRDE has no equilibrium points "at infinity" (under the assumption that L is nonnegative definite). We will now see that controllability actually has a much stronger implication for the phase portrait of the SRDE – namely that there are no invariant tori "at infinity." We reinvoke the assumptions about the eigenvalues of H made in the previous section and proceed as follows: Let S^- denote the invariant subspace of H associated with the n eigenvalues of H with negative real part. I.e.,

$$S^- = E_1 \oplus \cdots \oplus E_r$$

Similarly, let S^+ denote the invariant subspace of H associated with the n eigenvalues with positive real part. I.e.,

$$S^+ = E_{r+1} \oplus \cdots \oplus E_{2r}$$

Note that S^- and S^+ are 0-dimensional invariant tori (i.e., equilibrium points) for the ESRDE corresponding to

$$\begin{aligned} l^- &= (\dim E_1, \cdots, \dim E_r, 0, \cdots, 0) \\ l^+ &= (0, \cdots, 0, \dim E_{r+1}, \cdots, \dim E_{2r}) \end{aligned} \tag{4.15}$$

respectively. By Theorem 4.5, S^- and S^+ are contained in $\mathcal{L}_0(n)$ and hence correspond to equilibrium points $K^+ = \psi^{-1}(S^-)$ and $K^- = \psi^{-1}(S^+)$ of the SRDE. (Note that the signs are interchanged!) Let $\Delta = K^+ - K^-$. The following result is well-known [4.16].

Proposition 4.5. *Suppose that L is nonnegative definite and* (A, L) *is controllable. Then* Δ *is positive definite.*

Proof. From the algebraic Riccati equation which is satisfied by both K^+ and K^-, one obtains

$$(A - LK^+)'\Delta + \Delta(A - LK^+) = -\Delta L \Delta$$

It is easy to verify that $A - LK^+$ is the matrix for the restriction of H to S^- relative to the basis given by the columns of $\begin{pmatrix} I \\ K^+ \end{pmatrix}$. Thus, $A - LK^+$ is Hurwitz and Δ is nonnegative definite. If $\Delta x = 0$, then

$$\begin{pmatrix} x \\ K^+ x \end{pmatrix} = \begin{pmatrix} x \\ K^- x \end{pmatrix} \in S^- \cap S^+ = 0$$

showing that Δ is nonsingular. □

Proposition 4.6. *Suppose that L is nonnegative definite and* (A, L) *is controllable. Let* $S \in \mathcal{L}(n)$ *be such that* $S = (S \cap S^-) \oplus (S \cap S^+)$. *Then* $S \in \mathcal{L}_0(n)$.

Proof. Let $k = \dim S \cap S^-$. There exist full rank $n \times k$ and $n \times n - k$ matrices C, D such that

$$S = \mathrm{Sp}\begin{pmatrix} C & D \\ K^+C & K^-D \end{pmatrix}. \tag{4.16}$$

It follows easily from the requirement that $S^\# = S$ that $C'\Delta D = 0$. This in turn implies that the $n \times n$ matrix $(\,C \quad D\,)$ is nonsingular. □

Corollary 4.2. *Suppose that L is nonnegative definite and* (A, L) *is controllable. Let* $l_j + l_{2r-j+1} = \dim E_j$, $j = 1, \cdots, r$. *Then* $T(l) \cap \mathcal{L}(n) \subset \mathcal{L}_0(n)$.

Proof. Immediate consequence of Proposition 4.6. □

Theorem 4.6. *Suppose that L is nonnegative definite and (A, L) is controllable. The invariant tori for the SRDE on $S(n)$ are*

$$\{\psi^{-1}(T(l) \cap \mathcal{L}(n)) \mid l_j + l_{2r-j+1} = \dim E_j, \quad j = 1, \cdots, r\}$$

There are $\binom{q}{k} 2^{r-k}$ invariant tori of dimension k.

Proof. Immediate from Theorem 4.3 and Corollary 4.2 . □

Corollary 4.3. If L is nonnegative definite and (A, L) is controllable, then the SRDE has 2^r equilibrium points and $q2^{r-1}$ isolated periodic orbits.

We comment on the relationship between Theorem 4.6 and the eigenvector method for constructing the solutions of the algebraic Riccati equation. Each equilibrium point of the SRDE is of the form $\psi^{-1}(T(l) \cap \mathcal{L}(n))$ where $\dim T(l) \cap \mathcal{L}(n) = 0$. I.e., either $l_j = 0$ or $l_j = \dim E_j$, $j = 1, \cdots, r$. Thus, each equilibrium point is of the form

$$K = \psi^{-1}(F_1 \oplus \cdots \oplus F_r), \qquad F_j = E_j \text{ or } E_{2r-j+1}.$$

Choosing eigenvectors to span each chosen primary component F_j and using them as the columns of a $2n \times n$ matrix $\begin{pmatrix} X \\ Y \end{pmatrix}$, we have $K = \psi^{-1}(\mathrm{Sp} \begin{pmatrix} X \\ Y \end{pmatrix}) = YX^{-1}$, which is precisely the eigenvector method.

More generally, we can extend the eigenvector method to construct the invariant torus $\psi^{-1}(T(l) \cap \mathcal{L}(n))$. Instead of choosing F_j to be E_j or E_{2r-j+1}, we choose S_j to be an l_j-dimensional subspace of E_j and set $F_j = S_j \oplus S_j^{\#} \cap E_{2r-j+1}$. The $2n \times n$ matrix $\begin{pmatrix} X \\ Y \end{pmatrix}$ is constructed so that its columns span each of the subspaces F_j. Then $K = YX^{-1}$ belongs to the invariant torus.

Theorem 4.6 shows that under appropriate conditions, all of the invariant tori for the ESRDE actually correspond to invariant tori for the SRDE in the space of symmetric matrices. In addition to assuming that the eigenvalues of H are distinct and eigenvalues with the same real part are conjugate, we assumed that H has no imaginary axis eigenvalues, L is nonnegative definite and (A, L) is controllable. We briefly discuss the necessity of the last three assumptions. Suppose that H has no imaginary axis eigenvalues but (A, L) is uncontrollable. Since H has no imaginary axis eigenvalues, S^- and S^+ are 0-dimensional invariant tori. If they are both contained in $\mathcal{L}_0(n)$ (i.e., are not at infinity), then K^+ and K^- exist. $A - LK^+$ (respectively, $A - LK^-$) is a matrix representation for the restriction of H to S^- (respectively, S^+) and is Hurwitz (respectively, anti-Hurwitz). Thus, every eigenvalue of A can be shifted by state feedback, so (A, L) is controllable, a contradiction. Consequently, at least one of S^-, S^+ must not be in $\mathcal{L}_0(n)$.

Example 4.6. Suppose that $L = 0$ and A is Hurwitz. Thus, (A, L) is stabilizable but not controllable, and the SRDE is a Liapunov differential equation. The SRDE has a unique equilibrium point K^+ while the corresponding ESRDE has 2^r equilibrium points. Thus, $2^r - 1$ 0-dimensional invariant tori are at infinity. □

The next example is borrowed from Van Swieten [4.15] and shows that the conclusion of Theorem 4.6 can fail when L is indefinite – i.e., for the SRDE associated with a zero-sum differential game.

Example 4.7. Consider the SRDE associated with the Hamiltonian matrix

$$H = \begin{pmatrix} 1 & 0 & 0 & -1 \\ 0 & 3 & -1 & 0 \\ 0 & 5 & -1 & 0 \\ 5 & 0 & 0 & -3 \end{pmatrix}.$$

(A, L) is controllable and the eigenvalues of H are $\pm 1 \pm i$. However, L is indefinite. We have $E_1 = \mathrm{Sp}\{e_1, e_4\}$, $E_2 = \mathrm{Sp}\{e_2, e_3\}$. It follows that each of the 0-dimensional invariant tori ($S^- = E_1$, $S^+ = E_2$) is at infinity, while the single 1-dimensional invariant torus contains the point $\mathrm{Sp}\{e_3, e_4\}$ at infinity. Thus, the phase portrait of the SRDE contains no points corresponding to the two equilibrium points of the ESRDE and a trajectory which has finite escape in both positive and negative time instead of the periodic orbit of the ESRDE. □

The next example shows that the conclusion of Theorem 4.6 can fail when H has imaginary axis eigenvalues – even when L is nonnegative definite and (A, L) is controllable.

Example 4.8. Consider the scalar SRDE

$$\dot{K} = K^2 + 1$$

Then $L = 1$ and $A = 0$, so the nonnegativity and controllability requirements are satisfied. However, the Hamiltonian matrix

$$H = \begin{pmatrix} 0 & -1 \\ 1 & 0 \end{pmatrix}$$

has imaginary axis eigenvalues. The phase portrait of the ESRDE consists of a single periodic orbit, whereas every solution of the SRDE has a finite escape time. Thus, the 1-dimensional invariant torus for the ESRDE contains a point at infinity. □

In order to describe the role of observability in the phase portrait of the SRDE, we dualize Propositions 4.4, 4.6 and Corollary 4.2. Let

$$\mathcal{L}'_0(n) = \{S \in \mathcal{L}(n) \mid S \cap \mathrm{Sp}\begin{pmatrix} I_n \\ 0 \end{pmatrix} = 0\}$$

Proposition 4.7. *If Q is nonnegative definite and (Q, A) is observable, then every element $S \in \mathcal{L}(n)$ which is H-invariant belongs to the subset $\mathcal{L}'_0(n)$.*

Proof. Applying Proposition 4.4 with the Hamiltonian matrix H' in place of H, we obtain that every $\bar{S} \in \mathcal{L}(n)$ which is H'-invariant is contained in $\mathcal{L}_0(n)$. If $S \in \mathcal{L}(n)$ is H-invariant, then $S^\perp = J(S) \in \mathcal{L}(n)$ (since J is symplectic) and is H'-invariant. Thus, $J(S) \in \mathcal{L}_0(n)$ so $S \in \mathcal{L}_0'(n)$. □

Proposition 4.8. *Suppose that Q is nonnegative definite and (Q, A) is observable. Let $S \in \mathcal{L}(n)$ be such that $S = (S \cap S^-) \oplus (S \cap S^+)$. Then $S \in \mathcal{L}_0'(n)$.*

Proof. Let $\bar{S} = J(S)$ and let $\bar{S}^-$, $\bar{S}^+$ denote the spectral subspaces of H' associated with its left half-plane eigenvalues and right half-plane eigenvalues respectively. Then $\bar{S}^- = J(S^+)$ and $\bar{S}^+ = J(S^-)$. Thus,

$$\bar{S} = (\bar{S} \cap \bar{S}^+) \oplus (\bar{S} \cap \bar{S}^-).$$

Applying Proposition 4.6 with H' in place of H gives $\bar{S} \in \mathcal{L}_0(n)$ and hence $S \in \mathcal{L}_0'(n)$. □

There has been considerable interest in the signatures of solutions of the algebraic Riccati equation – i.e., of the equilibrium points of the SRDE. The following is a more general result which shows that (under appropriate assumptions) the signature of solutions is constant on each invariant torus. In addition, it gives a formula for this signature.

Theorem 4.7. *Suppose that L, Q are both nonnegative definite and that (A, L), (Q, A) are controllable and observable respectively. Then every $K \in \psi^{-1}(T(l) \cap \mathcal{L}(n))$ is nonsingular and has exactly $l_1 + \cdots + l_r$ positive eigenvalues.*

Proof. Let $S = \psi(K) \in T(l) \cap \mathcal{L}(n)$. By Proposition 4.8, $S \in \mathcal{L}_0'(n)$, so K is nonsingular. Express S as in (4.16). Let P denote the projection onto Sp C along Sp $D = \Delta^{-1}(\text{Sp } C)^\perp$. Then

$$S = \text{Sp}\begin{pmatrix} P & I-P \\ K^+P & K^-(I-P) \end{pmatrix} = \text{Sp}\begin{pmatrix} I \\ K^+P + K^-(I-P) \end{pmatrix}$$

which shows that

$$K = K^+P + K^-(I-P). \tag{4.17}$$

Under the present assumptions, it is well-known that $K^+ > 0$ and $K^- < 0$. To see this, note that

$$\begin{aligned} (A - LK^+)'K^+ + K^+(A - LK^+) &= -Q - K^+LK^+ \\ -(A - LK^-)'K^- - K^-(A - LK^-) &= Q + K^-LK^- \end{aligned}$$

Since $A - LK^+$ and $-(A - LK^-)$ are each Hurwitz, it follows that K^+ is positive definite and K^- is negative definite. It then follows from (4.17) that the quadratic form associated with K is positive definite on Sp C and negative definite on the complementary subspace Sp D, which implies that the number of positive eigenvalues of K is equal to dim Sp C = dim $S \cap S^- = l_1 + \cdots + l_r$. □

Theorem 4.7 can be viewed as a generalization of an "Inertia Theorem" due to Wimmer [4.17]. Wimmer's result states that under the assumptions of Theorem 4.7, every solution K of the *algebraic* Riccati equation is nonsingular and the number of positive eigenvalues of K is equal to the number of left half-plane eigenvalues of $A-LK$. Theorem 4.7 extends this result for the equilibrium points to all of the invariant tori of the SRDE. Note that in the special case where K is an equilibrium point, then $A-LK$ is a matrix for the restriction of H to the invariant subspace $\psi(K)$. Thus, the number of left half-plane eigenvalues of $A-LK$ is equal to the dimension of $\psi(K)\cap S^-$, namely $l_1+\cdots l_r$.

An extension of a result of J.C. Willems [4.16] is also implicit in the proof of Theorem 4.7. Willems shows that under the assumptions that L is nonnegative definite and (A,L) is controllable, every solution K of the *algebraic* Riccati equation can be expressed in the form (4.17) where P is the projection onto an $(A-LK^+)$-invariant subspace M along $\Delta^{-1}M^{\perp}$. Conversely, every such expression defines a solution of the algebraic Riccati equation. Under the same assumptions (without assuming that Q is nonnegative definite or that (Q,A) is observable), (4.17) shows that every K belonging to an invariant torus can be expressed in this same form, but where M is not necessarily $(A-LK^+)$-invariant.

More generally, for each ν, the $n\times n$ matrix $B:=A-LK^+$ generates an ERDE (nonsymplectic!) flow on $G^\nu(\mathbb{R}^n)$ via (4.7). The mapping

$$M\mapsto K^+P+K^-(I-P)$$

imbeds $G^\nu(\mathbb{R}^n)$ into $S(n)$, and the image consists of those symmetric matrices K such that

$$\psi(K)=(\psi(K)\cap S^-)\oplus(\psi(K)\cap S^+),\qquad \dim\psi(K)\cap S^-=\nu$$

Furthermore, the invariant tori for the ERDE on $G^\nu(\mathbb{R}^n)$ are mapped onto those invariant tori $\psi^{-1}(T(l)\cap\mathcal{L}(n))$ of the SRDE for which $l_1+\cdots+l_r=\nu$.

Finally, we turn to the question of convergence to the invariant tori. For the ESRDE on $\mathcal{L}(n)$, we know from Theorem 4.4 that every solution converges to an invariant torus in forward time and to an invariant torus in backward time. This implies that every solution of the SRDE *which does not have a finite escape in positive time* converges to an invariant torus as $t\to\infty$. Similarly, every solution of the SRDE *which does not have a finite escape in negative time* converges to an invariant torus as $t\to-\infty$. Under the assumptions that L is nonnegative definite, (A,L) is controllable and H has no imaginary axis eigenvalues, the conditions for finite escape times are well-known [4.1], [4.7]: $K(t,K_0)$ has no finite escape in positive time if and only if $K_0\le K^+$ and no finite escape in negative time if and only if $K_0\ge K^-$. The next result is then an immediate consequence of Theorem 4.4.

Theorem 4.8. *Suppose that L is nonnegative definite and (A,L) is controllable.*

(a) $K(t,K_0)$ *converges to the invariant torus* $\psi^{-1}(T(l)\cap\mathcal{L}(n))$ *as* $t\to\infty$ *if and only if* $K\le K^+$ *and* $\dim\psi(K_0)\cap M_j=\sum_{i=1}^{j}l_i$, $j=1,\cdots,r$.

(b) $K(t, K_0)$ *converges to the invariant torus* $\psi^{-1}(T(l) \cap \mathcal{L}(n))$ *as* $t \to -\infty$ *if and only if* $K_0 \geq K^-$ *and dim* $\psi(K_0) \cap N_j = \sum_{i=1}^{j} l_{2r-i+1}$, $j = 1, \cdots, r$.

A well-known convergence result is that $K(t, K_0) \to K^+$ as $t \to -\infty$ if and only if $K_0 > K^-$, and $K(t, K_0) \to K^-$ as $t \to \infty$ if and only if $K_0 < K^+$ [4.16]. This can be recovered as a special case of Theorem 4.8: $\{K^+\} = \psi^{-1}(T(l^-) \cap \mathcal{L}(n))$, where l^- is defined in (4.15). The second condition in Theorem 4.8(b) reduces to $\psi(K_0) \cap S^+ = 0$, or equivalently, $K_0 - K^-$ being nonsingular. Combined with the condition that $K_0 \geq K^-$, this yields $K_0 > K^-$. Similarly, Theorem 4.8(a) specializes to give $K(t, K_0) \to K^-$ as $t \to \infty$ if and only if $K_0 < K^+$.

4.6 Conclusion

We have described how a "geometric approach" leads to a rather complete understanding of the phase portrait of the symplectic Riccati differential equation on the space of symmetric matrices under mild assumptions. The strategy is to begin by extending the flow to the Grassmann manifold. On the Grassmann manifold, the flow is given by the projection of a linear flow, and the asymptotic behavior of all solutions can be characterized explicitly. There are invariant tori of various dimensions to which all solutions converge. The stable and unstable manifolds of the tori are unions of Schubert cells.

The phase portrait on the Grassmann manifold is then intersected with the Lagrange-Grassmann submanifold, an invariant manifold for the flow. Each invariant torus which has nonempty intersection with this submanifold intersects it in a torus of one half its dimension. Thus, the flow restricted to the Lagrange-Grassmann manifold is also characterized by invariant tori to which all solutions converge. The stable and unstable manifolds are unions of Schubert cells for the Lagrange-Grassmann manifold.

Finally, the phase portrait for the original Riccati equation on the space of symmetric matrices is recovered from that on the Lagrange-Grassmann manifold. This step requires system-theoretic assumptions such as nonnegativity of the quadratic term and controllability of the associated linear system. Under these assumptions, each invariant torus for the flow on the Lagrange-Grassmann manifold corresponds to an invariant torus for the original Riccati equation. Every nonescaping solution converges to an invariant torus.

The geometric approach also leads to reinterpretation and/or generalization of several of the well-known results in the theory of the Riccati equation. Examples include Radon's formula, the eigenvector method for constructing the solutions of the algebraic Riccati equation, Willems' classification of the solutions of the algebraic Riccati equation and Wimmer's inertia theorem for the solutions of the algebraic Riccati equation.

Similar techniques can be used to obtain corresponding results for the Riccati differential equation with periodic coefficients. Details may be found in [4.13].

Acknowledgement. This research was partially supported by the National Science Foundation under grants ECS-8301015 and ECS-8696108.

References

1. Bucy, R.S.; Rodriguez-Canabal, J.: A negative definite equilibrium and its induced cone of global existence for the Riccati equation. SIAM J. Math. Anal., 3 (1972), 644-646.
2. Čurilov, A.N.: On the solutions of quadratic matrix equations. Nonlinear Vibrations and Control Theory, 2 (1978), 24-33.
3. Hermann, R.: Cartanian Geometry, Nonlinear Waves, and Control Theory, Part A. Interdisciplinary Mathematics, Vol. XX, Math. Sci. Press, Brookline, MA, 1979.
4. Hermann, R.; Martin, C.: Lie and Morse theory for periodic orbits of vector fields and matrix Riccati equations, II. Math. Systems Theory, 16 (1983), 297-306.
5. Kučera, V.: A review of the matrix Riccati equation. Kybernetika, 9 (1973), 42-61.
6. Lancaster, P.; Rodman, L.: Existence and uniqueness theorems for the algebraic Riccati equation. Int. J. Control, 32 (1980), 285-309.
7. Martin, C.: Finite escape time for Riccati differential equations. Systems and Control Letters, 1 (1981), 127-131.
8. Martin, C.: Grassmannian manifolds, Riccati equations and feedback invariants of linear systems. Geometrical Methods for the Theory of Linear Systems, C. Byrnes and C. Martin, eds., Reidel, Dordrecht, 1980.
9. Potter, J.E.: Matrix quadratic solutions. SIAM J. Appl. Math., 14 (1966), 496-501.
10. Radon, J.: Zum problem von Lagrange. Abh. Math. Sem. Univ. Hamburg, 6 (1928), 273-299.
11. Schneider, C.R.: Global aspects of the matrix Riccati equation. Math. Systems Theory, 7 (1973), 281-286.
12. Shayman, M.A.: Geometry of the algebraic Riccati equation, I. SIAM J. Control Opt., 21 (1983), 375-394.
13. Shayman, M.A.: On the phase portrait of the matrix Riccati equation arising from the periodic control problem. SIAM J. Control Opt., 23 (1985), 717-751.
14. Shayman, M.A.: Phase portrait of the matrix Riccati equation. SIAM J. Control Opt., 24 (1986), 1-65.
15. Van Swieten, A.C.M.: Qualitative behavior of dynamical games with feedback strategies. Ph.D. Thesis, University of Groningen, The Netherlands, 1977.
16. Willems, J.C.: Least squares stationary control and the algebraic Riccati equation. IEEE Trans. Automat. Control, AC-16 (1971), 621-634.
17. Wimmer, H.K.: On the algebraic Riccati equation. Bull. Austral. Math. Soc., 14 (1976), 457-461.

5 The Geometry of the Matrix Riccati Equation and Associated Eigenvalue Methods

Clyde F. Martin and Gregory Ammar

5.1 Introduction

The mathematics used to study the matrix Riccati equation is as widely varied as its applications and occurrences. In this chapter we consider some of the geometric aspects of the equation and by necessity must consider some basic differential and algebraic geometry. This mathematics is at least as simple as the classical analysis used to study ordinary differential equations and is probably more intuitive. However a certain level of understanding is necessary. In the book of Doolin and Martin, [5] there is an introduction to the subject that is designed to appeal to ones intuition rather than to ones mathematical skills. For the considerations of length we refer the reader to that book and hope that this will encourage the reader to pursue more advanced monographs for the future study of the geometry of the matrix Riccati equation.

The matrix Riccati equation has intimate connections with the problem of factoring polynomial equations that are given in the form of characteristic polynomials of square matrices. This is easily seen by considering the standard two point boundary value problem

$$\begin{pmatrix} \dot{x}_1 \\ \dot{x}_2 \end{pmatrix} = \begin{pmatrix} A_1 & A_2 \\ A_3 & A_4 \end{pmatrix} \begin{pmatrix} x_1 \\ x_2 \end{pmatrix} \qquad x_1(0) = a_1 \; x_2(1) = a_2.$$

We transform variables by the time varying transformation

$$\begin{pmatrix} x_1 \\ x_2 \end{pmatrix} = \begin{pmatrix} I & O \\ P(t) & I \end{pmatrix} \begin{pmatrix} z_1 \\ z_2 \end{pmatrix}$$

to obtain the system

$$\begin{pmatrix} \dot{z}_1 \\ \dot{z}_2 \end{pmatrix} =$$

$$\left[\begin{pmatrix} O & O \\ -\dot{P}(t) & O \end{pmatrix} + \begin{pmatrix} I & O \\ -P(t) & I \end{pmatrix} \begin{pmatrix} A_1 & A_2 \\ A_3 & A_4 \end{pmatrix} \begin{pmatrix} I & O \\ P(t) & I \end{pmatrix} \right] \begin{pmatrix} z_1 \\ z_2 \end{pmatrix}.$$

Bittanti, Laub, Willems (Eds.)
The Riccati Equation

After simplification we obtain the standard

$$\begin{pmatrix} \dot{z}_1 \\ \dot{z}_2 \end{pmatrix} = \begin{pmatrix} A_1 - A_2P & A_2 \\ -\dot{P} + A_3 - PA_1 + A_4P - PA_2P & A_4 - PA_2 \end{pmatrix} \begin{pmatrix} z_1 \\ z_2 \end{pmatrix}$$

Now we choose the function of the transformation to satisfy the Riccati differential equation

$$\dot{P} = A_3 - PA_1 + A_4P - PA_2P, \quad P(1) = O.$$

However, if instead of the time varying solution of the differential equation we choose an equilibrium point then the matrix of the original system is reduced to a block upper triangular matrix and the characteristic equation of A is factored into the characteristic equations of the matrices $A_1 - A_2P$ and $A_4 - PA_2$. Thus, in principle, the eigenvalues of a matrix can be found by integrating the matrix Riccati equation to determine an equilibrium point.

However there is a better way to proceed. The equilibrium point of the Riccati equation can be determined by the one parameter group $\exp(At)$ acting on the grassmannian manifold of k-planes in Euclidean space. This would suggest the connection between Riccati equations and the so called power methods for determining eigenvalues. In this paper we explore some of the relations and refer the reader to [1] for proofs and additional details. We hope in this paper to acquaint the reader with some of the possibilities of the using Riccati equations as a theoretical numerical tool and to also point out the serious shortcomings of using the equations as a serious tool for computation.

5.2 The Classical Power Method

Let A be a real $n \times n$-matrix, and let $\Lambda(A) = \{\lambda_1, \ldots, \lambda_n\}$ denote the spectrum of A, with $|\lambda_1| \geq |\lambda_2| \geq \cdots \geq |\lambda_n|$. The power method applied to A is defined as the repeated action of A on the unit sphere S^{n-1} in $\mathbb{R}^n$: Given $x_0 \in S^{n-1}$, the *power iterates* x_i of A starting at x_0 are defined by the recursive formula

$$x_{i+1} = \frac{Ax_i}{\|Ax_i\|} \ (i = 0, 1, \ldots).$$

It is well known that if $\lambda_1 > |\lambda_2|$, then the power iterates will converge to the unit eigenvector v or $-v$ corresponding to λ_1 provided x_0 is not contained in an A-invariant subspace of $\mathbb{R}^n$ which excludes v. Moreover, convergence occurs at a linear rate with convergence factor $|\lambda_2/\lambda_1|$. This was classically studied as the Perron-Frobenius theory for positive matrices, [7].

The power iterates x_i defined above are representatives from the subspaces $X_i = \text{span}\{x_i\}$. Note, that if $-\lambda_1 > |\lambda_2|$, then the power iterates will oscillate between the eigenvectors v and $-v$ corresponding to λ_1. In this case the x_i will not converge while the subspaces $X_i = \text{span}\{x_i\}$ will converge. The reason for this ambiguity is the fact that every one-dimensional subspace of $\mathbb{R}^n$ corresponds to two antipodal points on S^{n-1} i.e., the vectors x and $-x$ determine the same subspace X. We can guarantee that each subspace has only one representative by constraining each power iterate to lie on a fixed hemisphere of S^{n-1}. This is usually accomplished by multiplying each x_i by ± 1 so that the first nonzero component of x_i is positive. In this way, we can assure that the power iterates will converge whenever $|\lambda_1| > |\lambda_2|$ and x_0 is not contained in an A-invariant subspace which excludes the eigenvector corresponding to λ_1. Of course, any hemisphere would work provided suitable case was taken with the boundary.

Thus, the classical power method can be regarded as an iteration on the set of one-dimensional subspaces of R^n, and the points of S^{n-1}, or of a hemisphere of S^{n-1}, serve as representatives of these subspaces. It is advantageous to define the power method as an iteration on subspaces without reference to a particular set of representatives. Questions of convergence can then be posed using any naturally defined metric on the set of one-dimensional subspaces, for example, the angle between spaces. This approach is used in [4] and [12] to obtain convergence criteria for the classical power method as well as its generalizations. We now consider the set of one-dimensional subspaces more closely. In particular, this set is a compact analytic manifold, and is also a homogeneous space of a Lie group of matrices. These ideas provide a natural geometric setting in which to study the classical power method.

Let $\mathrm{R}P^{n-1}$ denote the set of all one-dimensional subspaces of R^n; this set is a well-known $(n-1)$-dimensional real analytic manifold called *real projective space.* Any invertible linear operator $A; \mathrm{R}^n \to \mathrm{R}^n$ acts on $\mathrm{R}P^{n-1}$ in the natural manner: if $X = \mathrm{span}(x)$ with $x \in \mathrm{R}^n$, then $A \cdot X = \mathrm{span}\{Ax\}$. This definition is clearly independent of the representative vector x for the subspace X. The *power iterates* X_i of A starting at $X_0 \in \mathrm{R}P^{n-1}$ are then defined as the successive action of A on X_0, i.e.,

$$X_i = A \cdot X_{i-1} = A^i \cdot X_0 \, (i = 1, 2, \ldots),$$

In this way, we view the classical power method as an iteration on $\mathrm{R}P^{n-1}$. The fact that A must be invisible is a minor inconvenience; Y not perturb to $A + \in I$.

Riccati equations arise when the power iterates are considered in a particular set of representatives of $\mathrm{R}P^{n-1}$, namely, the set of all vectors in R^n with first component equal to one. Assume that each power iterate $X_i = A \cdot X_{i-1}$ can be written as $X_i = \mathrm{span}\{x_i\}$, where $x_i = \begin{pmatrix} 1 \\ K_i \end{pmatrix}$ and K_i ia an $(n-1)$-dimensional vector. Then the iteration on this set of representatives is given by

$$K_{i+1} = \frac{A_3 + A_4 K_i}{A_1 + A_2 K_i}, \tag{5.1}$$

where the matrix A is written as a partitioned matrix

$$A = \begin{pmatrix} A_1 & A_2 \\ A_3 & A_4 \end{pmatrix}$$

with A_1 a scalar and A_4 and $(n-1) \times (n-1)$ matrix. By writing the recursive relationship (5.1) in difference equation form, we obtain

$$K_{i+1} - K_i = (A_3 + A_4 K_i - K_i A_1 - K_i A_2 K_i)\left(\frac{1}{A_1 + A_2 K_i}\right) \tag{5.2}$$

which can be recognized as a discretization of the vector-valued Riccati differential equation

$$\frac{dK(t)}{dt} = A_3 + A_4 K(t) - K(t) A_1 - K(t) A_2 K(t). \tag{5.3}$$

Note that the fixed points of the iteration (5.1), or equivalently (5.2), are precisely the equilibrium points of the differential equation (5.3); these are the solutions of the algebraic Riccati equation

$$A_3 + A_4 K - K A_1 - K A_2 K = 0.$$

Thus, we see that there is a relationship between the classical power method and the Riccati equations. Note that we are defining a set of local coordinates in $\mathbb{R}P^{n-1}$ and the Riccati equation acts within the local coordinate system.

Now let Γ_1 be the subset of $\mathbb{R}P^{n-1}$ consisting of those subspaces that are spanned by a vector of the form $\begin{pmatrix} 1 \\ K \end{pmatrix}$ with $K \in \mathbb{R}^{n-1}$. It is evident that a subspace $X \in \mathbb{R}P^{n-1}$ is contained in Γ_1 if and only if every vector $x \in X$ has its first component not equal to zero. Hence if $X_i \in \Gamma_1$ and $X_{i+1} = A \cdot X_i$ is not in Γ_1, then the scalar $A_1 + A_2 K_i$ must be zero. Similarly, if $X_i \in \Gamma_1$ and the X_i converge to X not in Γ_1, then the scalars $A_1 + A_2 K_i$ will tend to zero, and the vectors $K_i \in \mathbb{R}^{n-1}$ will become unbounded. This possibility corresponds to the finite escape time phenomenon which is exhibited by Riccati equations. However, note that the action of A on $\mathbb{R}P^{n-1}$ is well defined.

The power method applied to and $n \times n$ complex matrix is defined exactly as in the real case: if $x_i \in \mathbb{C}^n$ with $\|x_i\| = 1$, then $x_{i+1} = Ax_i / \|Ax_i\|$. In this case each vector x_i represents a one-dimensional complex subspace of $\mathbb{C}^n$, and we can guarantee that each subspace has only one representative by multiplying each iterate x_i by a scalar so that its first nonzero component is equal to one. The set of one-dimensional subspaces of $\mathbb{C}^n$ is an $(n-1)$-dimensional complex analytic manifold called *complex projective space*, and denoted by $\mathbb{C}P^{n-1}$. The action of an invertible linear map $A : \mathbb{C}^n \to \mathbb{C}^n$ on $\mathbb{C}P^{n-1}$ is defined in the natural manner, and we can therefore consider the power iterates of A on $\mathbb{C}P^{n-1}$.

The projective spaces have a rich structure which can be exploited in the study of the local and global behavior of the classical power method. In particular, $\mathbb{R}P^{n-1}$ and $\mathbb{C}P^{n-1}$ can be realized as compact homogeneous spaces of a Lie group.

5.3 The Power Method on the Grassmann Manifold

We now consider a multi-dimensional generalization of the classical power method, which we will refer to simply as the power method. In its basic form, the power method is not fast enough for practical use; nevertheless, it provides the theoretical basis for understanding the more useful matrix eigenvalue methods, particularly the QR algorithm. In this section we formulate the power method as an iteration on the Grassmann manifold as a generalization of $\mathbf{R}P^{n-1}$. A set of local coordinates for the Grassmann manifold is the defined in a natural way, and in these coordinates the power method is seen to be directly related to both the continuous-time and discrete-time matrix Riccati equations. We can therefore study the power method using the same tools which have been applied to the study of Riccati equations. Refer to [4, 8, 12, 14] for background information on the power method and its relationship to the QR algorithm.

Let V be an n-dimensional vector space over the field $\Im$, where $\Im$ is either $\mathbf{R}$ or $\mathbf{C}$. The same methods work over any field and have applications in coding theory for finite fields, for example the works of Ryan and Ryan, [14]. The set of all k-dimensional $\Im$-linear subspaces of V will be denoted by $G^k(V)$; this is the *Grassmann manifold* of k-planes in V. The general linear group $Gl(V)$ acts on $G^k(V)$ in the natural manner; if $U \in G^k(V)$ and $A \in Gl(V)$, then

$$A \cdot U = \{Ax : x \in U \subseteq V\}.$$

The *power iterates* U_i of A starting at $U_0 \in G^k(V)$ are then given by $U_i = A \cdot U_{i-1}$ $(i = 1, 2 \ldots)$. To implement the power algorithm on $G^k(V)$, a basis $x_1, \ldots, x_k$ of U_i is chosen (usually orthonormal), and then

$$U_{i+1} = \text{span}\{Ax_1, \ldots, Ax_k\} \quad (i = 0, 1, \ldots).$$

Of course, one of the problems is that the resulting basis of U_{c+1} is not orthonormal.

The power method is studied as an iteration on the set of k-dimensional subspaces in Parlett and Poole [12] and Buurema [4]. We say that an invariant subspace $U \in G^k(V)$ of $A \in Gl(V)$ is *dominant subspace* if the eigenvalues of the restriction $A_1 = A\mid_U$ have moduli equal to or exceeding that of all other eigenvalues of A. The following is a basic convergence result for the power algorithm.

Theorem 1. *If $A \in Gl(V)$ has only one dominant k-dimensional invariant subspace U, and if there exists an $(n-k)$–dimensional A-invariant subspace W which is complementary to U (i.e., $U \cap W = \{0\}$), then the power iterates of A starting at $X_0 \in G^k(V)$ will converge to U, with a linear rate of convergence, provided X_0 is complementary to W (i.e., $X_0 \cap W = \{0\}$).*

For a proof of this result, see [12]. Note that if the eigenvalues of A satisfy

$$|\lambda_1| \geq |\lambda_2| \cdots \geq |\lambda_k| > |\lambda_{k+1}| \geq \cdots \geq |\lambda_n|,$$

then the hypotheses of the theorem are satisfied by letting U and W be the invariant subspaces corresponding to $\lambda_1, \ldots, \lambda_k$ and $\lambda_{k+1}, \ldots, \lambda_n$, respectively. In this case the linear convergence factor is equal to $\mid \lambda_{k+1}/\lambda_k \mid$, [12].

The Grassmann manifold is the natural space on which to study the power method and bears the same relation to the method of Parlett and Poole [12] and Buurema [4] as does $\mathrm{R}P^{n-1}$ to the classical power method. We define a system of coordinates (i.e., a chart structure) for $G^k(V)$ as follows [5, 9]. This is roughly the same as forcing a coordinate to be 1 in $\mathrm{R}P^{n-1}$. Fix $U \in G^k(V)$ and let $W \in G^{n-k}(V)$ be a complementary subspace to U in V. By choosing bases for U and W we can express any element of $V = U \oplus W$ as a partitioned vector $\begin{pmatrix} u \\ w \end{pmatrix}$, where $U \in \Im^k$ and $w \in \Im^{n-k}$. Similarly, a linear operator $A : V \to V$ takes the form $\begin{pmatrix} A_1 & A_2 \\ A_3 & A_4 \end{pmatrix}$, where A_1 and A_4 are order k and $n-k$ square matrices, respectively. For each linear map $K : U \to W$, we define

$$U_k = \left\{ \begin{pmatrix} u \\ Ku \end{pmatrix} : u \in \Im^k \right\}.$$

Equivalently, U_K can be defined as the result of acting on U by the invertible map

$$\begin{pmatrix} I & 0 \\ K & I \end{pmatrix} : U \oplus W \to U \oplus W;$$

it is evident that $U_K \in G^k(V)$ for each linear map $K : U \to W$.

Let $L(U, W)$ denote the set of all linear maps from U to W. We then have a mapping

$$\psi : L(U, W) \to G^k(V), \quad K \mapsto U_k$$

The following proposition indicates that the map ψ can be used to define a coordinate system in an open subset of $G^k(V)$.

Proposition 1. *The map ψ is injective, and its image $\Gamma(W)$ is open and dense in $G^k(V)$. Moreover, $\Gamma(W)$ consists of all k-dimensional subspaces of V which are complementary to W.*

Thus, corresponding to the direct sum decomposition $V = U \oplus W$ we have an open neighborhood $\Gamma(W)$ of $U \in G^k(V)$ and a map

$$\psi_w : \Gamma(W) \to L(U, W) = \Im^{k(n-k)}$$

given by $\psi_w(U_k) = K$. The pair $(\Gamma(W), \psi_w)$ will be referred to as the *canonical chart* of $G^k(V)$ corresponding to the composition $V = U \oplus W$, and for $U_k \in \Gamma(W)$, the entries of K (with respect to fixed bases for U and W) will be called the *coordinates* of U_k.

The Riccati equation arises when the power iterates are considered in the canonical charts of $G^k(V)$ just as it did in $\mathrm{R}P^{n-1}$. Let $V = U \oplus W$ with $U \in G^k(V)$, and let $A : V \to V$ be an invertible map. Also, let $U_k \in \Gamma(W)$, and write A as a partitioned matrix

$$A = \begin{pmatrix} A_1 & A_2 \\ A_3 & A_4 \end{pmatrix} : U \oplus W \to U \oplus W.$$

Then we have by an easy calculation

$$\begin{aligned} A \cdot U_k &= \left\{ \begin{pmatrix} A_1 & A_2 \\ A_3 & A_4 \end{pmatrix} \begin{pmatrix} u \\ Ku \end{pmatrix} : u \in \Im^K \right\} \subset U \oplus W \\ &= \left\{ \begin{pmatrix} (A_1 & + & A_2K)u \\ (A_3 & + & A_4K)u \end{pmatrix} : u \in \Im^K \right\} \\ &= \left\{ \begin{pmatrix} cu \\ (A_3 + A_4K)(A_1 + A_2K)^{-1}u \end{pmatrix} : u \in \Im^K \right\} \\ &= U_{(A_3+A_4K)(A_1+A_2K)^{-1}}. \end{aligned}$$

Thus, the action of A on U_k is given in local coordinates by the *generalized linear fractional transformation*

$$K \to (A_3 + A_4K)(A_1 + A_2K)^{-1},$$

provided the indicated inverse exists. The power iterates of A starting at $U_0 = U_{K_0} \in \Gamma(W)$ are therefore given in local coordinates by the recursive relation

$$K_{i+1} = (A_3 + A_4K_i)(A_1 + A_2K_i)^{-1},$$

or equivalently, by the difference equation

$$K_{i+1} - K_i = (A_3 + A_4K_i - K_iA_1 - K_iA_2K_i)(A_1 + A_2K_i)^{-1}. \tag{5.4}$$

Now (5.4) is immediately recognized as a discretization of the *Riccati differential equation*

$$\frac{d}{dt}K(t) = A_3 + A_4K(t) - K(t)A_1 - K(t)A_2K(t) \tag{5.5}$$

with $K(0) = K_0$. This type of Riccati equation arises in the continuous-time linear regulator problem of control theory [3] for specific values of the A_i and arises in general in the solution of two point boundary value problems. In fact, (5.4) can be interpreted as a generalized Euler approximation to (5.5) with the factor $(A_1 + A_2K_i)^{-1}$ serving as a variable generalized step size. Note that if the power iterates of A starting at $U_{K_0} \in \Gamma(W)$ converge to $U_{\bar{K}} \in \Gamma(W)$, then $\bar{K}$ will satisfy the algebraic Riccati equation

$$A_3 + A_4\bar{K} - \bar{K}A_1 - \bar{K}A_2\bar{K} = 0. \tag{5.6}$$

This reflects the well-known correspondence between the solutions of (5.6) and the invariant subspaces of the associated matrix $\begin{pmatrix} A_1 & A_2 \\ A_3 & A_4 \end{pmatrix}$, [13, 16]. Every solution of the algebraic Riccati equation corresponds to an invariant subspace of A. However, the converse need not be true because of the local nature of the algebraic Riccati equation i.e., there may be solutions at "infinity".

Equation (5.4) also arises in the solution of the discrete-time linear regulator problem of control theory. Consider the discrete linear system

$$x_i = Ax_{i-1} + Bu_{i-1},\ x_0 \text{ given}, i = 1, 2, \ldots, N,$$

where the state vectors x_i and control vectors u_i are elements of R^n and R^m, respectively. A sequence of controls $\{u_i^*\}$ is to be chosen so that the quadratic cost functional

$$\frac{1}{2} x_N' S x_N + \frac{1}{2} \sum_{i=0}^{N-1} (x_i' Q_{x_i} + u_i' R_{u_i})$$

is minimized. Here, S and Q are symmetric nonnegative-definite matrices, R is a symmetric positive-definite matrix, and $(')$ denotes transpose. See [3] or [6] for a complete treatment of the discrete-time regulator.

The optimal sequence $\{u_i^*\}$, found using the technique of dynamic programming, is given in closed-loop form as

$$u_i^* = -(R + B'K_{i+1}B)^{-1} B'K_{i+1}Ax_i,$$

where the matrices K_i satisfy the *discrete Riccati equation*

$$K_i = A'K_{i+1} + Q - (B'K_{i+1}A)'(R + B'K_{i+1}B)^{-1} B'K_{i+1}A,$$

with $K_N = S$. The calculus of variations provides another way to solve the linear regulator. Using this approach, the optimal sequence of controls is given by

$$u_i^* = -R^{-1} B' p_{i+1}^*,$$

where the partitioned second-vector $\begin{pmatrix} x_i^* \\ p_i^* \end{pmatrix}$ satisfies the linear Hamiltonian system

$$\begin{pmatrix} x_i \\ p_i \end{pmatrix} = \begin{pmatrix} A^{-1} & A^{-1}BR^{-1}B' \\ QA^{-1} & A' + QA^{-1}BR^{-1}B' \end{pmatrix} \begin{pmatrix} x_{i+1} \\ p_{i+1} \end{pmatrix}$$

with x_0 and p_N fixed. The two formulas for the optimal controls are equivalent, and so the discrete Riccati equation can be identified as the power iterates of the above matrix. Note that this implies that the theory of the discrete time Riccati equations is the same as the continuous time Riccati equations.

Note that the formula (5.4) is not valid when the $k \times k$ matrix $A_1 + A_2K_i$ is singular. Moreover, $A_1 + A_2K_i$ will be singular only when the next iterate $U_{i+1} = A \cdot U_{K_i}$ is not an element of $\Gamma(W)$. In other words, formula (5.4) is no longer valid if the power iterates leave the chart $(\Gamma(W), \phi_W)$. This phenomenon corresponds to the finite escape time exhibited by Riccati equations, in which a solution to (5.5) becomes unbounded as t tends to some t_0. We can therefore identify every subspace in the complement of $\Gamma(W)$ in $G^k(V)$ with 'point at infinity' for $L(U, W) \cong \Im^{(n-k)\times k}$; this is the well-known interpretation of $G^k(V)$ as a compactification of $\Im^{(n-k)\times k}$, [11].

Thus, we see a direct relationship between the power method and both the continuous-time and discrete-time matrix Riccati equations. This relationship stems from the fact that the power method and the Riccati equations can be described in terms of an action of a matrix on the Grassmann manifold. A detailed analysis of the convergence of the power methods using geometric and Riccati techniques is contained in [1].

5.4 The QR Algorithm and Flag Manifolds

The QR algorithm is the most widely used method for finding the eigenvalues and invariant subspaces of a matrix. In fact, it was the QR algorithm that led to the popularity of linear quadratic control. The method of Potter for solving the Riccati equation was only useful after the QR algorithm was published. It is an attractive algorithm for several reasons including that it is easily implemented and numerically stable. Also, several techniques exist which can be used to implement the QR algorithm efficiently, increase the rate of convergence, and guarantee convergence under general conditions. The QR algorithm is intimately related to the power method and, in fact, the well-known convergence criteria for the QR algorithm follow directly from those of the power method on the Grassmann manifold. This relationship leads us to consider the QR algorithm as an iteration on a compact manifold known as the *full flag manifold.* In fact, we can view the QR algorithm as the power iteration of a linear operator on the flag manifold. Moreover, the flag manifold is a homogeneous space of the general linear group. In this way, we obtain a natural geometric interpretation of the QR algorithm, and this interpretation is directly analogous to our previous formulation of the power method. One can think of QR algorithm as simultaneous power algorithms.

Let $A \in Gl(\mathbb{R}^n)$, give $\mathbb{R}^n$ a basis $e_1, \dots, e_n$ and let A_0 denote the matrix representation of A with respect to this basis. The QR algorithm generates a sequence of similar matrices A_i as follows. For each integer $I \geq 0$, let $A_i = Q_i \cdot R_i$ be the (unique) factorization of A_i as a product of an orthogonal matrix Q_i and an upper triangular matrix with positive diagonal entries R_i. Then A_{i+1} is given recursively by $A_{i+1} = R_i \cdot Q_i$, or equivalently, since Q_i ia an orthogonal matrix,

$$A_{i+1} = Q_i' \cdot A_i \cdot Q_i = P_i' \cdot A_0 \cdot P_i,$$

where $P_i = Q_0 \cdot Q_1, \dots Q_i$ and prime denotes transpose. Thus, the QR algorithm effects a sequence of orthogonal similarity transformations, and the orthogonal matrix P_i represents the composition of these transformations up to step i. In fact, each A_i is the representation of the same transformation of $\mathbb{R}^n$ with respect to the basis formed by the columns of P_i.

The QR algorithm is shown to be equivalent to a nested sequence of power iterations as follows [12]. Note that for any i,

$$\begin{aligned} A_0 \cdot P_i &= (Q_0 R_0) Q_0 Q_1 \cdots Q_i \\ &= Q_0 (Q_1 R_1) Q_1 \cdots Q_i \\ &= Q_0 Q_1 \cdots Q_i (R_i Q_i) \\ &= Q_0 Q_1 \cdots Q_{i+1} R_{i+1}; \end{aligned}$$

so that for each i,

$$A_0 P_i = P_{i+1} R_{i+1}.$$

Thus, since each R_i is upper triangular and invertible, the span of the first k columns of P_i is equal to the ith power iterate of A_0 starting at the subspace spanned by the first k columns of the $n \times n$-identity matrix for each k. In this way, the QR algorithm can be viewed as the simultaneous execution of n power iterations, on $G^k(\mathbb{R}^n)$ for each value of k.

The convergence properties of the QR algorithm can now be derived from those of the power method. This is the approach used in [4] and [12]. For each $k = 1, 2, \ldots, n$, let E_k denote the span of the first k columns of the identity matrix (i.e., $E_k = \text{span}\{e_1, \ldots, e_k\}$).

Theorem 2. *Suppose that the eigenvalues of A have distinct moduli, and let U_k denote the sole k-dimensional dominant subspace of $A (k = 1, \ldots, n)$. Then the matrices A_i generated by the QR algorithm will converge to an upper triangular matrix whose diagonal entries are the eigenvalues of A in order of descending modulus provided $U_k \cap E_k = \{0\}$ for each $k = 1, \ldots n-1$.*

More generally, we have the following (see Wilkinson [15]).

Theorem 3. *Let A have eigenvalues $\mid \lambda_1 \mid \geq \mid \lambda_2 \mid \geq \cdots \geq \mid \lambda_k \mid$. Then if $\mid \lambda_k \mid > \mid \lambda_{k+1} \mid$, the QR iterates A_i will converge to upper block triangular form with diagonal blocks of order k and $n-k$ provided $U_k \cap E_k = \{0\}$.*

Note that this theorem does not imply that the matrices A_i will converge, but that their entries in the $(n-k) \times k$ subdiagonal block will tend to zero.

Now let V be an n-dimensional vector space over the field $\Im$, where $\Im$ is either $\mathbb{R}$ or $\mathbb{C}$. For any ordered partition $\sigma = (\sigma_1, \ldots, \sigma_p)$ of the integer n, the *flag manifold* $\mathcal{F}(V; \sigma)$ of V corresponding to σ is the set of all p-tuples.

$$\gamma = (U_1, U_2, \ldots, U_p) \in G^{\sigma_1}(V) \times G^{\sigma_1+\sigma_2}(V) \times \cdots \times G^n(V)$$

which satisfy $U_1 \subset U_2 \subset \cdots \subset U_p = V$. The transitive action of the Lie group $Gl(V)$ on each Grassmann manifold induces a transitive action of $Gl(V)$ on $\mathcal{F}(V; \sigma)$, so the flag manifolds can also be identified with homogeneous space of $Gl(V)$. As indicated above, the QR algorithm applied to an operator $A \in Gl(\mathbb{R}^n)$ with respect to a basis $e_1, \ldots, e_n$ of $\mathbb{R}^n$ is equivalent to the power iteration of A

on $\mathcal{F}(\mathrm{R}^n; (1,1,\ldots,1))$ starting at the n-tuple

$$\lambda_0 = (\mathrm{span}\{e_1\}, \mathrm{span}\{e_1, e_2\}, \ldots, \mathrm{R}^n).$$

Consider now the isotropy subgroup of the action of $Gl(V)$ on $\mathcal{F}(V;\sigma)$ at an arbitrary point $\gamma_0 \in \mathcal{F}(V;\sigma)$. For the special case of $\sigma = (1,1,\ldots,1)$, let $\gamma_0 = (U_1, U_2, \ldots, U_n)$ and choose a basis $e_1, e_2, \ldots e_n$ of V such that U_k =span $\{e_1, \ldots, e_k\}$ for each $k = 1, 2, \ldots, n$. We say in this case that the basis is *compatible* with γ_0, or γ_0-*compatible.* Now with respect to such a basis, the isotropy subgroup H of $Gl(V)$ consists of all invertible upper triangular matrices, since $A \in H$ implies $AU_k = U_k$ for each k. By counting dimensions, we see that $\dim(H) = (n(n+1)/2)$, and therefore $\mathcal{F}(V; (1,1,\ldots,1))$ is an $(n(n-1)/2)$-dimensional manifold over the field $\Im$. Similarly, the isotropy subgroup of $Gl(V)$ at $\gamma_0 \in \mathcal{F}(V; (\sigma_1, \sigma_2, \ldots, \sigma_p))$ can be identified with a set of all invertible block upper triangular matrices whose ith diagonal block is a $\sigma_i \times \sigma_i$ submatrix; the dimension of this subgroup is therefore equal to

$$\frac{n^2 + \Sigma_{i=1}^p \sigma_i^2}{2}$$

and $\dim(\mathcal{F}(V;\sigma)) = \frac{1}{2}(n^2 - \Sigma_{i=1}^p \sigma_i^2)$. Note that each flag manifold can be viewed as a subset of $\mathcal{F}_0(V) = \mathcal{F}(V; (1,\ldots,1))$ in the sense that there is a canonical projection map of $\mathcal{F}_0(V)$ onto any flag manifold $\mathcal{F}_0(V;\sigma)$. This is the map which 'forgets' all subspaces which are not of dimension $k_i = \sigma_1 + \cdots \sigma_i$ for some $i = 1, \ldots, p$. $\mathcal{F}_0(V)$ is referred to as the *full flag manifold* of V.

Note that the orthogonal group $O(n)$ also acts transitively on $\mathcal{F}_0(\mathrm{R}^n;\sigma)$, and that the unitary group $U(n)$ acts transitively on $\mathcal{F}_0(\mathrm{C}^n;\sigma)$. The flag manifolds can therefore be interpreted as homogeneous spaces of these compact subgroups of $Gl(V)$; consequently, the flag manifolds are compact. In fact, $O(n)$ and $U(n)$ are maximal compact subgroups of $Gl(\mathrm{R}^n)$ and $Gl(\mathrm{C}^n)$, respectively.

Now consider the following interpretation of the QR algorithm. The general linear group $G = Gl(\mathrm{R}^n)$ can be written as the product of subgroups $G = K \cdot H$, where K is the orthogonal group $O(n)$, and H is the isotropy subgroup of G at some $\gamma_0 \in \mathcal{F}_0(\mathrm{R}^n)$. This is not a direct product of groups, but a product in the sense that any $A \in G$ can be written $A = Q \cdot R$, with $Q \in K$ and $R \in H$. Now give R^n an orthonormal basis compatible with γ_0, so that each $R \in H$ is represented by an upper triangular matrix, and let $A_0 = Q_0 \cdot R_0$ be the matrix representation of a QR decomposition of $A \in G$. Then since $R_0 \in H$, we have $\gamma_1 = A_0\gamma_0 = Q_0\gamma_0$. Thus, the action of A_0 on γ_0 is identical to that of Q_0, and we see that the orthogonal similarity transformation $A_1 = Q_0' A_0 Q_0$ represents a change of the orthonormal basis for R^n from one that is γ_0-compatible to one that is γ_1-compatible. The basis has been 'moved' from γ_0 to γ_1. Similarly,

$$A_2 = Q_1' A_1 Q_1 = P_1' A_0 P_1$$

is the representation of A with respect to a basis compatible with $\gamma_2 = A \cdot \gamma_1 = A^2 \cdot \gamma_0$. The QR algorithm is therefore a sequence of orthogonal basis changes of R^n such that the basis at the ith iteration is compatible with $\gamma_i = A^i\gamma_0 \in \mathcal{F}_0(\mathrm{R}^n)$.

If the power iterates $\gamma_i = A\gamma_{i-1}$ on $\mathcal{F}_0(\mathsf{R}^n)$ converge to $\bar{\gamma}$, then with respect to a basis compatible with $\bar{\gamma}$, the operator A takes upper triangular form.

Note that the decomposition $A = Q \cdot R$ given above is not unique, since the subgroup $H \cap K$ is nontrivial. In fact, for any $D \in H \cap K$, we see that if $A = Q \cdot R$ is a decomposition of A, then so is $A = (Q \cdot D)(D^{-1}R)$. Also that several orthonormal bases for R^n are compatible with a given $\gamma_0 \in \mathcal{F}_0(\mathsf{R}^n)$; if $e_1, \ldots, e_n$ is such a basis, then the other orthonormal γ_0-compatible bases are of the form $\{\pm e_1, \pm e_2, \ldots, \pm e_n\}$. If we fix an orthonormal γ_0-compatible basis $\{e_1, \ldots, e_n\}$, then $H \cap K$ can be identified with the set of unit diagonal matrices in G; these are the diagonal matrices with diagonal entries equal to 1 or -1. Thus, the action of $D \in H \cap K$ fixes γ_0, but changes the orientation of the chosen basis $e_1, \ldots, e_n$.

Now, we can write H as the product $H = (H \cap K) \cdot \tilde{H}$, where $\tilde{H}$ is the subgroup of all positive-definite elements of H. $\tilde{H}$ can be identified with the upper triangular matrices with positive diagonal entries; these are precisely those matrices which preserve the orientation of a given γ-compatible basis. Since the intersection $\tilde{H} \cap K$ is trivial, we see that the decomposition $A = Q \cdot R$ of $A \in Gl(\mathsf{R}^n)$ with respect of the product $Gl(\mathsf{R}^n)$ with respect of the product $Gl(\mathsf{R}^n) = K \cdot \tilde{H}$ is unique. This guarantees that convergence of the QR iterates A_i is equivalent to convergence of the Q_i to the identity matrix.

The QR algorithm generates a sequence of basis changes Q_i, and the P_i represent the composition of these basis changes; at stage i, R^n has a basis compatible with $\gamma_i = A^i\gamma_0 = P_i\gamma_0$. Each P_i effects the same transformation on γ_0 as A^i, and so P_i represents a change from a γ_0-compatible basis for R^n to a γ_i-compatible basis.

If the power iterates $\gamma_i = A\gamma_{i-1}$ converge to $\bar{\gamma}$, then the matrices Q_i will tend to the identity matrix, and the P_i will converge to an orthogonal matrix $\bar{P}$. Convergence of the Q_i and P_i is guaranteed because of the uniqueness of the factorization $A_i = Q_iR_i$, where R_i is positive-definite. Since $\bar{\gamma}$ is fixed by the action of A_0, the operator A takes the form of an upper triangular matrix $\bar{A}$ with respect to a $\bar{\gamma}$-compatible basis of R^n.

Thus, the QR algorithm can be regarded as the successive action of a linear operator on the full flag manifold. Since this manifold is a compact homogeneous space of the general linear group, the same ideas as those used in the study of the power methods can be used to study the convergence properties of the QR algorithm. The QR algorithm can be viewed as a discretization of a nested sequence of Riccati equations.

5.5 Riccati Equations and Roots of Polynomials

The problem of finding the zeros of equations has been successfully attacked using differential equations. The continuation methods that have enjoyed so much popularity in the last two decades are methods of this sort. We have seen that Riccati equations can be used to factor polynomial equations that are given as the characteristic polynomials of matrices. The question arises of whether or not the Riccati differential equation can be used effectively for this purpose.

The major difficulty associated with integrating a Riccati differential equation is that it may have finite escape time. Unfortunately it is very difficult, in general, to determine what sets of initial data will allow a particular matrix Riccati equation to be integrated from zero to infinity without finite escape. Only in special conditions is an answer known, [11].. The problem is equivalent to the problem of existence and uniqueness of linear two point boundary value problems, [2].

It is worth noting that the problem of factoring a given polynomial into the product of two polynomials reduces to the problem of solving a system of quadratic equations. The algebraic Riccati equation is likewise a system of quadratic equations. The Riccati equation has the potential of being competitive with continuation methods for this particular problem. However the problem will always remain of the finite escape time. The continuation methods have much the same problem due to non ivertibility of the Jacobian.

5.6 Conclusion

We have seen that the numerical methods for finding the eigenvalues and invariant subspaces of a matrix are intimately related to the matrix Riccati equations. This relationship stems from the fact that the Riccati equations and the iterative matrix eigenvalue methods are all naturally formulated in terms of an action of a linear operator on a compact homogeneous space of the general linear group. Thus, the tools of Lie theory and the theory of dynamical systems on compact manifolds which have been useful in the study of Riccati equations can also be applied to the numerical matrix eigenvalue methods. This approach provides a unified mathematical setting in which to study the matrix eigenvalue methods as well as the Riccati equations.

References

1. Ammar, G. and Martin, C., 'The geometry of matrix eigenvalue methods', *Acta Applicandae Mathematicae*, 5, (1986), 239–278.
2. Balakumar, S. and Martin, C., 'Two point boundary value problems and the matrix riccati equation'. In *Operator Methods for Optimal Control Problems*, Sung J. Lee (ed.), Vol. 108, Lecture Notes in Pure and Applied Mathematics. New York: Dekker Publishing Co., 1987 pp. 31–56.
3. Anderson, B.D.O. and Moore, J.B., *Linear Optimal Control*, Prentice-Hall, Englewood Cliffs, 1971.

4. Buurema, Hendrik Jan, 'A Geometric Proof of Convergence for the QR Method', PhD dissertation, Rijksuniversiteit te Groningen, 1970.
5. Doolin, B.F. and Martin, C.F., 'Introduction to Differential Geometry for Engineers', Marcel Dekker, Inc, New York, 1990.
6. Dorato, Peter and Levis, Alexander H., 'Optimal Linear Regulators: The Discrete-Time Case', IEEE *Trans. Aut. Cont. AC-16* (1971), 613-620.
7. Gantmacher, F., Matrix Theory, Vol II, Chelsea Publishing Co, NY, 1977.
8. Golub, Gene H. and Van Loan, Charles F., *Matrix Computations*, Johns Hopkins University Press, Baltimore, 1983.
9. Hermann, Robert and Martin, Clyde,'Lie Theoretic Aspects of the Riccati Equation', *Proceedings of the 1977 IEEE Conference on Decision and Control*, New Orleans, 1977, pp. 265-270.
10. Hermann, Robert and Martin, Clyde, 'Lie and Morse Theory of Periodic Orbits of Vector Fields and Matrix Riccati Equations, I: General Lie-Theoretic Methods', *Math. Systems Theory 15* (1982), 277-284.
11. Martin, Clyde F., 'Finite Escape Time for Riccati Differential Equations', *Systems Control Lett. 1* (1981), 127-131.
12. Parlett, B.M. and Poole, W. G. Jr., 'A Geometric Theory for the QR. LU, and Power Iterations', *SIAM J. Numer. Anal. 10* (1973), 389-412.
13. Potter, James E., 'Matrix Quadratic Solutions', *J. SIAM Appl. Math. 14* (1966), 496-501.
14. Ryan, Charles T. and Ryan, Kevin M., 'An Appleication of Geometry to the Calculation of Weight Enumerators', *Congr. Numer. 67* (1988), 77-89.
15. Wilkinson, J. H., 'The Algebraic Eigenvalue Problem', Clarendon Press, Oxford, 1965.
16. Willems, Jan C., 'Least Squares Stationary Optimal Control and the Algebraic Riccati Equation', IEEE *Trans. Aut. Cont.* AC-16 (1971), 621-634.

6 The Periodic Riccati Equation

Sergio Bittanti, Patrizio Colaneri and Giuseppe De Nicolao

6.1 Introduction

The history of the time-varying Riccati equation can be traced back to Riccati's original manuscripts of 1715-1725. Indeed, the major concern of Count Riccati was to study the problem of the separation of variables in quadratic and *time-varying* scalar differential equations [1]. The equation has been the subject of several contributions in the subsequent centuries. In recent times, the importance of the Riccati equation in Control, Systems, and Signals has led to the development of a considerable research activity on the subject, see e.g., [2], [3], [4] for the time-varying matrix Riccati equation.

Spurred by the periodic characteristics of many phenomena encountered in the natural and artificial world , the Riccati equation with periodic coefficients has been also extensively dealt with. Roughly, the research activity on this subject consists of three major periods. The first one refers to the pioneering works dealing with the geometrical analysis, the general integration methods and the second variation techniques for periodic optimization, see [5], [6], [7], [8], and references quoted there. The middle period is characterized by the attempt of analyzing the Riccati equation in terms of the structural properties of the underlying periodic system. The origin of this phase can be traced back to the work of Hewer [9], where a Quasi-Linearization approach applied to the periodic Riccati equation led to a necessary and sufficient existence condition for the symmetric, periodic and positive semidefinite solution. Analogous results were derived in [10] by an approach based on the Hamiltonian matrix analysis. The present period is characterized by a full clarification of the problems evolving around the structural properties of periodic systems along with the related issue of canonical decomposition. The role played by these properties in the analysis of the Riccati equation has been studied in many papers. The interested reader is referred to [11–12] for the Hamiltonian matrix approach, [13] for the iterative Quasi-Linearization methods, [14–15] for the geometric aspects, [16–17] for the Inertia theorems and [18–24] for the optimal control and filtering problems. Extensions to the infinite-dimensional and to the quasi-periodic cases can be found in [25–27] and [28], respectively.

Bittanti, Laub, Willems (Eds.)
The Riccati Equation

This chapter provides a unified survey on the main results on the Periodic Riccati Equation (PRE):

$$-\dot{P}(t) = A(t)'P(t) + P(t)A(t) - P(t)B(t)B(t)'P(t) + C(t)'C(t) \tag{PRE}$$

where $A(t) \in R^{n\times n}$, $B(t) \in R^{n\times m}$, $C(t) \in R^{p\times n}$. Matrix functions $A(\cdot)$, $B(\cdot)$, $C(\cdot)$ are periodic of period T:

$$A(t+T) = A(t),\ B(t+T) = B(t),\ C(t+T) = C(t),$$

and integrable over $[0, T]$.

We will also be concerned with the discrete-time version of the PRE (DPRE), defined as

$$\begin{aligned} P(t) &= A(t)'P(t+1)A(t) + C(t)'C(t) \\ &\quad - A(t)'P(t+1)B(t)(I + B(t)'P(t+1)B(t))^{-1}B(t)'P(t+1)A(t). \end{aligned} \tag{DPRE}$$

Here $A(\cdot)$,$B(\cdot)$ and $C(\cdot)$, which depend upon a discrete time index t, are again periodic matrices of period T, where T is an integer.

The PRE and DPRE play a major role in the optimal periodic control problem, whereas their dual equations

$$\dot{P}(t) = A(t)P(t) + P(t)A(t)' - P(t)C(t)'C(t)P(t) + B(t)B(t)',$$

and

$$\begin{aligned} P(t+1) &= A(t)P(t)A(t)' + B(t)B(t)' \\ &\quad - A(t)P(t)C(t)'(I + C(t)P(t)C(t)')^{-1}C(t)P(t)A(t)' \end{aligned}$$

arise in the optimal periodic filtering problem.

In this chapter, we will focus on the PRE and DPRE; however a few nonstandard periodic Riccati equations, relative to game theory problems, H_∞ control, etc., will also be introduced and briefly discussed. Parallel results for the equations relative to the dual problems can be easily obtained via duality theory, and are therefore omitted.

The chapter is organized as follows. Some basic concepts relative to linear continuous-time periodic systems are introduced in Section 6.2. In the same Section the Lyapunov lemma and the inertia theorems for the Periodic Lyapunov Equation (PLE) are given. Section 6.3 is devoted to the PRE. Preliminarily, we present the main theoretical tools for the subsequent analysis, i.e., the Hamiltonian matrix properties, the Inertia Theorems, the Quasi-Linearization approach and the decomposition of the PRE induced by the Kalman canonical decomposition of the underlying system. The core of the Section is constituted by the analysis of the Symmetric Periodic Positive Semidefinite (SPPS) solutions. We will prove a number of existence and uniqueness conditions with special emphasis on stabilizing, maximal, strong and positive definite solutions. The various results are then translated into parallel ones for the Symmetric Periodic Negative Semidefinite (SPNS) solutions. The issue of convergence to periodic equilibria is also touched upon. Finally, some remarks on the nonstandard periodic Riccati equation and the monotonicity property are given. Sections 6.1 to 6.3 deal with the continuous-time case only. Discrete-time periodic systems and the DPRE are analyzed in Section 6.4. The analysis of the DPRE differs from that of the PRE in two basic issues. First, discrete-time systems may be nonreversible, so that some intriguing issues are to be coped with. The second one is the existence of a natural isomorphism between discrete-time periodic systems and (finite-dimensional) time-invariant ones, which reflects in a correspondence between the periodic solutions of the DPRE and those of a suitable algebraic Riccati equation. A few important applications are reviewed in Section 6.5. They refer to LQG and H_∞ periodic filtering and control, and multirate sampled-data control systems.

For easy reference, we end this Section by listing the set of acronyms of extensive use in this Chapter:

PRE: continuous-time Periodic Riccati Equation
DPRE: Discrete-time Periodic Riccati Equation
ARE: continuous-time Algebraic Riccati Equation
DARE: Discrete-time Algebraic Riccati Equation
PLE: continuous-time Periodic Lyapunov Equation
SP: Symmetric and Periodic
SPPS: Symmetric, Periodic and Positive Semidefinite
SPPD: Symmetric, Periodic and Positive Definite
SPNS: Symmetric, Periodic and Negative Semidefinite.

6.2 Analysis of Linear Periodic Systems – An Overview

6.2.1 Preliminaries and Stability

The following periodic system can be associated with the PRE:

$$\dot{x}(t) = A(t)x(t) + B(t)u(t)$$
$$y(t) = C(t)x(t)$$

where $u(t) \in R^m$, $x(t) \in R^n$, $y(t) \in R^p$, are the input, state and output vector variables respectively; as already stated, $A(\cdot)$, $B(\cdot)$, $C(\cdot)$, are periodic matrices of period T:

$$A(t+T) = A(t),\ B(t+T) = B(t),\ C(t+T) = C(t).$$

It is easy to see that the system transition matrix, denoted by $\Phi_A(t,\tau)$, enjoys the property

$$\Phi_A(t+T,\tau+T) = \Phi_A(t,\tau).$$

$\Phi_A(t+T,t)$ is named monodromy matrix at t. Its eigenvalues are independent of t, and named *characteristic multipliers*. In the sequel, attention will be focussed on the monodromy matrix at a given time point, say t_0, which will be denoted by $\Psi_A(t_0)$, i.e.,

$$\Psi_A(t_0) = \Phi_A(t_0+T,t_0).$$

The system (or $A(\cdot)$) is stable [antistable]iff the characteristic multipliers of $A(\cdot)$ belong to the open unit disk [do not belong to the closed unit disk]. Notice that the characteristic multipliers of $\tilde{A}(\cdot)$, with $\tilde{A}(t) = A(-t)'$ coincide with those of $A(\cdot)$, whereas the characteristic multipliers of $\bar{A}(\cdot)$, with either $\bar{A}(t) = -A(t)'$ or $\bar{A}(t) = -A(-t)$ are given by the reciprocals of those of $A(\cdot)$. For more details on the basic properties of periodic systems, the reader is referred to [29].

6.2.2 Structural Properties

6.2.2.1 Reachability, Observability and Canonical Decomposition. The structural properties of periodic systems have been extensively investigated in the last decade. Among the various characterizations of reachability and observability now available, see e.g. [29], we will focus on the modal notions only.

Definition 6.1. A characteristic multiplier λ of $A(\cdot)$ is said to be $(A(\cdot),B(\cdot))$-unreachable if

$$\Psi_A(t_0)'x = \lambda x\,, x \neq 0, \Longrightarrow B(t)'\Phi_A(t_0,t)'x = 0, \; a.e. \; t \in [t_0, t_0+T].$$

A characteristic multiplier of $A(\cdot)$ is said to be $(A(\cdot),B(\cdot))$-reachable if it is not $(A(\cdot),B(\cdot))$ unreachable. The system, or the pair $(A(\cdot),B(\cdot))$, is said to be reachable if all the characteristic multipliers af $A(\cdot)$ are $(A(\cdot),B(\cdot))$-reachable.

Definition 6.2. A characteristic multiplier λ of $A(\cdot)$ is said to be $(A(\cdot),C(\cdot))$-unobservable if

$$\Psi_A(t_0)x = \lambda x\,, x \neq 0, \Longrightarrow C(t)\Phi_A(t,t_0)x = 0, \; a.e. \; t \in [t_0, t_0+T].$$

A characteristic multiplier of $A(\cdot)$ is said to be $(A(\cdot),C(\cdot))$-observable if it is not $(A(\cdot),C(\cdot))$-unobservable. The system, or the pair $(A(\cdot),C(\cdot))$, is said to be observable if all the characteristic multipliers of $A(\cdot)$ are $(A(\cdot),C(\cdot))$-observable. □

It can be shown, [29], that the notions of unreachable and unobservable multiplier previously introduced are independent of the particular time point t_0.

The Kalman canonical decomposition can be extended to periodic systems, as indicated in [30]. In particular, the observable/unobservable decomposition takes the form

$$A(t) = \begin{bmatrix} A_o(t) & 0 \\ \tilde{A}_o(t) & \bar{A}_o(t) \end{bmatrix}, \; B(t) = \begin{bmatrix} B_o(t) \\ \bar{B}_o(t) \end{bmatrix}, \; C(t) = [C_o(t)\; 0].$$

Obviously, the characteristic multipliers of $A_o(\cdot)$, resp. $\bar{A}_o(\cdot)$, are $(A(\cdot),C(\cdot))$-observable, resp.$(A(\cdot),C(\cdot))$-unobservable, characteristic multipliers of $A(\cdot)$.

Finally, by means of the Kalman canonical decomposition, it is easy to show that the $(A(\cdot),B(\cdot))$-unreachable characteristic multipliers of $A(\cdot)$ are invariant under periodic state feedback. To be precise, λ is an $(A(\cdot),B(\cdot))$-unreachable characteristic multiplier of $A(\cdot)$ iff it is an $(F(\cdot),B(\cdot))$-unreachable characteristic multiplier of $F(\cdot)$, where $F(t)=A(t)+B(t)K(t)$, $K(\cdot)$ being any T-periodic feedback matrix.

6.2.2.2 Stabilizability and Detectability. Thanks to the concepts of the previous subsection, we are now in a position to introduce three different, yet equivalent, characterizations of stabilizability and detectability of periodic systems.

The system is stabilizable, resp. detectable, if the unreachable part, resp. unobservable part, is stable. We will also say that $(A(\cdot),B(\cdot))$, resp. $(A(\cdot),C(\cdot))$, is stabilizable, resp. detectable.

As usual, stabilizability is linked with the existence of a stabilizing state-feedback, whereas detectability is linked with the existence of a stable state-observer. To be precise, $(A(\cdot),B(\cdot))$, resp. $(A(\cdot),C(\cdot))$, is stabilizable, resp. detectable, iff there exists a periodic matrix $K(\cdot)$, resp. $L(\cdot)$, such that $A(\cdot)+B(\cdot)K(\cdot)$, resp. $A(\cdot)+L(\cdot)C(\cdot)$, is stable.

Finally, the modal definitions of reachability and observability yield the corresponding modal notions of stabilizability [antistabilizability] and detectability [antidetectability]. For instance, the system, or the pair $(A(\cdot),C(\cdot))$, is detectable [antidetectable] if all the $(A(\cdot),C(\cdot))$-unobservable characteristic multipliers of $A(\cdot)$ [do not] belong to the open [closed] unit disk.

6.2.3 The Periodic Lyapunov Equation

In this section, preliminary results on the differential Periodic Lyapunov Equation (PLE) are recalled.

6.2.3.1 The Periodic Lyapunov Lemma. As is well known, the Lyapunov equation plays an important role in the analysis of Riccati equations. Here, we will present the basic results on the PLE, which will be useful in the following. For an extensive analysis of the PLE, the interested reader is referred to [31]. The PLE is defined as

$$-\dot{P}(t)=F(t)'P(t)+P(t)F(t)+Q(t)$$

where $F(\cdot)$ and $Q(\cdot)$ are T-periodic matrices of suitable dimensions. The solution of the PLE is given by the celebrated formula

$$P(t)=\Phi_F(t_f,t)'P(t_f)\Phi_F(t_f,t)+\int_t^{t_f}\Phi_F(\sigma,t)'Q(\sigma)\Phi_F(\sigma,t)\mathrm{d}\sigma \tag{6.1}$$

In particular, if $F(\cdot)$ is asymptotically stable, there exists a unique T-periodic solution $\bar{P}(\cdot)$, whose periodic generator at time t, $\bar{P}(t)$, is the unique solution of the discrete-time algebraic Lyapunov equation

$$P(t) = \Psi_F(t)'P(t)\Psi_F(t) + \int_t^{t+T} \Phi_F(\sigma,t)'Q(\sigma)\Phi_F(\sigma,t)\mathrm{d}\sigma.$$

Moreover, if $Q(t)$ is positive semidefinite, $\forall t$, it can be seen that the periodic generator is positive semidefinite, whereby $P(t)$ is positive semidefinite, $\forall t$. Conversely, if $Q(\cdot)$ is positive semidefinite $\forall t$ and $(A(\cdot), Q(\cdot))$ is observable, then the existence of an SPPS solution of the PLE implies the stability of $A(\cdot)$. This gives the statement of the periodic *Lyapunov lemma*. When observability is replaced by detectability, the previous result can be extended as follows, see [32] .

Extended Lyapunov lemma. *$A(\cdot)$ is stable iff, for any SPPS $Q(\cdot)$ such that $(A(\cdot), Q(\cdot))$ is detectable, there exists an SPPS solution of the PLE.*

6.2.3.2 Periodic Inertia Theorems. In general, even if $Q(\cdot)$ is SPPS, a periodic solution of the PLE may be sign-indefinite. Under weak assumptions, however, its inertia (i.e., the triple of integers constituted by the number of positive, null and negative eigenvalues) is time invariant, and is determined by the position in the complex plane of the characteristic multipliers of $A(\cdot)$. To precisely state the inertia theorems, a few symbols, relative to a square matrix S, are needed.

- $\nu_c(S)$ = number of eigenvalues of S with negative real part
- $\delta_c(S)$ = number of eigenvalues of S with null real part
- $\pi_c(S)$ = number of eigenvalues of S with positive real part
- $\nu_d(S)$ = number of eigenvalues of S with modulus lower than 1
- $\delta_d(S)$ = number of eigenvalues of S with modulus equal to 1
- $\pi_d(S)$ = number of eigenvalues of S with modulus greater than 1.

Among the various inertia results, the following ones, given in [16] and [17] are stated here.

Inertia theorem. *Assume that $Q(\cdot)$ is SPPS and suppose that there exists an SP solution of the PLE. Then, if $(A(\cdot), Q(\cdot))$ is observable,*

$$\delta_d(\Psi_A(t_0)) = 0 = \delta_c(P(t))$$

$$\nu_d(\Psi_A(t_0)) = \pi_c(P(t)).$$

Extended inertia theorem. *Assume that $Q(\cdot)$ is SPPS and suppose that there exists an SP solution of the PLE. Then, if $(A(\cdot), Q(\cdot))$ is detectable,*

$$\delta_d(\Psi_A(t_0)) = 0$$

$$\nu_d(\Psi_A(t_0)) = \pi_c(P(t)) + \delta_c(P(t)).$$

6.3 The Periodic Differential Riccati Equation

6.3.1 Preliminaries

In this section, the terminology for the solutions of the PRE is introduced. Furthermore, we present the basic approaches used in the analysis, namely the Hamiltonian theory, the canonical decomposition, and the Quasi-Linearization technique. In addition, some useful results on the inertia of the PRE are derived as straightforward consequences of the inertia theorems for the PLE.

6.3.1.1 Classification of Periodic Solutions. We will consider the real Symmetric and Periodic (SP) solutions of the PRE. Among all SP solutions, particular attention will be paid to the following ones:

- $P_M(t)$: maximal solution, i.e., $P_M(t) \geq P(t)$, $\forall t$, for any SP solution $P(\cdot)$ of the PRE
- $P_m(t)$: minimal solution, i.e., $P_m(t) \leq P(t)$, $\forall t$, for any SP solution $P(\cdot)$ of the PRE
- $P_S(t)$: strong solution, i.e., the characteristic multipliers of $F_S(t) = A(t) - B(t)B(t)'P_S(t)$ belong to the closed unit disk.
- $P_a(t)$: antistrong solution, i.e., the characteristic multipliers of $F_a(t) = A(t) - B(t)B(t)'P_a(t)$ do not belong to the open unit disk.
- $P_+(t)$: stabilizing solution, i.e., $F_+(t) = A(t) - B(t)B(t)'P_+(t)$ is stable.
- $P_-(t)$: antistabilizing solution, i.e., $F_-(t) = A(t) - B(t)B(t)'P_-(t)$ is antistable.
- $P_o(t)$: minimal nonnegative solution, i.e., an SPPS solution $P_o(\cdot)$ s.t. $P_o(t) \leq P(t)$, $\forall t$, for any SPPS solution $P(\cdot)$.

These definitions are the generalization of analogous definitions given for the algebraic Riccati equation. From the definitions of maximal and minimal solution, one can introduce the so-called *gap* as

$$\Delta(t) = P_M(t) - P_m(t).$$

Finally, the value $P(t_0)$ taken by a periodic solution $P(\cdot)$ at time t_0, is named periodic generator at t_0.

6.3.1.2 Hamiltonian Matrix. The Hamiltonian matrix associated with the PRE is

$$H(t) = \begin{bmatrix} A(t) & -B(t)B(t)' \\ -C(t)'C(t) & -A(t)' \end{bmatrix}.$$

Matrix $H(t)$ is Hamiltonian, namely $H(t)'J + JH(t) = 0$, $\forall t$, where J is the symplectic matrix:

$$J = \begin{bmatrix} 0 & I \\ -I & 0 \end{bmatrix}.$$

Consequently, the corresponding transition matrix $\Phi_H(t,t_0)$ is symplectic, i.e.,

$$\Phi_H(t,t_0)' J \Phi_H(t,t_0) = J.$$

Partition matrix $\Phi_H(t,t_0)$ into four $n \times n$ blocks as follows

$$\Phi_H(t,t_0) = \begin{bmatrix} \Phi_{11}(t,t_0) & \Phi_{12}(t,t_0) \\ \Phi_{21}(t,t_0) & \Phi_{22}(t,t_0) \end{bmatrix}.$$

Then, as is well known, the solution of the differential Riccati equation with initial condition P_0 at time t_0, is given by

$$P(t) = (\Phi_{21}(t,t_0) + \Phi_{22}(t,t_0)P_0)(\Phi_{11}(t,t_0) + \Phi_{12}(t,t_0)P_0)^{-1} \tag{6.2}$$

The solution can also be given a backward expression, easily deducible from (6.2), as follows:

$$P(t) = (P_f \Phi_{12}(t_f,t) - \Phi_{22}(t_f,t))^{-1} (\Phi_{21}(t_f,t) - P_f \Phi_{11}(t_f,t)),$$

where t_f is the final time point and P_f is the final condition, i.e., $P(t_f) = P_f$. We will not dwell upon the case when $\Phi_{11}(t,t_0) + \Phi_{12}(t,t_0)P_0$ or $P_f\Phi_{12}(t_f,t) - \Phi_{22}(t_f,t)$ are singular, which corresponds to the existence of a finite escape time, see [15] for a discussion on this topic.

The solution given by eq. (6.2) will be periodic if $P(t_0+T) = P_0$, i.e., letting $\Psi_{ij}(t_0) = \Phi_{ij}(t_0+T,t_0)$, $i,j = 1,2$:

$$P_0\,(\Psi_{11}(t_0) + \Psi_{12}(t_0)P_0) = (\Psi_{21}(t_0) + \Psi_{22}(t_0)P_0) \tag{6.3}$$

Conversely, any solution of this algebraic equation is a periodic generator for the differential Riccati equation.

Eq. (6.2) can be given a more standard expression by resorting to a Floquet representation of the Hamiltonian matrix, see [8]. In such a way, one can associate to the original differential Riccati equation an algebraic Riccati equation whose unknown is a periodic generator. This reformulation approach proved particularly powerful in the integration of the periodic Riccati equation.

The properties of the Hamiltonian matrix are also strictly related to the characteristics of the so called closed-loop matrix, i.e.,

$$F(t) = A(t) - B(t)B(t)'P(t)$$

where $P(\cdot)$ is a solution of the PRE. It is readily seen from (6.2) that

$$\Phi_F(t,t_0) = \Phi_{11}(t,t_0) + \Phi_{12}(t,t_0)P_0. \tag{6.4}$$

In particular, if $P(\cdot)$ is periodic, then $F(\cdot)$ is periodic too, and eq. (6.2) over a period can be rewritten as

$$\Psi_H(t_0)\begin{bmatrix} I \\ P_0 \end{bmatrix} = \begin{bmatrix} I \\ P_0 \end{bmatrix}(\Psi_{11}(t_0) + \Psi_{12}(t_0)P_0). \tag{6.5}$$

Note that (6.5) expresses the invariance of the range of the matrix $[I\, P_0']'$ with respect to $\Psi_H(t_0)$, see [15] for more details and a discussion on the interpretation of this condition from a geometric viewpoint. This formula entails that the characteristic multipliers of $F(\cdot)$ constitute a subset of the characteristic multipliers of $H(\cdot)$. Indeed, consider the matrix

$$S(t) = \begin{bmatrix} I & 0 \\ -P(t) & I \end{bmatrix}.$$

and perform the change of basis in R^{2n} leading to:

$$\tilde{H}(t) = S(t)H(t)S(t)^{-1} + \dot{S}(t)S(t)^{-1} = \begin{bmatrix} F(t) & -B(t)B(t)' \\ 0 & -F(t)' \end{bmatrix}$$

and

$$\Psi_{\tilde{H}}(t_0) = S(t_0)\Psi_H(t_0)S(t_0)^{-1} = \begin{bmatrix} \Psi_F(t_0) & * \\ 0 & \Psi_F(t_0)^{-1'} \end{bmatrix}, \qquad (6.6)$$

where * is a block of no specific interest. From (6.6), it appears that the characteristic multipliers of $H(\cdot)$ are those of $F(\cdot)$ together with their reciprocal ones.

Among the characteristic multipliers of the Hamiltonian matrix, one may find some characteristic multipliers of $A(\cdot)$. Indeed, the following results hold, (see [10–12] for the proof).

Lemma 6.1.

(i) *The characteristic multipliers of* $A(\cdot)$ *which are* $(A(\cdot), B(\cdot))$*-unreachable or/and* $(A(\cdot), C(\cdot))$*-unobservable are also characteristic multipliers of* $H(\cdot)$.

(ii) *The unit-modulus characteristic multipliers of* $H(\cdot)$ *are* $(A(\cdot), B(\cdot))$*-unreachable or/and* $(A(\cdot), C(\cdot))$*-unobservable characteristic multipliers of* $A(\cdot)$. □

Lemma 6.1(ii) points out that all the unit-modulus closed loop characteristic multipliers stem from the unreachable or unobservable open loop characteristic multipliers. As for Lemma 6.1(i), some further elaboration shows that the open loop unreachable multipliers are preserved under the feedback $u(t) = K(t)x(t)$, $K(t) = -B(t)'P(t)$, whereas the open loop unobservable multipliers may be either preserved or commuted into their reciprocal ones.

6.3.1.3 Inertia Theorems for the PRE The inertia results relative to the periodic Lyapunov equation can be transferred to the periodic Riccati equation. Indeed, as is well known, any time-varying Riccati equation can be written as

$$-\dot{P}(t) = F(t)'P(t) + P(t)F(t) + H(t)'H(t), \qquad (6.7)$$

where

$$F(t) = A(t) - B(t)B(t)'P(t)$$

$$H(t)'H(t) = C(t)'C(t) + P(t)B(t)B(t)'P(t).$$

In particular, eq. (6.7) holds in the periodic case. Therefore, if $P(\cdot)$ is an SP solution of the PRE, then $P(\cdot)$ is also a solution of the Lyapunov-type eq. (6.7).

The structural properties of the pair $(F(\cdot), H(\cdot))$ can be related to the ones of the pair $(A(\cdot), C(\cdot))$ by means of the following lemma, which goes back to [16], [17]:

Lemma 6.2. *If λ is an $(F(\cdot), H(\cdot))$-unobservable characteristic multiplier of $F(\cdot)$, then λ is an $(A(\cdot), C(\cdot))$-unobservable characteristic multiplier of $A(\cdot)$.* □

In particular, the observability [detectability] of $(A(\cdot), C(\cdot))$ implies the observability [detectability] of $(F(\cdot), H(\cdot))$. Consequently, from the extended inertia theorem (Section 6.2.3.2) the result below follows.

Theorem 6.1. *Let $P(\cdot)$ be an SP solution of the PRE. Then*

1. *If $(A(\cdot), C(\cdot))$ is observable, then*

   $\delta_d(\Psi_F(t_0)) = \delta_c(P(t)) = 0, \forall t.$

   $\nu_d(\Psi_F(t_0)) = \pi_c(P(t)), \forall t$

2. *If $(A(\cdot), C(\cdot))$ is detectable, then*

   $\delta_d(\Psi_F(t_0)) = 0, \forall t$

   $\nu_d(\Psi_F(t_0)) = \pi_c(P(t)) + \delta_c(P(t)), \forall t.$ □

For an extensive presentation of the inertia theory for time-invariant and periodic systems, the interested reader is referred to [33].

6.3.1.4 Canonical Decomposition of the PRE If a change of basis is performed in the state space, say $\tilde{x}(t) = T(t)x(t)$, $det T(t) \neq 0, \forall t$, the Riccati equation becomes

$$-\dot{\tilde{P}}(t) = \tilde{A}(t)'\tilde{P}(t) + \tilde{P}(t)\tilde{A}(t) + \tilde{C}(t)'\tilde{C}(t) - \tilde{P}(t)\tilde{B}(t)\tilde{B}(t)'\tilde{P}(t), \quad (6.8)$$

where

$$\tilde{A}(t) = T(t)A(t)T(t)^{-1} + \dot{T}(t)T(t)^{-1}$$
$$\tilde{B}(t) = T(t)B(t)$$
$$\tilde{C}(t) = C(t)T(t)^{-1}.$$

The solutions of the original Riccati equation and those of eq. (6.8) are in a one-to-one correspondence: $P(t) = T(t)'\tilde{P}(t)T(t)$. Notice that such a correspondence preserves the inertia of the solution. In particular, if $P(t) \geq 0$, then $\tilde{P}(t) \geq 0$, and viceversa. Therefore, for the analysis of the solutions, there is no loss of generality in considering the system in the Kalman canonical decomposition, see Section (6.2.2.1). Correspondingly, a symmetric solution $P(\cdot)$ can be partitioned as

$$P(t) = \begin{bmatrix} P_o(t) & \tilde{P}_o(t) \\ \tilde{P}_o(t)' & \bar{P}_o(t) \end{bmatrix},$$

where $P_o(t)$ is a square matrix of dimension equal to the dimension of the observability subspace. A cumbersome computation shows that the Riccati equation decomposes into the three subequations:

$$\begin{aligned}-\dot{P}_o(t) &= P_o(t)A_o(t) + \tilde{P}_o(t)\tilde{A}_o(t) + A_o(t)'P_o(t) + \tilde{A}_o(t)'\tilde{P}_o(t)'+\\ &+ C_o(t)'C_o(t) - P_o(t)B_o(t)B_o(t)'P_o(t) - P_o(t)B_o(t)\bar{B}_o(t)'\tilde{P}_o(t)'+\\ &- \tilde{P}_o(t)\bar{B}_o(t)B_o(t)'P_o(t) - \tilde{P}_o(t)\bar{B}_o(t)\bar{B}_o(t)'\tilde{P}_o(t)'\end{aligned} \tag{6.9}$$

$$\begin{aligned}-\dot{\tilde{P}}_o(t) &= \tilde{P}_o(t)\bar{A}_o(t) + A_o(t)'\tilde{P}_o(t) + \tilde{A}_o(t)'\bar{P}_o(t)+\\ &- P_o(t)B_o(t)B_o(t)'\tilde{P}_o(t) - P_o(t)B_o(t)\bar{B}_o(t)'\bar{P}_o(t)+\\ &- \tilde{P}_o(t)\bar{B}_o(t)B_o(t)'\tilde{P}_o(t) - \tilde{P}_o(t)\bar{B}_o(t)\bar{B}_o(t)'\bar{P}_o(t)\end{aligned} \tag{6.10}$$

$$\begin{aligned}-\dot{\bar{P}}_o(t) &= \bar{P}_o(t)\bar{A}_o(t) + \bar{A}_o(t)'\bar{P}_o(t)+\\ &- \tilde{P}_o(t)'B_o(t)B_o(t)'\tilde{P}_o(t) - \tilde{P}_o(t)'B_o(t)\bar{B}_o(t)'\bar{P}_o(t)\\ &- \bar{P}_o(t)\bar{B}_o(t)B_o(t)'\tilde{P}_o(t) - \bar{P}_o(t)\bar{B}_o(t)\bar{B}_o(t)'\bar{P}_o(t).\end{aligned} \tag{6.11}$$

Notice that, if $\tilde{P}_o(t) = 0$ and $\bar{P}_o(t) = 0, \forall t$, then the first equation reduces to

$$\begin{aligned}-\dot{P}_o(t) &= P_o(t)A_o(t) + A_o(t)'P_o(t) + C_o(t)'C_o(t)+\\ &- P_o(t)B_o(t)B_o(t)'P_o(t),\end{aligned} \tag{6.12}$$

which is a Riccati equation for the observable part. Thus, if $P_o(\cdot)$ is a periodic solution of eq. (6.12), then

$$\begin{bmatrix} P_o(\cdot) & 0 \\ 0 & 0 \end{bmatrix}, \tag{6.13}$$

is a periodic solution of the canonically-decomposed PRE.

6.3.1.5 Quasi-Linearization of the PRE In order to find the solution of the PRE, one can resort to a Newton-type algorithm. Consider the operator

$$Ric : P(\cdot) \longmapsto \dot{P}(\cdot) + A(\cdot)'P(\cdot) + P(\cdot)A(\cdot) + C(\cdot)'C(\cdot) - P(\cdot)B(\cdot)B(\cdot)'P(\cdot).$$

A symmetric periodic solution of the PRE satisfies the operator equation

$$Ric(P(\cdot)) = 0. \tag{6.14}$$

Suppose that $P_i(\cdot)$ is a symmetric periodic function matrix approximating a solution of eq. (6.14). The Newton method consists in iteratively solving the first-order approximation of eq. (6.14) evaluated in $P_i(\cdot)$:

$$Ric(P_i(\cdot)) + \mathrm{d}Ric(P_i(\cdot), \Delta P_i(\cdot)) = 0,$$

where $\mathrm{d}Ric(.,.)$ is the differential of Ric evaluated in $P_i(\cdot)$ along $\Delta P_i(\cdot)$. From $\Delta P_i(\cdot)$, the new approximation is obtained as

$$P_{i+1}(\cdot) = P_i(\cdot) + \Delta P_i(\cdot).$$

After some manipulations, it turns out that $P_{i+1}(\cdot)$ can be computed from $P_i(\cdot)$ by solving the following PLE:

$$-\dot{P}_{i+1}(t) = A_i(t)'P_{i+1}(t) + P_{i+1}(t)A_i(t) + C(t)'C(t) + K_i(t)'K_i(t), \tag{6.15.a}$$

where

$$A_i(\cdot) = A(\cdot) - B(\cdot)K_i(\cdot) \tag{6.15.b}$$

$$K_i(\cdot) = B(\cdot)'P_i(\cdot). \tag{6.15.c}$$

Suppose that, $\forall i$, eq.(6.15.a) admits a symmetric periodic solution. If the sequence of these solutions $\{P_i(\cdot) | \, i = 1, 2, ...\}$ converges,

$$\lim_{i \to \infty} P_i(t) = P_\infty(t),$$

then it can be easily verified that $P_\infty(\cdot)$ is a symmetric periodic solution of the PRE. As discussed in the subsequent section, by introducing suitable assumptions, one can ensure the global convergence of this method with respect to any feasible initialization. As for any Newton method, the convergence rate is quadratic in a neighbourhood of $P_\infty(\cdot)$.

6.3.2 Positive Semidefinite Solutions

This Section is devoted to the analysis of the Symmetric Periodic Positive Semidefinite (SPPS) solutions of the PRE. We will develop a consistent picture of results by a chain of logical steps, the most important of which will be given a formal proof. The results are gleaned out from various papers, such as [9–15] and [18], [20].

As a first step, we prove that the stabilizability of $(A(\cdot), B(\cdot))$ is sufficient for the existence of an SPPS solution. There are various ways to attain this result: the Hamiltonian matrix analysis [10], the Optimal Control Theory [20], or the Quasi-Linearization technique [9], [13]. The proof given herein relies on the last approach, which, as already seen in Section 6.3.1.5., is an iterative linearization scheme.

Theorem 6.2. *Suppose that $(A(\cdot), B(\cdot))$ is stabilizable and consider the sequence of periodic Lyapunov equations defined in eq. (6.15). Let $K_0(\cdot)$ be a T-periodic matrix such that $A_0(\cdot)$ is stable. Then:*

(i) *For each $i \geq 0$, there exists a unique SPPS solution $P_{i+1}(\cdot)$ of eq. (6.15.a) and $A_{i+1}(\cdot)$ is stable.*
(ii) *The sequence $\{P_i(\cdot)\}$ is a monotonically nonincreasing sequence of SPPS matrices, i.e., $0 \leq P_{i+1}(t) \leq P_i(t)$.*
(iii) *The sequence $\{P_i(\cdot)\}$ is such that $\lim_{i \to \infty} P_i(t) = P_M(t)$, where $P_M(t)$ is a maximal and strong solution of the PRE.*

Proof. The proof is by induction. First of all, note that since $(A(\cdot), B(\cdot))$ is stabilizable, there exists a T-periodic matrix $K_0(\cdot)$ such that the feedback matrix $A_0(\cdot)$ is stable. Suppose now that $A_i(\cdot)$ is stable. Then, in view of the extended Lyapunov lemma (Section 6.2.3.1), eq. (6.15.a) has a unique SPPS solution $P_{i+1}(\cdot)$ After some computations, eq. (6.15.a) can be rewritten as follows:

$$-\dot{P}_{i+1}(t) = A'_{i+1}P_{i+1} + P_{i+1}A_{i+1} + C'_{i+1}C_{i+1} \tag{6.16}$$

where $C_{i+1}(\cdot)$ is any matrix such that

$$\begin{aligned} C'_{i+1}C_{i+1} = {} & C(t)'C(t) + K_{i+1}(t)'K_{i+1}(t) + \\ & + [K_i(t) - K_{i+1}(t)]'[K_i(t) - K_{i+1}(t)]\,. \end{aligned}$$

Eq. (6.16) is a Lyapunov equation, the dynamics of which is determined by the matrix $A_{i+1}(\cdot)$ in place of $A_i(\cdot)$. $P_{i+1}(\cdot)$ is an SPPS solution of this equation. Therefore, by showing that $(A_{i+1}(\cdot), C_{i+1}(\cdot))$ is detectable, from the extended Lyapunov Lemma, it follows that $A_{i+1}(\cdot)$ is stable and the induction is complete. To this purpose, let $\bar{C}_{i+1}(\cdot)$ be any matrix such that

$$\bar{C}'_{i+1}\bar{C}_{i+1} = C(t)'C(t) + K_{i+1}(t)'K_{i+1}(t)$$

Since $A_i(\cdot)$ is stable, the pair $(A_i(\cdot), \bar{C}_{i+1}(\cdot))$ is detectable. Then resorting to Lemma 6.2 (replace there $A(t)$ with $A_i(t)$, $C(t)$ with $\bar{C}_{i+1}(t)$ and $B(t)'P(t)$ with $K_{i+1}(t) - K_i(t)$) it turns out that $((A_i(\cdot) - B(\cdot)[K_{i+1}(\cdot) - K_i(\cdot)], C_{i+1}(\cdot))$ is detectable too. On the other hand, in view of (6.15.b),

$$A_i(\cdot) - B(\cdot)[K_{i+1}(\cdot) - K_i(\cdot)] = A_{i+1}.$$

Therefore, $(A_{i+1}(\cdot), C_{i+1}(\cdot))$ is detectable, so that the stability of $A_{i+1}(\cdot)$ follows from the Lyapunov lemma applied to eq. (6.16). The induction step is now completed, thus leading to the conclusion of point (i).

So far, let $X_i(t) = P_i(t) - P_{i+1}(t)\,, \forall i \geq 1$. Then, by suitable manipulations of eq. (6.15.a) it follows that

$$-\dot{X}_i(t) = A_i(t)'X_i(t) + X_i(t)A_i(t) + [K_{i-1} - K_i(t)]'[K_{i-1}(t) - K_i(t)].$$

Viewing the last term as a known term, this is again a Lyapunov equation, so that the extended Lyapunov lemma implies that $X_i(\cdot)$ is positive semidefinite. Hence, the monotonic pattern of $P_i(\cdot)$ is proven (point (ii)). In turn, this leads to the conclusion that

$$\lim_{i\to\infty} P_i(t) = \bar{P}(t)$$

exists, and it is obviously SPPS. $\bar{P}(\cdot)$ is, in fact, an SPPS solution of the PRE. Indeed, compute the limit as $i \to \infty$ of the two sides of eq. (6.15.a). Then, the following equation is obtained

$$-\dot{\bar{P}}(t) = \bar{A}(t)'\bar{P}(t) + \bar{P}(t)\bar{A}(t) + \bar{C}(t)'\bar{C}(t) + \bar{K}(t)'\bar{K}(t) \tag{6.17}$$

where $\bar{A}(t)$ and $\bar{K}(t)$ are matrices (6.15.b) and (6.15.c) with $P_i(\cdot)$ replaced by $\bar{P}(\cdot)$. Now, simple computations show that eq. (6.17) coincides with the PRE.

We finally show that $\bar{P}(\cdot)$ is the maximal solution. For any symmetric and T-periodic solution $P(\cdot)$ of the PRE, let $Y_i(t) = P_{i+1}(t) - P(t)$. Based on the original formula of the PRE and eq. (6.15.a), it follows that $Y_i(\cdot)$ is a symmetric and T-periodic solution of

$$-\dot{Y}_i(t) = A_i(t)'Y_i(t) + Y_i(t)A_i(t) + Y_{i-1}(t)B(t)B(t)'Y_{i-1}(t)$$

In view of the extended Lyapunov lemma, the already proven stability of $A_i(\cdot)$ implies that $Y_i(t)$ is positive semidefinite $\forall t$. Consequently, $\bar{P}(t) - P(t) = \lim_{i\to\infty} Y_i(t)$ is also positive semidefinite $\forall t$. Finally, since $A_i(\cdot)$ is stable, $\bar{A}(\cdot) = \lim_{i\to\infty} A_i(\cdot)$ has characteristic multipliers belonging to the closed unit disk. Hence $P_M(\cdot) = \bar{P}(\cdot)$ is both maximal and strong (point (iii)). □

Notice that, under the sole assumption of stabilizability, the maximal solution $P_M(\cdot)$ need not be stabilizing. The least restrictive assumption to be added to stabilizability in order for $P_M(\cdot)$ to be stabilizing is provided in the following

Theorem 6.3. *There exists a stabilizing symmetric periodic solution $P_+(\cdot)$ of the PRE if and only if $(A(\cdot), B(\cdot))$ is stabilizable and no unit-modulus characteristic multiplier of $A(\cdot)$ is $(A(\cdot), C(\cdot))$-unobservable.*

Proof. Under the stabilizability assumption, Theorem 6.2 ensures the existence of the strong SPPS solution $P_S(\cdot)$. All the characteristic multipliers of $A(\cdot) - B(\cdot)B(\cdot)'P_S(\cdot)$ have modulus lower than or equal to one. On the other hand, the characteristic multipliers of $A(\cdot) - B(\cdot)B(\cdot)'P_S(\cdot)$ are characteristic multipliers of the associated Hamiltonian matrix as well (see Section 6.3.1.2). In particular the unit-modulus characteristic multipliers are $(A(\cdot), B(\cdot))$ unreachable or/and $(A(\cdot), C(\cdot))$ unobservable characteristic multipliers of $A(\cdot)$, see Lemma 6.1. This is inhibited by the assumptions of the Theorem, so that $P_S(\cdot) = P_+(\cdot)$.

Conversely, if $P_+(\cdot)$ exists, $(A(\cdot), B(\cdot))$ is obviously stabilizable. Moreover, since all the characteristic multipliers of $A(\cdot) - B(\cdot)B(\cdot)'P_+(\cdot)$ have modulus less than one, the Hamiltonian matrix cannot have unit-modulus characteristic multipliers. Therefore, in view of Lemma 6.1, $A(\cdot)$ cannot have unit-modulus characteristic multipliers which are $(A(\cdot), C(\cdot))$-unobservable. □

In Theorem 6.2 it has been shown that the stabilizability of $(A(\cdot), B(\cdot))$ is a sufficient condition for the existence of an SPPS solution. A natural question is to find the least restrictive condition ensuring the existence of an SPPS solution. The answer is provided in

Theorem 6.4. *There exists an SPPS solution of the PRE if and only if the observable and unreachable part of system $(A(\cdot), B(\cdot), C(\cdot))$ is stable.*

Remark. The previous statement can be given the equivalent form: There exists an SPPS solution of the PRE if and only if any $(A(\cdot), B(\cdot))$-unreachable characteristic multiplier of $A(\cdot)$ with modulus greater than or equal to one is also $(A(\cdot), C(\cdot))$-unobservable.

Proof. For the 'if part' suppose that the system is already given in the Kalman canonical decomposition. As discussed in Section 6.3.1.4, this does not imply any loss of generality. Then, the PRE can be partitioned into the three subequations (6.9)-(6.11). Since the observable part is stabilizable, the periodic Riccati equation relative to the observable part only (eq. (6.12)) admits an SPPS solution, say $P_o(\cdot)$, (Theorem 6.2). Therefore, as observed in Section 6.3.1.4, matrix (6.13) is an SPPS solution of the PRE. As for the 'only if' part, suppose that there exists an SPPS solution $\bar{P}(\cdot)$. Let μ, with $|\mu| \geq 1$, be an $(A(\cdot), B(\cdot))$-unreachable characteristic multiplier of $A(\cdot)$. The problem is to show that this multiplier is also unobservable. Since μ is unreachable, it is also a characteristic multiplier of $F(\cdot) = A(\cdot) - B(\cdot)B(\cdot)'\bar{P}(\cdot)$, see Section 6.2.2.1, i.e., for some $y \neq 0$, $\Psi_F(0)y = \mu y$. Moreover, since $\bar{P}(\cdot)$ is an SPPS solution of the periodic Lyapunov equation (6.7), $\bar{P}(0)$ solves the discrete-time algebraic Lyapunov equation

$$\Psi_F(0)'\bar{P}(0)\Psi_F(0) - \bar{P}(0) + \int_0^T \Phi_F(t,0)'H(t)'H(t)\Phi_F(t,0)\mathrm{d}t = 0 \qquad (6.18)$$

Pre and post-multiplying (6.18) by y^* and y, respectively, it is easy to see that

$$(|\mu|^2 - 1)y^*\bar{P}(0)y + \int_0^T \|H(t)\Phi_F(t,0)y\|^2\mathrm{d}t = 0$$

Since $|\mu| \geq 1$ and $\bar{P}(0) \geq 0$, it follows that $H(t)\Phi_F(t,0)y = 0$, a.e. $t \in [0,T]$. This means that μ, $|\mu| \geq 1$, is an $(F(\cdot), H(\cdot))$-unobservable characteristic multiplier of $F(\cdot)$. Then, Lemma 6.2 entails that μ is an $(A(\cdot), C(\cdot))$-unobservable characteristic multiplier of $A(\cdot)$. □

In the 'if part' of the theorem, it has been proven the existence of a particular SPPS solution $P_0(\cdot)$, starting from the Kalman canonical decomposition of the system into the observable/unobservable parts. It can be easily shown that such a solution is in fact the minimal SPPS solution of the PRE.

When the uniqueness property of the SPPS solution and the stability of the corresponding closed-loop system are jointly imposed, one arrives at the extension of the well known Wonham-Kucera theorem.

Theorem 6.5. *There exists a unique SPPS solution of the PRE and the corresponding closed-loop system is stable if and only if $(A(\cdot), B(\cdot))$ is stabilizable and $(A(\cdot), C(\cdot))$ is detectable.*

Proof. Suppose that $(A(\cdot), B(\cdot))$ is stabilizable and $(A(\cdot), C(\cdot))$ is detectable. Then, Theorem 6.2 implies that there exists an SPPS solution $\bar{P}(\cdot)$. In view of Theorem 6.1, detectability implies that $\bar{P}(\cdot)$ is stabilizing, i.e., $F(\cdot) = A(\cdot) - B(\cdot)B(\cdot)'\bar{P}(\cdot)$ is stable. Let's now prove the uniqueness of the solution. Suppose by contradiction that $P_1(\cdot)$ is a second stabilizing solution and denote by $F_1(\cdot)$ the associated closed-loop matrix. It is a simple fact of matrix manipulation to show that $X(\cdot) = \bar{P}(\cdot) -$

$P_1(\cdot)$ satisfies the following linear periodic equation

$$-\dot{X}(t) = F_1(t)'X(t) + X(t)F(t).$$

Since $F(\cdot)$ and $F_1(\cdot)$ are both stable, the unique T-periodic solution of such an equation is $X(\cdot) = 0$.

Conversely, suppose that there exists a unique SPPS solution $\bar{P}(\cdot)$ and that $F(\cdot) = A(\cdot) - B(\cdot)B(\cdot)'\bar{P}(\cdot)$ is stable. Obviously this implies that $(A(\cdot), B(\cdot))$ is stabilizable. To prove detectability, assume, without any loss of generality, that the system is canonically decomposed into the observable/unobservable parts. Hence the unique solution of the PRE must take the form given by (6.13). Correspondingly,

$$F(\cdot) = \begin{bmatrix} ? & 0 \\ ? & \bar{A}_0(\cdot) \end{bmatrix}$$

The stability of $F(\cdot)$ entails the stability of $\bar{A}_0(\cdot)$, i.e., the detectability of $(A(\cdot), C(\cdot))$. □

So far, only positive semidefinite solutions of the PRE have been considered. Turning now to the Symmetric Periodic *Positive Definite* (SPPD) solutions, the analogous of Theorems 6.3 and 6.5 can be established. Interestingly enough, these results can be jointly formulated as follows.

Theorem 6.6. *The following statements are equivalent:*

(i) *There exists a stabilizing SPPD solution*
(ii) *There exists a unique SPPD solution $\bar{P}(\cdot)$ and $F(\cdot) = A(\cdot) - B(\cdot)B(\cdot)'\bar{P}(\cdot)$ is asymptotically stable*
(iii) *$(A(\cdot), B(\cdot))$ is stabilizable and $(A(\cdot), C(\cdot))$ is antidetectable.*

Proof. (iii) $\Longleftrightarrow$ (i). Suppose that $(A(\cdot), B(\cdot))$ is stabilizable and $(A(\cdot), C(\cdot))$ is antidetectable. Notice that the latter hypothesis entails that $A(\cdot)$ has no $(A(\cdot), C(\cdot))$-unobservable characteristic multipliers with modulus equal to one. In view of Theorem 6.3 there exists the SPPS stabilizing solution $P_+(\cdot)$. In order to prove the positive definiteness of $P_+(\cdot)$, it is preliminarily shown that the pair $(F_+(\cdot), H_+(\cdot))$ is observable, where

$$F_+(\cdot) = A(\cdot) - B(\cdot)B(\cdot)'P_+(\cdot),$$
$$H_+(\cdot)'H_+(\cdot) = C(\cdot)'C(\cdot) + P_+(\cdot)B(\cdot)B(\cdot)'P_+(\cdot)$$

For, assume by contradiction that λ is an $(F_+(\cdot), H_+(\cdot))$-unobservable characteristic multiplier of $F_+(\cdot)$. Since $F_+(\cdot)$ is stable, $|\lambda| < 1$. Now, in view of Lemma 6.2, λ is also an $(A(\cdot), C(\cdot))$-unobservable characteristic multiplier of $A(\cdot)$. This contradicts antidetectability. Finally, the observability of $(F_+(\cdot), H_+(\cdot))$ implies that the solution $P_+(\cdot)$ of the Lyapunov equation

$$-\dot{P}_+(t) = F_+(t)'P_+(t) + P_+(t)F_+(t) + H_+(t)'H_+(t)$$

is positive definite, see the extended Lyapunov Lemma of Section 6.2.3.1.

Conversely, suppose that there exists an SPPD solution $P_+(\cdot)$ such that $F_+(\cdot) = A(\cdot) - B(\cdot)B(\cdot)'P_+(\cdot)$ is stable. This obviously entails that $(A(\cdot), B(\cdot))$ is stabilizable. The antidetectability of the pair $(A(\cdot), C(\cdot))$ is proven by showing that the dual pair $(\tilde{A}(\cdot), \tilde{B}(\cdot))$ is antistabilizable, where $\tilde{A}(t) = A(-t)'$ and $\tilde{B}(t) = C(-t)'$. Actually let $\tilde{P}(t) = -P_+(-t)^{-1}$ and $\tilde{C}(t) = B(-t)'$. It is easy to show that $\tilde{P}(\cdot)$ is an SP negative definite solution of

$$-\dot{\tilde{P}}(t) = \tilde{P}(t)\tilde{A}(t) + \tilde{A}(t)'\tilde{P}(t) - \tilde{P}(t)\tilde{B}(t)\tilde{B}(t)'\tilde{P}(t) + \tilde{C}(t)'\tilde{C}(t).$$

Moreover, the pair $(A(\cdot), B(\cdot))$ being stabilizable, the pair $(\tilde{A}(\cdot), \tilde{C}(\cdot))$ is detectable. The inertia Theorem 6.1 for this equation entails that $\tilde{A}(\cdot) - \tilde{B}(\cdot)\tilde{B}(\cdot)'\tilde{P}(\cdot)$ is antistable, i.e., $(\tilde{A}(\cdot), \tilde{B}(\cdot))$ is antistabilizable.
$(iii) \Longleftrightarrow (ii)$. Let assume that $(A(\cdot), B(\cdot))$ is stabilizable and $(A(\cdot), C(\cdot))$ is antidetectable. Then, we already proved that there exists an SPPD solution $P(\cdot)$ such that $F_+(\cdot) = A(\cdot) - B(\cdot)B(\cdot)'P_+(\cdot)$ is stable. We have now to prove that such a solution is the *unique* SPPD solution of the PRE. This is done by showing that, under the antidetectability assumption, any SPPD solution is stabilizing, a fact which implies uniqueness (see the proof of Theorem 6.5). Actually, let $P(\cdot)$ be any SPPD solution. With the symbols used in the previous proof, it is easy to show that $\hat{P}(t) = -P(-t)$ is a periodic negative definite solution of the periodic Riccati equation

$$-\dot{\hat{P}}(t) = -\hat{P}(t)\tilde{A}(t) - \tilde{A}(t)'\hat{P}(t) - \hat{P}(t)\tilde{C}(t)'\tilde{C}(t)\hat{P}(t) + \tilde{B}(t)\tilde{B}(t)'$$

Moreover, the antidetectability of $(A(\cdot), C(\cdot))$ is equivalent to the detectability of $(-\tilde{A}(\cdot)', \tilde{B}(\cdot)')$. The inertia Theorem 6.1 applied to this last equation entails that $\tilde{F}(\cdot) = -\tilde{A}(\cdot)' - \tilde{C}(\cdot)'\tilde{C}(\cdot)\hat{P}(\cdot)$ is antistable. Since $F(t) = -\tilde{F}(-t)$, it follows that $F(\cdot) = A(\cdot) - B(\cdot)B(\cdot)'P(\cdot)$ is stable.

Conversely, suppose that there exists a unique SPPD solution $P_+(\cdot)$ and that $F_+(\cdot) = A(\cdot) - B(\cdot)B(\cdot)'P_+(\cdot)$ is stable. Then, $(A(\cdot), B(\cdot))$ is obviously stabilizable. The antidetectability of $(A(\cdot), C(\cdot))$ can be proven with the same line of reasoning adopted to derive implication $(i) \Rightarrow (iii)$. □

Many other properties of the periodic Riccati equation have been pointed out. Among them, the following ones, proven in [11], are here reported.

Theorem 6.7.

(i) *The strong solution exists and is unique if and only if $(A(\cdot), B(\cdot))$ is stabilizable*
(ii) *The strong solution coincides with the stabilizing solution if and only if $(A(\cdot), B(\cdot))$ is stabilizable and the unit-modulus characteristic multipliers of $A(\cdot)$ are $(A(\cdot), C(\cdot))$-observable*
(iii) *The strong solution is the the only SPPS solution if and only if $(A(\cdot), B(\cdot))$ is stabilizable and no $(A(\cdot), C(\cdot))$-unobservable characteristic multiplier of $A(\cdot)$ lies outside the unit circle.*

6.3.3 Negative Semidefinite Solutions and Gap

The properties of the Symmetric Periodic Negative Semidefinite (SPNS) solutions of the PRE are parallel to those of the SPPS solutions. This is a consequence of the following observation. If $P(\cdot)$ is a solution of the PRE, then $\hat{P}(\cdot)$, with $\hat{P}(t) = -P(-t)$, is a solution of the following periodic Riccati equation:

$$-\dot{\hat{P}}(t) = \hat{A}(t)'\hat{P}(t) + \hat{P}(t)\hat{A}(t) + \hat{C}(t)'\hat{C}(t) - \hat{P}(t)\hat{B}(t)\hat{B}(t)'\hat{P}(t),$$

where $\hat{A}(t) = -A(-t)$, $\hat{B}(t) = B(-t)$, $\hat{C}(t) = C(-t)$. Moreover, it is easily seen that the controllability and observability properties are preserved when passing from $(A(\cdot), B(\cdot), C(\cdot))$ to $(\hat{A}(\cdot), \hat{B}(\cdot), \hat{C}(\cdot))$, whereas stability, stabilizability and detectability of $(A(\cdot), B(\cdot), C(\cdot))$ translate into the corresponding "anti" properties of $(\hat{A}(\cdot), \hat{B}(\cdot), \hat{C}(\cdot))$. In this way, all theorems of the previous section give rise to analogous results for the SPNS solutions. In particular, Theorem 6.2 becomes

Theorem 6.8. *Suppose that $(A(\cdot), B(\cdot))$ is antistabilizable and consider the sequence of periodic Lyapunov equations defined in (6.15). Let $K_0(\cdot)$ be a T-periodic matrix such that $A_0(\cdot)$ is antistable. Then:*

(i) *For each $i \geq 0$, there exists a unique SPNS solution $P_{i+1}(\cdot)$ of (6.15.a) and $A_{i+1}(\cdot)$ is antistable.*

(ii) *The sequence $\{P_i(\cdot)\}$ is a monotonically nondecreasing sequence of SPNS matrices, i.e., $0 \geq P_{i+1}(t) \geq P_i(t)$.*

(iii) *The sequence $\{P_i(\cdot)\}$ is such that $\lim_{i\to\infty} P_i(t) = P_m(t)$, where $P_m(t)$ is a minimal and antistrong solution of the PRE.* □

From Theorems 6.2 and 6.8, it follows that, if $(A(\cdot), B(\cdot))$ is controllable, then the PRE admits both the maximal and minimal solutions. The difference $P_M(t) - P_m(t)$ is known as *gap*. Obviously, the gap is positive semidefinite for each time point. As a matter of fact, it turns out that, under the additional hypothesis that no unit-modulus characteristic multiplier of $A(\cdot)$ is $(A(\cdot), C(\cdot))$-unobservable, the gap is positive definite:

Theorem 6.9. *Assume that $(A(\cdot), B(\cdot))$ is controllable, and no unit-modulus characteristic multiplier of $A(\cdot)$ is $(A(\cdot), C(\cdot))$-unobservable. Then the gap $\Delta(t) = P_M(t) - P_m(t)$ is positive definite for each t.*

Proof. Consider the minimal solution $P_m(z)$ of the PRE, and let $Z(t) = P(t) - P_m(t)$, where $P(\cdot)$ is any SP solution of the PRE. It is easy to verify that

$$-\dot{Z}(t) = F_m(t)'Z(t) + Z(t)F_m(t) - Z(t)B(t)B(t)'Z(t) \tag{6.19}$$

with $F_m(t) = A(t) - B(t)B(t)'P_m(t)$.

Thanks to the controllability assumption, the minimal solution $P_m(\cdot)$ exists and is antistrong. Moreover, since no unit-modulus characteristic multiplier is $(A(\cdot), C(\cdot))$-unobservable, the minimal solution is also antistabilizing. Therefore,

$F_m(\cdot)$ is antistable, so that $(F_m(\cdot), C(\cdot))$ is antidetectable. Furthermore, $(F_m(\cdot), B(\cdot))$ is controllable.

From Theorem 6.6 (implication $(iii) \Rightarrow (i)$), it follows that the stabilizing solution of eq. (6.19) is positive definite. Due to the controllability assumption, this is the maximal solution $Z_M(\cdot)$ of eq. (6.19). Since $Z_M(\cdot) = \Delta(\cdot)$, the statement is proven. □

As a matter of fact, under the assumption of the last theorem, it has been shown [15] that the set of SP solutions of the PRE constitutes a complete lattice with the respect to the usual partial ordering of symmetric matrices, with $P_M(\cdot)$ and $P_m(\cdot)$ playing the role of maximum and minimum elements of the lattice.

6.3.4 Convergence to Periodic Equilibria

As is well known, the backward solution of the PRE with a given final condition plays a key role in the design of optimal controllers for periodic systems. However such a solution is not necessarily periodic. Should the periodic solution be attractive, then one could employ it to implement a sub-optimal periodic controller. The attractiveness of the periodic solution of the PRE has been analyzed in a number of papers, such as [9], [14] and [15]. Under controllability and observability assumptions, the original results of Kalman [2], relative to the time-varying Riccati equation, can be applied to periodic systems to conclude that all solutions of the PRE with positive semidefinite final condition converge to the unique SPPS solution. When the observability assumption is dropped, this global convergence property does not hold any more. Precisely, depending on the final condition, a variety of situations may occur, including convergence to the strong solution, convergence to a generic SPPS solution, or even convergence to an almost periodic solution, see [15]. The convergence analysis with stabilizability only has been recently treated in [11]. However, the derivation of the main theorem [11, Theorem 4.2] relies on a result of [9] the proof of which should be revisited under the light of the comments provided in [34].

6.3.5 Monotonicity Property and the Nonstandard Periodic Riccati Equation

A classical issue in the analysis of the Riccati equation is to study the behaviour of the solution when the 'true' parameters are different from the assumed ones. In this connection, a result of interest is the so-called monotonicity property, which will be now briefly recalled, see [35] for more details. Letting

$$M(t) = \begin{bmatrix} Q(t) & A(t)' \\ A(t) & -B(t)B(t)' \end{bmatrix},$$

the PRE can be written as

$$-\dot{P}(t) = [I_n \, P(t)]M(t) \begin{bmatrix} I_n \\ P(t) \end{bmatrix}. \tag{6.20}$$

The SP solutions of the PRE will be compared with the solutions of another PRE defined as follows

$$-\dot{\tilde{P}}(t) = [I_n \, \tilde{P}(t)]\tilde{M}(t) \begin{bmatrix} I_n \\ \tilde{P}(t) \end{bmatrix} \tag{6.21}$$

where

$$\tilde{M}(t) = \begin{bmatrix} \tilde{Q}(t) & \tilde{A}(t)' \\ \tilde{A}(t) & -\tilde{B}(t)\tilde{B}(t)' \end{bmatrix}.$$

Here, $\tilde{A}(\cdot)$, $\tilde{B}(\cdot)$, $\tilde{Q}(\cdot)$ are T-periodic matrices of the same dimensions of $A(\cdot)$,$B(\cdot)$ and $Q(\cdot)$, respectively. The comparison will be made under the basic monotonicity assumption that is:

$$M(t) \geq \tilde{M}(t), \; \forall t \in [0, T].$$

This assumption reflects into a monotonicity property of the solutions of the two periodic Riccati equations. Precisely, the following result holds true.

Theorem 6.10. *Suppose that* $(A(\cdot), B(\cdot))$ *is stabilizable and* $M(t) \geq \tilde{M}(t)$, $\forall t \in [0, T]$. *Then* $P_S(t) \geq \tilde{P}(t)$, $\forall t$, *where* $P_S(\cdot)$ *is the strong solution of eq. (6.20) and* $\tilde{P}(\cdot)$ *is any SP solution of eq. (6.21).* □

As a matter of fact, the monotonicity property is valid even in the *nonstandard* case, i.e., the known term $Q(t)$ is not necessarily positive semidefinite. In the sign-indefinite case, however, the existence of the strong solution is not guaranteed by the stabilizability assumption. For such an existence it is sufficient to complement stabilizability with the assumption that the nonstandard PRE admits at least an SP solution. Indeed, as discussed in [36], denoting by $\bar{P}(\cdot)$ the particular periodic solution and by $P(\cdot)$ a generic periodic solution of (6.20), it is easy to see that $Z(\cdot) = P(\cdot) - \bar{P}(\cdot)$ satisfies the standard periodic Riccati equation (with zero known term)

$$-\dot{Z}(t) = \bar{F}(t)'Z(t) + Z(t)\bar{F}(t) - Z(t)B(t)B(t)'Z(t) \tag{6.22}$$

with $\bar{F}(t) = A(t) - B(t)B(t)'\bar{P}(t)$.

Since the stabilizability of $(A(\cdot), B(\cdot))$ is equivalent to the stabilizability of $(F(\cdot), B(\cdot))$, in view of Theorem 6.2, eq. (6.22) admits the strong solution $Z_S(\cdot)$. A simple computation shows that $P_S(\cdot) = \bar{P}(\cdot) + Z_S(\cdot)$ is the strong solution of the nonstandard PRE.

6.4 The Discrete-Time Periodic Riccati Equation

The interest for the theory of discrete-time periodic systems has at least two major motivations. The first one regards the possibility of implementing controllers and filters for periodic systems by means of digital devices. The second one has to do with multirate control and filtering of time-invariant systems, see Section 6.5. In both cases, the design of optimal controllers and filters calls for the solution of the Difference Periodic Riccati Equation (DPRE).

Aside from some difficulties deriving from the possible nonreversibility of discrete-time systems, the analysis of the DPRE may be developed by following the tracks of the analysis of the continuous-time PRE, see [37–40]. Herein, however, we will follow a different approach, which goes under the name of Time-Invariant Reformulation (TIR) [41]. Precisely, a one-to-one correspondence is established between the solutions of the DPRE and the solutions of a suitable time-invariant difference Riccati equation. In this way, it is possible to "translate" several results concerning the time-invariant DRE into corresponding results for the DPRE. Moreover, the parallelism between the properties of the time-invariant and periodic Riccati equation is completely clarified.

The present section will be organized as follows. First, some elements of the theory of discrete-time linear periodic systems are concisely recalled. The theory of continuous-time periodic systems having been already treated in Section 6.2, we will focus only on the main differences between continuous and discrete-time periodic systems. Then, the time-invariant reformulation for discrete-time periodic systems is introduced. Finally, the correspondence between the periodic solutions of the DPRE and the solutions of a suitable ARE is established and used to derive the main results of the section.

6.4.1 Discrete-Time Linear Periodic Systems

If it were not for the possible system nonreversibility, the theory of discrete-time linear systems would not consistently differ from its continuous-time counterpart. In general, the class of reversible discrete-time linear systems shares the same properties as continuous time linear systems and the proofs can be carried out in a very similar fashion. On the contrary, when passing to nonreversible systems, new, more intriguing, properties arise, while even the properties that remain unchanged may require more complex proofs. Herein, we will not dwell upon the features that are shared by continuous and discrete-time systems but only on the differences that stem from nonreversibility. As is shown below, such differences concern the relationships between controllability and reachability (and, by duality, between reconstructibility and observability), the dimensions of the corresponding subspaces and the canonical decomposition.

Consider the discrete-time periodic system

$$x(t+1) = A(t)x(t) + B(t)u(t) \tag{6.23.a}$$

$$y(t) = C(t)x(t) + w(t), \tag{6.23.b}$$

where $A(t) = A(t+T)$, $B(t) = B(t+T)$, and $C(t) = C(t+T)$ are periodic matrices of suitable dimensions, with integer period T.

As in continuous-time, $\Phi_A(t,\tau)$ denotes the transition matrix, and $\Psi_A(\tau)$ the monodromy matrix at τ, i.e., $\Psi_A(\tau) = \Phi_A(\tau+T,\tau)$. It can be shown that the spectrum of $\Psi_A(\tau)$ is independent of τ, see e.g. [29]. Contrary to the continuous-time case, in discrete-time the characteristic multipliers, defined as the eigenvalues of $\Psi_A(\tau)$, may be equal to zero. The symbols $X_r(\tau,t)$, $X_c(\tau,t)$, $X^\omega(\tau,t)$ and $X^\rho(\tau,t)$ are used to denote the reachability, controllability, observability and reconstructibility subspaces over the interval (τ,t). $X_r(t)$, $X_c(t)$, $X^\omega(t)$ and $X^\rho(t)$ are the reachability, controllability, observability and reconstructibility subspaces at time t. Differently from continuous periodic systems, the reachability and controllability subspaces need not coincide. Indeed, for a nonreversible periodic system, in general, $X_r(t) \neq X_c(t)$ and $X^\omega(t) \neq X^\rho(t)$. Moreover, as discussed in [29], the dimensions of the controllability and reconstructibility subspaces are constant, whereas, in general, the dimensions of the reachability and observability subspaces are periodically time-varying. Therefore, when dealing with a nonreversible system, the only canonical decomposition that can always be performed is the one in terms of controllability and reconstructibility, [42–44]. This allows to define the notion of stabilizability [detectability] as the stability of the uncontrollable [unreconstructible] part [45]. A modal notion of detectability can also be introduced by making reference to the stability of the so-called unobservable characteristic multipliers. Precisely, a characteristic multiplier λ is said to be $(A(\cdot),C(\cdot))$-unobservable at time τ, if there exists $\eta \neq 0$, such that

$$\Psi_A(\tau)\eta = \lambda\eta$$

$$C(j)\Phi_A(j,\tau)\eta = 0, \qquad \forall j \in [\tau, \tau+T-1].$$

It can be shown that, if $\lambda \neq 0$ is $(A(\cdot),C(\cdot))$-unobservable at τ, it is $(A(\cdot), C(\cdot))$-unobservable at any time point, i.e., the unobservability notion is time independent for the nonzero characteristic multipliers [41]. As for the null multipliers, they may well be unobservable at certain time points, and observable at others.

6.4.2 Time-Invariant Reformulation

The attempt of analyzing discrete-time periodic systems by means of a suitable time-invariant representation goes back to the pioneer paper [46] and was subsequently developed in the seventies and eighties [47–49]. Below, the time-invariant representation is introduced and the relationship between the structural properties of the original system and the time-invariant reformulation is discussed.

Once a time point τ has been selected, a time-invariant system can be associated with system (6.23). Precisely, define

$$\bar{A}_\tau = \Phi_A(\tau), \qquad \bar{A}_\tau \in R^{n\times n}, \tag{6.24.a}$$

$$\bar{B}_\tau = [\Psi_A(\tau+T,\tau+1)B(\tau) \quad \Psi_A(\tau+T,\tau+2)B(\tau+1) \ \ldots \ B(\tau+T-1)], \qquad \bar{B}_\tau \in R^{n\times mT}, \tag{6.24.b}$$

$$\bar{C}_\tau = [C(\tau)' \quad \Psi_A(\tau+1,\tau)'C(\tau+1)' \ \ldots \ \Psi_A(\tau+T-1,\tau)'C(\tau+T-1)']', \qquad \bar{C}_\tau \in R^{pT\times n}, \tag{6.24.c}$$

$$\bar{D}_\tau = [(\bar{D}_\tau)_{ij}], i,j = 1,2,\ldots,T, \qquad \bar{D}_\tau \in R^{pT\times mT}, \tag{6.24.d.1}$$

$$(\bar{D}_\tau)_{ij} = \begin{cases} 0, & \text{if } i \le j, \\ C(\tau+i-1)\Psi_A(\tau+i-1,\tau+j)B(\tau+j-1), & \text{if } i > j, \end{cases} \tag{6.24.d.2}$$

and consider the time-invariant system

$$\bar{x}_\tau(k+1) = \bar{A}_\tau\bar{x}_\tau(k) + \bar{B}_\tau\bar{u}_\tau(k), \tag{6.25.a}$$

$$\bar{y}_\tau(k) = \bar{C}_\tau\bar{x}_\tau(k) + \bar{D}_\tau\bar{u}_\tau(k), \tag{6.25.b}$$

where

$$\bar{u}_\tau(k) = [u(\tau+kT)' \ u(\tau+kT+1)' \ \ldots \ u(\tau+(k+1)T-1)']', \tag{6.26}$$

Note that (6.25) defines a stationary system that is no more proper because the output $\bar{y}_\tau(\cdot)$ depends directly on the input variable $\bar{u}_\tau(\cdot)$. Moreover, if the input $\bar{u}_\tau(\cdot)$ of system (6.25) is constructed from the input $u(\cdot)$ of system (6.23) according to (6.26), and if the initial condition is taken as $\bar{x}_\tau(0) = x(\tau)$, then

$$\bar{x}_\tau(k) = x(\tau+kT), \tag{6.27.a}$$

$$\bar{y}_\tau(k) = [y(\tau+kT)' \ y(\tau+kT+1)' \ \ldots \ y(\tau+(k+1)T-1)']', \qquad \forall k \ge 0. \tag{6.27.b}$$

Therefore, the time-invariant system (6.25) can be seen as a state-sampled representation of system (6.23), fed by an augmented input vector and producing an augmented output vector. From the very definition of $\bar{A}_\tau$, it is obvious that system (6.25) is stable if and only if system (6.23) is asymptotically stable. As for the structural properties, the results stated in the theorem below at points (i) and (ii) can be found in [48] and [50]. Statement (iii) is straightforward.

Lemma 6.3

(i) *Let $\bar{X}_{r\tau}(0,k)$, $\bar{X}_{c\tau}(0,k)$, $\bar{X}^{\omega}_{\tau}(0,k)$ and $\bar{X}^{\varrho}_{\tau}(0,k)$ be the reachability, controllability, observability and reconstructibility subspaces over the interval $(0,k)$ of system (6.25). Then*

$$X_a(\tau,\tau+kT) = \bar{X}_{a\tau}(0,k), \ a = r, \ c,$$

$$X^\alpha(\tau, \tau + kT) = \bar{X}_\tau^\alpha(0, k), \; \alpha = \omega, \; \rho.$$

(ii) *The periodic pair* $(A(\cdot), B(\cdot))$ $[A(\cdot), C(\cdot)]$ *is stabilizable [detectable] if and only if the constant pair* $(\bar{A}_\tau, \bar{B}_\tau)$ $[(\bar{A}_\tau, \bar{C}_\tau)]$ *is stabilizable [detectable].*

(iii) *A characteristic multiplier* λ *of system (6.23) is* $(A(\cdot), C(\cdot))$*-unobservable at* τ *if and only if it is* $(\bar{A}_\tau, \bar{C}_\tau)$*-unobservable.*

6.4.3 DPRE Analysis via Time-Invariant Reformulation

In this section a one-to-one correspondence between the solutions of the DPRE relative to the periodic system (6.23) and the solutions of the difference Riccati equation relative to the time-invariant system (6.25) will be established by means of optimal control theory. Consider the problem of minimizing the performance index

$$J = x(\tau + NT)' S x(\tau + NT) + \sum_{t=\tau}^{\tau+NT-1} \{y(t)'y(t) + u(t)'u(t)\}. \tag{6.28}$$

It is well known that the optimal control that attains the minimum is

$$u(t) = -(I + B(t)'P(t+1)B(t))^{-1} B(t)'P(t+1)A(t)x(t),$$

where $P(\cdot)$ is the backward solution of the following DPRE

$$\begin{aligned} P(t) &= A(t)'P(t+1)A(t) + C(t)'C(t) \\ &\quad - A(t)'P(t+1)B(t)(I + B(t)'P(t+1)B(t))^{-1}B(t)'P(t+1)A(t), \end{aligned} \tag{6.29}$$

with final condition $P(\tau + NT) = S$. Moreover, the optimal value of the performance index is equal to $x(\tau)'P(\tau)x(\tau)$. It is worth noticing that, in general, the solution $P(\cdot)$ of eq. (6.29) need not be periodic, unless a suitable final condition is chosen. Such a final condition will be referred to as a periodic generator.

Consider now the following performance index associated with system (6.25) with initial condition $\bar{x}_\tau(0) = x(\tau)$:

$$\bar{J}_\tau = \bar{x}_\tau(N)' S \bar{x}_\tau(N) + \sum_{k=0}^{N-1} \{\bar{y}_\tau(k)'\bar{y}_\tau(k) + \bar{u}_\tau(k)'\bar{u}_\tau(k)\}. \tag{6.30}$$

Letting

$$\tilde{A}_\tau = \bar{A}_\tau - \bar{B}_\tau(I + \bar{D}_\tau'\bar{D}_\tau)^{-1}\bar{D}_\tau'\bar{C}_\tau, \tag{6.31.a}$$

$$\tilde{C}_\tau'\tilde{C}_\tau = \bar{C}_\tau'(I + \bar{D}_\tau\bar{D}_\tau')^{-1}\bar{C}_\tau, \tag{6.31.b}$$

the control minimizing such a performance index turns out to be [4]

$$\bar{u}_\tau(k) = -(I + \bar{D}_\tau'\bar{D}_\tau + \bar{B}_\tau'\bar{P}_\tau(k+1)\bar{B}_\tau)^{-1}\bar{B}_\tau'\bar{P}_\tau(k+1)\tilde{A}_\tau\bar{x}_\tau(k),$$

where $\bar{P}_\tau(\cdot)$ is the backward solution of the time-invariant difference Riccati equation:

$$\begin{aligned}\bar{P}_\tau(k) &= \tilde{A}'_\tau \bar{P}_\tau(k+1)\tilde{A}_\tau + \tilde{C}'_\tau\tilde{C}_\tau \\ &\quad - \tilde{A}'_\tau \bar{P}_\tau(k+1)\bar{B}_\tau(I + \bar{D}'_\tau\bar{D}_\tau + \bar{B}'_\tau\bar{P}_\tau(k+1)\bar{B}_\tau)^{-1}\bar{B}'_\tau\bar{P}_\tau(k+1)\tilde{A}_\tau,\end{aligned} \tag{6.32}$$

with final condition $\bar{P}_\tau(N) = S$. The optimal value of the performance index is $\bar{x}_\tau(0)'\bar{P}_\tau(0)\bar{x}_\tau(0)$. The relationship between the symmetric positive semidefinite solutions of the Riccati equations (6.29) and (6.32) is stated in:

Lemma 6.4. *If* $\bar{P}_\tau(N) = P(\tau + NT) = S$, *then*

$$\bar{P}_\tau(k) = P(\tau + kT), \forall k \leq N.$$

Proof. It can be seen that, in view of (6.24)-(6.26) the optimization problems (6.28) and (6.30) are in fact coincident. Therefore, the optimal values of the performance indices must be the same, so that $\bar{P}_\tau(k) = P(\tau + kT)$. □

In particular, Lemma 6.4 entails that the constant positive semidefinite solutions of (6.32) supply the periodic generators for the original DPRE. In conclusion, the reformulation procedure introduced above enables one to establish a correspondence between the SPPS solutions of the DPRE (6.29) and the positive semidefinite solutions of the Discrete-time Algebraic Riccati Equation (DARE):

$$\begin{aligned}\bar{P}_\tau =& \tilde{A}'_\tau \bar{P}_\tau \tilde{A}_\tau + \tilde{C}'_\tau\tilde{C}_\tau \\ & - \tilde{A}'_\tau \bar{P}_\tau \bar{B}_\tau(I + \bar{D}'_\tau\bar{D}_\tau + \bar{B}'_\tau\bar{P}_\tau\bar{B}_\tau)^{-1}\bar{B}'_\tau\bar{P}_\tau\tilde{A}_\tau,\end{aligned} \tag{6.33}$$

As a consequence of the previous discussion, we can state the following

Proposition 6.1. *If* $P(\cdot)$ *is an SPPS solution of the DPRE (6.29), then* $\bar{P}_\tau = P(\tau)$ *is a positive semidefinite solution of the DARE (6.33). Conversely, if* $\bar{P}_\tau$ *is a positive semidefinite solution of (6.33), then the solution of (6.29) with* $P(\tau) = \bar{P}_\tau$ *is SPPS.*

Remark. So far, we have demonstrated that there exists a one to one correspondence between the symmetric positive semidefinite solutions of the DPRE (6.29) and the symmetric positive semidefinite solutions of the time-invariant difference Riccati equation (6.32). Actually, by a rationale based on the notion of analytic continuation for analytic functions [41], it can be shown that such a correspondence extends to all (non necessarily positive semidefinite) symmetric solutions. Therefore, the eq. (6.32) can be seen as the T-sampled representation of the DPRE (6.29). □

Prop. 6.1 above establishes a connection between the algebraic and periodic Riccati equation. For such a connection to be effective, one has to clarify the relationships between the structural properties of the original periodic system (6.23) and those of the time-invariant system underlying eq. (6.32). Lemma 6.3 points out the one-to-one correspondence between the structural properties of the original triple

$(A(\cdot), B(\cdot), C(\cdot))$ and those of the time-invariant triple $(\bar{A}_\tau, \bar{B}_\tau, \bar{C}_\tau)$, for any τ. In turn, $(\tilde{A}_\tau, \bar{B}_\tau, \tilde{C}_\tau)$ shares the same structural properties of $(\bar{A}_\tau, \bar{B}_\tau, \bar{C}_\tau)$. For instance, with reference to the reachability, controllability and stabilizability notions, note that $\tilde{A}_\tau$ is the closed-loop matrix stemming from a state feedback on system (6.25). On the other hand, this state feedback does not affect the mentioned properties. As for observability, reconstructibility and detectability, it is apparent from (6.31.b) that $\tilde{C}_\tau$ and $\bar{C}_\tau$ have the same range, so that one can focus on $(\tilde{A}_\tau, \bar{C}_\tau)$ in place of the pair $(\tilde{A}_\tau, \tilde{C}_\tau)$. Observe now that $\tilde{A}_\tau$ can be seen as the closed loop matrix resulting from an output feedback on the pair $(\tilde{A}_\tau, \bar{C}_\tau)$. Therefore, the pair $(\tilde{A}_\tau, \bar{C}_\tau)$ has the same observability properties as $(\bar{A}_\tau, \bar{C}_\tau)$. Finally, it is apparent that an $(\bar{A}_\tau, \bar{C}_\tau)$-unobservable eigenvalue is also $(\tilde{A}_\tau, \tilde{C}_\tau)$-unobservable.

A further significant property relates the dynamics of the periodic and time-invariant optimal closed-loop systems relative to the periodic and time-invariant systems (6.23) and (6.25), respectively. Precisely, consider an SPPS solution $P(\cdot)$ of the DPRE (6.29) and let

$$K(t) = (I + B(t)'P(t)B(t))^{-1}B(t)'P(t+1)A(t)$$
$$\hat{A}(t) = A(t) - B(t)K(t) \quad ,$$

be the optimal gain and the closed-loop matrix of the periodic optimal controller; the corresponding transition matrix will be denoted by $\Phi_{\hat{A}_\tau}(t, \tau)$, while $\Psi_{\hat{A}}(\tau) = \Psi_{\hat{A}}(\tau + T, \tau)$ is the closed-loop monodromy matrix. Moreover, consider the constant solution $\bar{P}_\tau = P(\tau)$ of eq. (6.32) and define

$$\hat{A}_\tau = \tilde{A}_\tau - \bar{B}_\tau \tilde{K}_\tau \quad ,$$

where

$$\tilde{K}_\tau = (I + \bar{D}_\tau'\bar{D}_\tau + \bar{B}_\tau'\bar{P}_\tau\bar{B}_\tau)^{-1}\bar{B}_\tau'\bar{P}_\tau\tilde{A}_\tau \quad .$$

Then, the following result holds.

Proposition 6.2. $\Psi_{\hat{A}}(\tau) = \hat{A}_\tau$.

Proof. The dynamics of the optimal closed-loop system is described by the equation

$$x(t+1) = \hat{A}(t)x(t) \quad ,$$

while the dynamics of the stationary optimal closed-loop system is given by

$$\bar{x}_\tau(k+1) = \hat{A}_\tau x_\tau(k) \quad .$$

Now, as already observed in the proof of Lemma 6.4, the optimization problems (6.28) and (6.30) are in fact coincident. Then, we have $x(\tau + kT) = \bar{x}_\tau(k)$, for any initial condition $x(\tau) = \bar{x}_\tau(0)$, so that $\Phi_{\hat{A}}(\tau) = \hat{A}_\tau$. □

The preservation of the structural properties when passing from the triple $(A(\cdot), B(\cdot), C(\cdot))$ to the triple $(\tilde{A}_\tau, \bar{B}_\tau, \tilde{C}_\tau)$, together with the statement of Prop. 6.2, enables one to transfer to the DPRE various results relative to the DARE. For example, if the assumption is made that $(A(\cdot), B(\cdot), C(\cdot))$ is stabilizable and detectable, then $(\tilde{A}_\tau, \bar{B}_\tau, \tilde{C}_\tau)$ turns out to be stabilizable and detectable as well. Then, it is well known that the DARE (6.32) has a unique positive semidefinite solution and the associated closed-loop system is asymptotically stable. Correspondingly, thanks to Props. 6.1 and 6.2, the DPRE (6.29) has a unique SPPS solution and the associated periodic closed-loop system is asymptotically stable. The converse statement that, if the DPRE (6.29) has a unique SPPS solution and the associated periodic closed-loop system is asymptotically stable, then $(A(\cdot), B(\cdot), C(\cdot))$ is stabilizable and detectable, can be derived in an analogous way. Similarly, the following four necessary and sufficient conditions follow directly from the results relative to the time-invariant DRE that can be found in Theorem 3.2 of [51].

Theorem 6.11.

(i) *The strong solution of the DPRE exists and is unique if and only if $(A(\cdot), B(\cdot))$ is stabilizable*
(ii) *The strong solution coincides with the periodic stabilizing solution of the DPRE if and only if $(A(\cdot), B(\cdot))$ is stabilizable and no $(A(\cdot), C(\cdot))$-unobservable characteristic multiplier lies on the unit circle.*
(iii) *The periodic stabilizing solution of the DPRE is positive definite at τ if and only if $(A(\cdot), B(\cdot))$ is stabilizable and no $(A(\cdot), C(\cdot))$-unobservable characteristic multiplier at time τ lies inside or on the unit circle.*
(iv) *The strong solution is the only SPPS solution of the DPRE if and only if $(A(\cdot), B(\cdot))$ is stabilizable and no $(A(\cdot), C(\cdot))$-unobservable characteristic multiplier lies outside the unit circle.* □

In the same way, the following convergence theorem can be easily proven from Corollary 4.1 in [51].

Theorem 6.12. *Subject to either $S_\tau > 0$ or $S_\tau \geq P_S(\tau)$, then the stabilizability of $(A(\cdot), B(\cdot))$ and the nonexistence of $(A(\cdot), C(\cdot))$-unobservable characteristic multipliers on the unit circle are necessary and sufficient conditions for the asymptotic convergence of $P(\cdot)$ towards $P_S(\cdot)$, where $P(\cdot)$ is the solution of the DPRE with initial condition $P(\tau) = S_\tau$, and $P_S(\cdot)$ is the unique periodic stabilizing solution of the DPRE.*

6.5 Applications

The purpose of this Section is to show how the results on the PRE presented in this Chapter can be applied to a few important problems of periodic control. Besides the classical LQG problem, we will cover more recent developments concerning H_∞ and multirate control, which are still subject of active research.

6.5.1 Optimal Control

6.5.1.1 Linear Quadratic Optimal Control. The linear quadratic (LQ) optimal control problem of periodic systems has a long history in Periodic Control Theory, see [52]. As in the time-invariant case, one has to cope with the possible discontinuity arising when passing from a finite horizon to an infinite horizon problem. This important issue has been addressed only recently in [53], see [54] and [55] for the time-invariant case. With reference to the periodic system

$$S : \dot{x}(t) = A(t)x(t) + B(t)u(t)$$

define the loss functionals:

$$J(t_0, t_f) = \int_{t_0}^{t_f} \{x(t)'C(t)'C(t)x(t) + u(t)'u(t)\}\mathrm{dt}$$

$$\bar{J}(t_0, t_f) = J(t_0, t_f) + x(t_f)'x(t_f) \tag{6.34}$$

where $C(\cdot)$ is T-periodic. By short, $J(t_0, \infty)$ will indicate $\lim_{t_f \to \infty} J(t_0, t_f)$, and analogously for similar cases. The formulation of the periodic LQ (PLQ) Control Problem will be posed by resorting to the class of linear, dynamic, periodic and stabilizing regulators C with complete information ($C : x(\cdot) \to u(\cdot)$). Such a class will be indicated as $R_{PS}^{(x)}$

***PLQ* Control Problem.** Find a regulator in the class $R_{PS}^{(x)}$ such that $J(t_0, \infty)$ exists and is minimal. □

In order to provide the solution of this problem, we are well advised to define two auxiliary control problems in one of which the class $R_P^{(x)}$ of linear, dynamic and periodic regulators with complete information enters the game. Obviously, $R_{PS}^{(x)} \subset R_P^{(x)}$.

$\overline{PLQ}$ Control Problem. Find a regulator in the class $R_P^{(x)}$ such that $\bar{J}(t_0, \infty)$ exists and is minimal.

$\overline{PLQ}^o$ Control Problem. Let $\bar{J}_o(t_0, t_f) = \min_{u(\cdot)} \bar{J}(t_0, t_f)$. Find $\bar{J}^o(t_0, \infty)$, if such a limit exists. □

In order to illustrate the subtle difference among these problems, consider the scalar time-invariant system: $\dot{x} = u$ and let $C(t) = \epsilon$. Then, the optimal control minimizing $\bar{J}(t_0, t_f)$ is $u(t) = p(t, t_f)x(t)$, where $p(t, t_f)$ is such that

$$\left|\frac{p(t, t_f) + \epsilon}{p(t, t_f) - \epsilon}\right| = \left|\frac{1 + \epsilon}{1 - \epsilon}\right| e^{2\epsilon(t_f - t)}.$$

By taking the limit for $t_f \to \infty$, one obtains $\bar{J}^o(t_0,\infty) = |\epsilon|x(t_0)^2$ and $u(t) = -|\epsilon|x(t)$ as optimal control law for the receding horizon control problem $\overline{PLQ}^o$. On the basis of the Lemmas below, it turns out that, if $\epsilon \neq 0$, $u(t) = -|\epsilon|x(t)$ is the solution of the $\overline{PLQ}$ and **PLQ** problems too. On the contrary, if $\epsilon = 0$, then $\bar{J}^o(t_0,\infty) = 0$ is the limiting value of the cost functional for the receding horizon control problem; however, neither $\overline{PLQ}$ nor **PLQ** has a solution. Indeed, with a feedback law $u(t) = -cx(t)$, $c > 0$, the value of the cost functional of $\overline{PLQ}$ or **PLQ** is $cx(t_0)^2$, so that it becomes closer and closer to 0, as $c \to 0$; however, there is no periodic control law attaining 0 as value of the cost functionals $J(t_0,\infty)$ or $\bar{J}(t,\infty)$.

The relationships among the three control problems defined above are clarified by the following Lemmas, the proofs of which are straightforward and therefore omitted.

Lemma 6.5. *If $\bar{J}^o(t_0,\infty)$ exists, then the cost functional $\bar{J}(t_0,\infty)$ associated with any regulator in the class R_P^x is such that $\bar{J}^o(t_0,\infty) \leq \bar{J}(t_0,\infty)$.*

Lemma 6.6. *If the regulator $C \in R_P^x$ is a solution of $\overline{PLQ}$ and is stabilizing, then C is a solution of PLQ too.* □

A sufficient condition for the solution of the PLQ problem can now be provided.

Theorem 6.12. *Suppose that $(A(\cdot),B(\cdot))$ is stabilizable and no unit-modulus characteristic multiplier of S is $(A(\cdot),C(\cdot))$-unobservable. Then, the optimal control law for the **PLQ** problem is $u(t) = -B(t)'\bar{P}(t)x(t)$, where $\bar{P}(\cdot)$ is the periodic stabilizing solution of the PRE.*

Proof. In [11], it is shown that, under the stated assumptions, the solution $P(t,t_f)$ of the PRE with the final condition $P(t_f,t_f) = I$ converges as $t_f \to \infty$ to the unique periodic stabilizing solution $\bar{P}(\cdot)$ of the PRE. Moreover, with reference to problem $\overline{PLQ}^o$, $\bar{J}^o(t_0,\infty)$ is finite and given by $x(t_0)'\bar{P}(t_0)x(t_0)$. Turning now to $\overline{PLQ}$, note that the periodic controller $u(t) = -B(t)'\bar{P}(t)x(t)$ gives rise to a cost functional $\bar{J}(t_0,\infty) = \bar{J}^o(t_0,\infty)$. In view of Lemma 6.5, such a control law is the optimal one for the $\overline{PLQ}$ problem. Considering that $\bar{P}(\cdot)$ is stabilizing, Lemma 6.6 leads to the conclusion.

6.5.1.2 Linear Quadratic and Gaussian Optimal Control. We now consider stochastic systems with incomplete information:

$$S_S : \begin{cases} \dot{x}(t) = A(t)x(t) + B(t)u(t) + G(t)w(t) \\ y(t) = D(t)x(t) + v(t) \end{cases},$$

where $w(\cdot)$, $v(\cdot)$ are independent zero-mean white noises with identity spectral density matrices. Reference will be made to the class $R_{PS}^{(y)}$ of linear, dynamic, periodic regulators C with incomplete information ($C : y(\cdot) \to u(\cdot)$) such that the linear periodic system $S_S\&C$ is stable. As for the cost functional, let

$$J_S(t_0,t_f) = \frac{1}{t_f - t_0} E\{J(t_0,t_f)\}.$$

PLQG Control Problem. Find a regulator in the class $R_{PS}^{(y)}$ such that $J_S(-\infty,+\infty)$ exists and is minimal. □

As discussed in [53], the PLQG problem can be reduced to an LQ problem by replacing the state x(t) with its estimate $\hat{x}(t)$ supplied by the optimal periodic Kalman filter. In this regard, note that, by dualizing Theorem 6.12 above, one obtains a sufficient condition for the periodic optimal filtering problem. Along this line of reasoning, the following result is derived in [53].

Theorem 6.13. *Suppose that*

(i) $(A(\cdot), B(\cdot))$ *is stabilizable*
(ii) $(A(\cdot), D(\cdot))$ *is detectable*
(iii) *no unit-modulus characteristic multiplier of* $A(\cdot)$ **is** $(A(\cdot), C(\cdot))$*-unobservable*
(iv) *no unit-modulus characteristic multiplier of* $A(\cdot)$ *is* $(A(\cdot), G(\cdot))$*-unreachable*

Then, the PLQG problem is solved by $u(t) = -B(t)'P(t)\hat{x}(t)$, *where* $\hat{x}(t)$ *is the state estimate at time t supplied by the periodic Kalman filter*

$$\dot{\hat{x}}(t) = A(t)\hat{x}(t) + B(t)u(t) + \Pi(t)D(t)'(y(t) - D(t)\hat{x}(t)),$$

where $P(\cdot)$ *is the periodic stabilizing solution of the PRE:*

$$-\dot{P}(t) = A(t)'P(t) + P(t)A(t) - P(t)B(t)B(t)'P(t) + C(t)'C(t)$$

and $\Pi(\cdot)$ *is the periodic stabilizing solution of*

$$\dot{\Pi}(t) = A(t)\Pi(t) + \Pi(t)A(t)' - \Pi(t)D(t)'D(t)\Pi(t) + G(t)G(t)'.$$

6.5.2 H_∞ Periodic Filtering and Control

The recent development of H_∞ control theory in the state-space domain, see [56], has set the basis for the extension of the theory to time-varying systems. Based on the results of [57] for the finite-horizon H_∞ filtering problem of time-varying systems, in [58] the infinite horizon H_∞ state-estimation problem for periodic systems is studied. Specifically, consider system S_S with $x(0) = 0$, $v(t) = E(t)w(t)$ and

$$E(t)\begin{bmatrix} G(t)' \\ E(t)' \end{bmatrix} = \begin{bmatrix} 0 \\ I \end{bmatrix}.$$

The only assumption on the noise $w(\cdot)$ is that it has bounded energy, i.e., $w(\cdot) \in L_2[0,\infty]$. The estimation problem consists in determining an unbiased linear periodic filter such that, for a given $\gamma > 0$, $J < \gamma^2$, where

$$J = sup\frac{\|x(\cdot) - \hat{x}(\cdot)\|^2_{L_2}}{\|w(\cdot)\|^2_{L_2}}$$

To deal with such a problem, consider the periodic sign-indefinite Riccati equation

$$\dot{P}(t) = A(t)P(t) + P(t)A(t)' + P(t)W(t)P(t) + G(t)G(t)' \quad (6.35.a)$$

$$W(t) = \frac{1}{\gamma^2}I - D(t)'D(t). \quad (6.35.b)$$

Then, the main result can be stated as follows.

Theorem 6.14. *Suppose that the pair* $(A(\cdot), D(\cdot))$ *is detectable and that* $(A(\cdot), G(\cdot))$ *has no uncontrollable unit-modulus characteristic multiplier. If, for a given* $\gamma > 0$, *the periodic Riccati equation (6.35) has an SPPS strong solution, then*

$$\dot{\hat{x}}(t) = (A(t) - P(t)D(t)'D(t))\hat{x}(t) + B(t)u(t) + P(t)D(t)y(t) \quad (6.36)$$

is a periodic state estimator such that $J < \gamma^2$. □

Interestingly enough, in analogy with the time-invariant case, as $\gamma \to \infty$, equation (6.35) tends to the standard PRE for the filtering problem and equation (6.36) tends to the periodic Kalman filter.

6.5.3 Multirate Sampled-Data Systems

Research in multirate sampled-data systems has been motivated by two important reasons. First, it was proven that the use of multirate and periodically time-varying controllers can significantly improve the closed-loop performance of linear time-invariant plants, see [59]. Second, in several industrial applications, specific technological or economical reasons require the use of control environments, where the inputs updating and the output sampling mechanisms are performed at different time instants and rates, see [60]. The distinctive feature of multirate sampled-data systems is that they present an inherently periodic (or quasi periodic) behavior. Depending on the control objectives, a number of control schemes have been introduced, see e.g. [61], [62]. In particular, starting from a discrete-time model of the process under control, the LQG problem has been treated in [63]. For the sake of simplicity, it is assumed that the input variables are updated at a unique and fast rate whereas the output variables can be measured with different sampling rates and/or at different sampling instants. Despite its simplicity, this situation allows to throw into sharp relief some peculiar aspects of multirate systems analysis and control. The analysis of more general situations is currently underway.

Consider the *time-invariant* stochastic system:

$$x(t+1) = Ax(t) + Bu(t) + v_1(t) \quad (6.37.a)$$

$$y(t) = Cx(t) + v_2(t). \quad (6.37.b)$$

where v_1 and v_2 are uncorrelated zero-mean Gaussian white noises, $v_1 \sim WGN(0, V_1)$, $V_1 \geq 0$ and $v_2 \sim WGN(0, V_2)$, $V_2 > 0$. Also assume that the initial state $x(t_0)$ is uncorrelated with v_1 and v_2. Suppose now that the i-th output variable $y_i(t)$ is sampled each, say, T_i time instants, with $\tau_i \in [0, T_i - 1]$ as initial sampling time. To each variable $y_i(\cdot)$ one can then associate a variable $z_i(\cdot)$ which is zero at each time point but the sampling instants, at which it coincides with $y_i(\cdot)$. This leads to the definition of $z(t) = N(t)y(t)$, where

$$N(t) = diag\{\nu_i(t)\},\ \nu_i(t) = \begin{cases} 1, & t = kT_i + \tau_i \\ 0, & t \neq kT_i + \tau_i \end{cases} \tag{6.38}$$

Notice that $N(\cdot)$ is a T-periodic matrix, where $T = l.c.m.\,\{T_i\}$. Hence, in view of (6.37) and (6.38), we obtain the following periodic system:

$$x(t+1) = Ax(t) + Bu(t) + v_1(t) \tag{6.39.a}$$

$$z(t) = N(t)Cx(t) + N(t)v_2(t) \tag{6.39.b}$$

With reference to (6.39), consider the problem of minimizing

$$J = \lim_{\substack{t_0 \to -\infty \\ t_f \to +\infty}} E[\sum_{t_0}^{t_f} \{x(t)'Qx(t) + u(t)'u(t)\}] \tag{6.40}$$

where $Q \geq 0$ and $R > 0$.

Due to the periodicity of matrix $N(\cdot)$, this is in fact a periodic LQG problem. If compared with the one dealt with in Section 6.5.1, however, it has the peculiarity that the output noise has a singular covariance matrix. This difficulty can be circumvented in various ways. As an alternative to the one suggested in [63], the following procedure can be adopted. Note that, at each time point t, a number, say $r(t)$, of entries of vector $z(t)$ are equal to zero. Therefore, without any loss of generality, the output equation can be given the form

$$z(t) = \begin{bmatrix} \zeta(t) \\ 0 \end{bmatrix} = \begin{bmatrix} H(t) \\ 0 \end{bmatrix} x(t) + \begin{bmatrix} w(t) \\ 0 \end{bmatrix},$$

where, denoting by p the dimension of $y(t)$, $\zeta \in R^{(p-r(t))}$, $H(t) \in R^{(p-r(t)) \times n}$ and $w(t) \sim WGN(0, W(t))$, $W(t) \in R^{(p-r(t)) \times (p-r(t))}$. Then, the periodic one step ahead predictor is given by

$$\hat{x}(t+1) = A\hat{x}(t) + Bu(t) + L(t)(\zeta(t) - H(t)\hat{x}(t)),$$

where

$$L(t) = A\Pi(t)H(t)'(W(t) + H(t)\Pi(t)H(t)')^{-1}$$

and $\Pi(t)$ an SPPS solution of the following DPRE

$$\begin{aligned} \Pi(t+1) &= A\Pi(t)A' + V_1 \\ &\quad - A\Pi(t)H(t)'(W(t) + H(t)\Pi(t)H(t)')^{-1}H(t)\Pi(t)A'. \end{aligned} \tag{6.41}$$

Notice that the dimension of the innovation vector $\zeta(t) - H(t)\hat{x}(t)$ is periodically time-varying. One can eliminate such a time dependence of the dimensions, by resorting to the following equivalent predictor:

$$\hat{x}(t+1) = A\hat{x}(t) + Bu(t) + K_F(t)(z(t) - \begin{bmatrix} H(t) \\ 0 \end{bmatrix} \hat{x}(t)),$$

with

$$K_F(t) = A\Pi(t) \begin{bmatrix} H(t) \\ 0 \end{bmatrix}' \Sigma(t)^{-1},$$

where, $\Pi(\cdot)$ is an SPPS solution of

$$\begin{aligned} \Pi(t+1) &= A\Pi(t)A' + V_1 \\ &\quad - A\Pi(t) \begin{bmatrix} H(t) \\ 0 \end{bmatrix}' \Sigma(t)^{-1} \begin{bmatrix} H(t) \\ 0 \end{bmatrix} \Pi(t)A', \end{aligned} \tag{6.42}$$

with

$$\Sigma(t) = \begin{bmatrix} W(t) & 0 \\ 0 & \Lambda(t) \end{bmatrix} + \begin{bmatrix} H(t) \\ 0 \end{bmatrix} \Pi(t) \begin{bmatrix} H(t) \\ 0 \end{bmatrix}'$$

and $\Lambda(t) \in R^{r(t)\times r(t)}$ is an arbitrary positive definite matrix. Observe that eq. (6.42) is a purely formal reformulation of eq. (6.41). As for matrix $\Lambda(t)$, an easy choice is the identity matrix. With such a choice, the previous procedure is equivalent to forcing the nonsingularity of the original covariance matrix $N(t)V_2N(t)$ by replacing all the 0's along the diagonal with 1's. The minimization of the cost functional (6.40) can be performed by the usual approach based on the separation principle, thanks to which the optimal control is based on the optimal one-step ahead prediction $\hat{x}$ of the state x.

The detectability of the periodic system (6.39) can be related to the detectability of the original time-invariant system (6.37), as indicated in [63], where the following result is proven.

Theorem 6.15. *Suppose that*

(i) (A, B) *and* (A, V_1) *are stabilizable,*
(ii) (A, C) *and* (A, Q) *are detectable*
(iii) *do not exists two distinct eigenvalues of* A, λ_i *and* λ_j, $|\lambda_i| \geq 1$, $|\lambda_j| \geq 1$, *such that* $\lambda_i^T = \lambda_j^T$.
(iv) *do not exist eigenvalues* λ *of* A, $\lambda \neq 1$, *such that* $\lambda^T = 1$.

Then, the solution of the LQG problem with cost given by eq. (6.40) exists and is given by

$$\begin{aligned} u(t) &= -[R + B'PB]^{-1} B'PA\hat{x}(t), \\ \hat{x}(t+1) &= A\hat{x}(t) + Bu(t) + K_F(t)(z(t) - N(t)C\hat{x}(t)), \\ K_F(t) &= A\Pi(t)C'N(t)(\tilde{V}_2(t) + N(t)C\Pi(t)C'N(t))^{-1}, \end{aligned}$$

where $\tilde{V}_2(t)$ *is given by* $N(t)V_2N(t)$ *with* 0*'s replaced by* 1*'s along the diagonal, P is the unique positive semidefinite solution of the algebraic Riccati equation*

$$P = A'PA + Q - A'PB[R + B'PB]^{-1}B'PA$$

and $\Pi(t)$ *is the unique SPPS solution of the periodic Riccati equation*

$$\begin{aligned}\Pi(t+1) &= A\Pi(t)A' + V_1 \\ &\quad - A\Pi(t)C'N(t)[\tilde{V}_2(t) + N(t)C\Pi(t)C'N(t)]^{-1}N(t)C\Pi(t)A'.\end{aligned}$$

Moreover, the T-periodic closed-loop system is asymptotically stable. Its characteristic multipliers are those of $A(\cdot) - K_F(\cdot)N(\cdot)C$ *along with the eigenvalues of* $(A + BK)^T$. □

To prove the result above, in [63] it is first shown that conditions (iii) and (iv) together with the detectability of the original pair (A,C) imply the detectability of $(A, N(\cdot)C)$. The theorem above shows how the classical LQG theory can be generalized to a multirate control scheme. Different and more complex situations have been recently considered in the literature, in order to take into account integral actions or holding mechanisms on the input variables, see [61] and [64] and references quoted there.

Acknowledgement. The partial support of Consiglio Nazionale delle Ricerche and Ministero dell'Universita' e della Ricerca Scientifica is gratefully acknowledged.

References

1. Riccati, J.F.: Animadversiones in aequationes differentiales secundi gradus. Acta Eruditorum Lipsiae, Supplementa, VIII, II (1724) 66-73
2. Kalman, R.E.: New methods in Wiener filtering. Proc. of First Symp. on Eng. Applications of Random Function Theory and Probability (J. Bogdanoff and F. Kozin eds.), J. Wiley 1963, 270–388
3. Kwakernaak, H; Sivan R.: Linear Optimal Control Systems. J. Wiley 1972
4. Anderson, B.D.O.; Moore, J.B.: Linear Optimal Control. Prentice Hall 1971
5. Yakubovich, V.A.; Starzhinskii, V.M.: Linear Differential Equations with Periodic Coefficients. Wiley 1972
6. Canabal J.R.: Periodic geometry of the Riccati equation. Stochastics, 1 (1974) 432–440
7. Bittanti, S.; Locatelli, A.; Maffezzoni, C.: Second-variation methods in periodic optimization. J. Opt. Th. Appl. 14 (1974) 31–49
8. Bekir, E.; Bucy, R.S.: Periodic equilibria for matrix Riccati equations. Stochastics 2 (1976) 1–104
9. Hewer G.A.: Periodicity, detectability and the matrix Riccati equation, SIAM J. Control 13 (1975) 1235–1251
10. Kano H.; Nishimura, T.: Periodic solutions of matrix Riccati equations with detectability and stabilizability. Int. J. Control 29 (1979) 471–487
11. de Souza, C.E.: Riccati differential equation in optimal filtering of periodic non-stabilizable systems. Int. J. Control 46 (1987) 1235–1250
12. Kano H.; Nishimura, T.: Periodic strong solutions of periodically time-varying matrix Riccati equations. Int. J. Control 49 (1989) 193–205

13. Bittanti, S.; Colaneri, P.; De Nicolao, G.: A note on the maximal solution of the periodic Riccati equation. IEEE Trans. on Automat. Control 34 (1989) 1316–1319
14. Shayman M. A.: On the periodic solutions of the matrix Riccati equation. Math. Systems Theory 16 (1983) 267–287
15. Shayman M. A.: On the phase portrait of the matrix Riccati equation arising from the periodic control problem. SIAM J. Contr. and Opt. 23 (1985) 717–751
16. Shayman M. A.: Inertia theorems for the periodic Lyapunov equation and periodic Riccati equation. Syst. & Control Letters 4 (1984) 27–32
17. Bittanti, S.; Colaneri, P.: Lyapunov and Riccati equations: Periodic inertia theorems". IEEE Trans. Automat.Control 31 (1986) 659–661
18. Bittanti, S.; Colaneri, P.; Guardabassi, G.: Periodic solutions of periodic Riccati equations. IEEE Trans. Automat. Control 29 (1984) 665–667
19. Speyer, J.L.; Evans, R.T.: A second variational theory for optimal periodic processes. IEEE Trans. on Automat. Control 39 (1984) 138–148
20. Bittanti, S.; Colaneri, P.; Guardabassi, G.: Analysis of the periodic Lyapunov and Riccati equations via canonical decomposition. SIAM J. Contr. and Opt. 24 (1986) 1138–1149
21. Yakubovich, V.A.: Linear-quadratic optimization problem and frequency theorem for periodic systems. Siberian Math. J. 27 (1986)
22. Bittanti, S.; Guardabassi, G.: Optimal periodic control and periodic systems analysis: An overview. Proc. 25th Conf. Decision Contr.. Athens, Greece (1986) 1417–1423
23. Wang, Q.; Speyer, J.L.: The periodic Riccati differential equation and the periodic regulator. Proc. 26th Conf. Decision Contr.. Los Angeles, USA (1987) 288–292
24. Wang, Q.; Speyer, J.L.: Necessary and sufficient conditions for local optimality of a periodic process. Proc. 11th IFAC World Congress. Munich, W. Germany 8 (1987) 188–193
25. Colonius, F.: Optimal Periodic Control. Habilitationsschrift, Universität Bremen 1986
26. Colonius, F.: A global maximum principle for periodic control of functional differential equations. Proc. 25th Conf. Decision Contr.. Athens, Greece (1986) 1424–1427
27. Da Prato, G.: Synthesis of optimal control for an infinite dimensional periodic problem. SIAM J. Contr. and Opt. 25 (1987) 706–714
28. Halanay, A.; Morozan, T.; Tudor, C.: Tracking almost periodic signals under white noise perturbations. Stochastics 21 (1988) 257–301
29. Bittanti, S.: Deterministic and Stochastic Linear Periodic systems. Time Series and Linear Systems (S. Bittanti Ed.), Springer 1986, 141–182
30. Bittanti, S.; Bolzern, P.: Stabilizability and detectability of linear periodic systems. Syst. & Control Letters 6 (1985) 141–145
31. Bolzern, P.; Colaneri, P.: The periodic Lyapunov equation. SIAM J. Matrix Anal. Appl. 9 (1988) 46–512
32. Bittanti, S.; Bolzern, P.; Colaneri, P.: The extended periodic Lyapunov lemma. Automatica 21 (1985) 603–605
33. Bittanti, S.; Bolzern, P.; Colaneri, P.: Inertia theorems for Lyapunov and Riccati equations-An updated view. Proc. SIAM Conf. on Linear Algebra in Signals, Systems, and Control, Boston, USA (1986) 11–35
34. Bittanti, S.; Guardabassi G.; Maffezzoni C.; Silverman L.: Periodic systems: controllability and the matrix Riccati equation. SIAM J. Contr. and Opt. 16 (1978) 37–40
35. Bittanti, S.; Colaneri, P.: A monotonicity result for the periodic Riccati equation. Robust Control of Linear Systems and Nonlinear Control (M.A. Kaashoek, J.H. Van Schuppen, A.C.M Ran, eds.), Birkhäuser, Boston 1990, 83–93.
36. Bittanti, S.; Colaneri, P.: Some remarks on the nonstandard periodic Riccati equation. Analysis and Control of Nonlinear Systems (C.I. Byrnes, C.F. Martin, R.E. Saeks eds.), North Holland 1988, 517–525
37. Bittanti, S.; Colaneri, P.; De Nicolao, G.: Analysis of the discrete-time periodic Riccati equation by a Kleinman procedure. Proc. 25th Conf. Decision Contr.. Athens, Greece (1986) 1444–1449
38. de Souza, C.E.; Goodwin, G.C.: Periodic solutions of matrix Riccati equation in optimal filtering of nonstabilizable periodic systems. Proc. 10th IFAC World Congress, Munich, W. Germany 9 (1987) 249–254

39. de Souza, C.E.: Maximal periodic solutions of periodic Riccati difference equations. Proc. Workshop on the Riccati Equation in Control, Systems and Signals. Como, Italy (1989) 177–182
40. Bittanti, S.; Colaneri, P.; De Nicolao, G.: The difference periodic Riccati equation for the periodic prediction problem. IEEE Trans. Autom. Control 33 (1988) 706–712
41. Bittanti, S.; Colaneri, P.; De Nicolao, G.: An algebraic Riccati equation for the discrete-time periodic prediction problem. Syst. & Control Letters 14 (1990) 71–78
42. Bittanti, S.; Bolzern, P.: Can the Kalman canonical decomposition be performed for a discrete-time linear periodic system?. Proc. Primero Congresso Latino Americano de Automatica. Campina Grande, Brazil (1984) 449–453
43. Grasselli, O.M.: A canonical decomposition of linear periodic discrete-time systems. Int. J. Contr. 40 (1984) 201–214
44. Bittanti, S.; Bolzern, P.: On the structure theory of discrete-time linear systems. Int. J. Syst Sci. 17 (1986) 33–47
45. Bolzern, P.: Criteria for reachability, controllability and stabilizability of discrete-time linear periodic systems. Proc. 5th Polish-English Seminar on Real-Time Process Control, Radzejowice, Poland (1986) 69–83
46. Jury, E.I.; Mullin, F.J.: The analysis of sampled-data control systems with a periodically time-varying sampling rate. IRE Trans. Autom. Control 5 (1959) 15–21
47. Meyer, R.A.; Burrus, C.S.:A unified analysis of multirate and periodically time-varying digital filters. IEEE Trans. Circuits Syst. 22 (1975) 162–167
48. Bittanti, S.; Colaneri, P.; De Nicolao, G.: Discrete-time linear periodic systems: A note on the reachability and controllability interval length. Syst. & Control Letters 8 (1986) 75–78
49. Bolzern, P.; Colaneri, P.; Scattolini, R.: Zeros of discrete-time linear periodic systems. IEEE Trans. Automat. Control 31 (1986) 1057–1058
50. Bolzern, P.; Colaneri, P.: Inertia theorems for the periodic Lyapunov difference equation and the periodic Riccati difference equation. Linear Algebra and Its Applications 85 (1987) 249–265
51. de Souza, C.E.; Gevers, M.R.; Goodwin, G.C.: Riccati equations in optimal filtering of non-stabilizable systems having singular state transition matrices. IEEE Trans. Automat. Control 31 (1986) 831–838
52. Marzollo, A. (ed.): Periodic Optimization. Springer 1982
53. Bittanti, S.; Colaneri, P.; De Nicolao, G.: Periodic regulators with incomplete and noisy measurements. Proc. 29th Conf. Decision Contr.. Honolulu, USA (160) to appear
54. Willems, J.L.; Callier, F.M.: Large finite horizon and infinite horizon LQ-optimal control problems. Optimal Control Appl. & Methods 4 (1983) 31-45
55. Callier, F.M.; Willems, J.L.: Criterion for the convergence of the solution of the Riccati differential equation. IEEE Trans. on Automat. Control 26 (1981) 1232–1242
56. Doyle, J.C.; Glover, K.; Khargonekar, P.P.; Francis, B.A.: State-space solutions to standard H_2 and H_∞ control problems. IEEE Trans. on Automat. Control 34 (1989) 831–847
57. Khargonekar, P.P.; Nagpal, K.M.: Filtering and smoothing in an H_∞ setting. Proc. 28th Conf. Decision Contr.. Tampa, USA (1989) 415–420
58. de Souza, C.E.: Private communication
59. Khargonekar, P.P.; Poolla, K.; Tannenbaum, A.: Robust control of linear time-invariant plants using periodic compensation. IEEE Trans. on Automat. Control 30 (1985) 1088-1096
60. Glasson, D.R.: Research in Multirate Estimation and Control. The Analytic Science Corp. 1980, Rep. TR1356-2
61. Al Rahmani, H.M.; Franklin, G.F.: A new optimal multirate control of linear periodic and time-invariant systems. IEEE Trans. on Automat. Control 35 (160) 406-415
62. Araki, M.; Hagiwara, T.: Pole-assignment by piecewise constant output feedback. Int. J. Control 29 (1979) 31-38
63. Colaneri, P.; Scattolini, R.; Schiavoni, N.: The LQG problem for multirate sampled-data systems. Proc. 28th Conf. Decision Contr.. Tampa, USA (1989) 469–474
64. Colaneri, P.; Scattolini, R.; Schiavoni, N.: LQG optimal control of multirate sampled-data systems. Dipartimento di Elettronica, Politecnico di Milano 1989, Int. Rep. 89-052

7 Invariant Subspace Methods for the Numerical Solution of Riccati Equations

Alan J. Laub

7.1 Introduction

In this tutorial paper, an overview is given of progress over the past ten to fifteen years towards reliable and efficient numerical solution of various types of Riccati equations. Our attention will be directed primarily to matrix-valued algebraic Riccati equations and numerical methods for their solution based on computing bases for invariant subspaces of certain associated matrices. Riccati equations arise in modeling both continuous-time and discrete-time systems in a wide variety of applications in science and engineering. One can study both algebraic equations and differential or difference equations. Both algebraic and differential or difference equations can be further classified according to whether their coefficient matrices give rise to so-called symmetric or nonsymmetric equations. Symmetric Riccati equations can be further classified according to whether or not they are definite or indefinite.

The rest of the paper is organized as follows. A brief review of "classical" methods is followed by a summary of the now-standard Schur method, introduced in 1978, for solving algebraic Riccati equations. Extensions of the basic Schur method, by means of associated generalized eigenvalue problems, are then described together with some applications. Next, some powerful new numerical results relating to Riccati equations in general will be described. These include a thorough analysis of iterative refinement via Newton's method (including a computable estimate of the region of convergence), a theorem on the relation of error estimates to residuals, estimation of the condition of algebraic Riccati equations, and promising new scaling strategies. Newton's method for computing the matrix sign function is then described and its implementation for parallel algorithms for Riccati equations (on a message-passing hypercube computer) is outlined. This method is particularly well suited to parallelization and vectorization and has been used successfully to solve fairly large order (several hundred) problems. A number of generalizations of this basic iteration have extended its applicability to a broader range of problems. For example, generalizations of the matrix sign function to the case of matrix pencils allows straightforward solution of discrete-time Riccati equations. Furthermore, the Newton iteration itself has been generalized considerably and found to be but a special case of a general family of iterations for the matrix

Bittanti, Laub, Willems (Eds.)
The Riccati Equation

sign function. Iteration formulas in this family can be chosen to be of arbitrary order convergence in exchange for, naturally, an increased computational burden. The effect of this increased computation can, however, be ameliorated by parallelization. Global convergence results for the iteration family have been proved. These issues are discussed together with applications and implementation results on vector supercomputers such as the Cray. Finally, after a few closing remarks, some further research directions are outlined.

7.2 Review of Invariant Subspace Methods

The numerical solution of Riccati equations by invariant subspace methods (eigenvectors), including the use of the matrix sign function, has been an extremely active research area for a long time and much of the relevant literature is included in an extensive References section at the end of this paper. Other solution techniques are also available for the solution of Riccati equations and most can be classified under the following headings:

- methods based on special canonical forms
- doubling and other direct integration techniques
- Newton's method
- parameter imbedding methods
- Chandrasekhar-type algorithms
- spectral factorization techniques
- "square root" formulations.

Apart from Newton's method, which will be discussed in detail later in the paper, we will not review any of these other methods in this paper. It suffices to note that, apart from fairly trivial examples, most of these methods encounter severe numerical difficulties when implemented in finite arithmetic. However, since the late 1970's, methods based on invariant subspaces (and, as we shall see, a significant generalization based on deflating subspaces) have emerged as a generally reliable family of methods, particularly for Riccati equations with relatively small dense coefficient matrices (say, of order a few hundred or less). We shall briefly review these methods here.

7.2.1 Methods Based on Invariant Subspaces

One of the simplest types of Riccati equations arises from the so-called linear-quadratic optimal control problem:

$$\min_u \frac{1}{2} \int_0^{+\infty} \left[x^T Q x + u^T R u\right] dt$$

$$\text{s.t.} \quad \dot{x} = Ax + Bu$$

Here $A, Q \in \mathbb{R}^{n \times n}, B \in \mathbb{R}^{n \times m}$, and $R \in \mathbb{R}^{m \times m}$. With a few natural assumptions related to controllability and observability, the solution to this problem is well known to be in the feedback form

$$u^*(t) = -R^{-1}B^T X x(t)$$

where X is the unique symmetric nonnegative definite solution of the symmetric Riccati equation

$$A^T X + XA - XBR^{-1}B^T X + Q = 0\,.$$

The Riccati solution X can be computed from a Jordan form reduction of an associated $2n \times 2n$ Hamiltonian matrix M. Specifically, compute a matrix of eigenvectors T to perform the following reduction:

$$\begin{pmatrix} T_{11} & T_{12} \\ T_{21} & T_{22} \end{pmatrix}^{-1} \begin{pmatrix} A & -BR^{-1}B^T \\ -Q & -A^T \end{pmatrix} \begin{pmatrix} T_{11} & T_{12} \\ T_{21} & T_{22} \end{pmatrix} = \begin{pmatrix} -\Lambda & 0 \\ 0 & \Lambda \end{pmatrix}$$

where $-\Lambda$ is composed of Jordan blocks corresponding to eigenvalues in the open left-half plane only and each T_{ij} is $n \times n$. Standard assumptions on the problem guarantee no eigenvalues on the imaginary axis and it is then easily seen that M has precisely n left-half-plane eigenvalues and n right-half-plane eigenvalues. Finally, the Riccati solution X is found by solving a system of linear equations:

$$X = T_{21}T_{11}^{-1}\,.$$

That this is true is easily seen by manipulating the invariant subspace equation:

$$\begin{pmatrix} A & -BR^{-1}B^T \\ -Q & -A^T \end{pmatrix} \begin{pmatrix} T_{11} \\ T_{21} \end{pmatrix} = \begin{pmatrix} T_{11} \\ T_{21} \end{pmatrix} (-\Lambda)$$

The connection between an apparently nonlinear matrix Riccati equation of order n and a linear eigenvalue problem of order $2n$ is classical and dates back at least to Von Escherich [221] in 1898 according to Reid [188]. The "eigenvector solution method" was popularized in the control literature by MacFarlane [152] and Potter [183]. However, there are severe numerical difficulties with this approach when the Hamiltonian matrix M has multiple or near-multiple eigenvalues. For a cogent discussion of the numerical difficulties associated with the numerical determination of Jordan forms, the reader is urged to consult the by-now-classic 1976 paper of Golub and Wilkinson in [97].

To ameliorate these difficulties, Schur methods were proposed in 1978 [139], [140]. The procedure is essentially the same as above except that instead of a nonsingular matrix T of eigenvectors, an orthogonal matrix U of "Schur vectors" is computed so that

$$\begin{pmatrix} U_{11} & U_{12} \\ U_{21} & U_{22} \end{pmatrix}^T \begin{pmatrix} A & -BR^{-1}B^T \\ -Q & -A^T \end{pmatrix} \begin{pmatrix} U_{11} & U_{12} \\ U_{21} & U_{22} \end{pmatrix} = \begin{pmatrix} S_{11} & S_{12} \\ 0 & S_{22} \end{pmatrix}.$$

where now S_{11} is a quasi-upper triangular matrix (in so-called real Schur form) with open left-half-plane eigenvalues and S_{22} is a quasi-upper triangular matrix with open right-half-plane eigenvalues. The key observation is that $\begin{pmatrix} T_{11} \\ T_{21} \end{pmatrix}$ and $\begin{pmatrix} U_{11} \\ U_{21} \end{pmatrix}$ span the same invariant subspace and it is easily seen that X can again be computed from

$$X = U_{21}U_{11}^{-1} .$$

The principal difficulty with the Schur approach is the ordering of the eigenvalues on the (block) diagonal of S so that the stable eigenvalues appear in the upper left $n \times n$ block, a problem that clearly does not arise in the eigenvector case. However, that problem can be overcome in a numerically stable way [209, 78] and the net result is an algorithm that is now considered one of the more reliable for Riccati equations and has the virtues of being simultaneously:

- efficient,
- numerically robust, and
- flexible enough to handle a broad range of problems.

7.2.2 Methods Based on Deflating Subspaces

Another significant advance took place in 1978 in the numerical solution of Riccati equations and that was the introduction of the idea of using generalized eigenvalue problems and deflating subspaces rather than eigenvalue problems and invariant subspaces. This turns out to be crucial for the case of Riccati equations arising from discrete-time models and, in fact, was motivated by such a problem arising in a discretized delay system [139, 175, 176, 177]. The difficulty with such systems is that the "A matrix" is singular. Unfortunately, the direct analog of the Hamiltonian matrix employed above is a symplectic matrix and such a matrix uses the explicit inverse of A. However, use of a factored form of the symplectic matrix M obviates the need to compute A^{-1}. We shall illustrate with a particular discrete-time Riccati equation.

Specifically, consider the equation

$$\begin{aligned} E^TXE &= A^TXA - (A^TXB + S)(B^TXB + R)^{-1}(A^TXB + S)^T + Q \\ &\equiv \hat{A}^TX\hat{A} - \hat{A}^TXB(B^TXB + R)^{-1}B^TX\hat{A} + Q - SR^{-1}S^T \end{aligned}$$

where $\hat{A} := A - BR^{-1}S^T$. Notice several extra matrices here. Specifically, we can include an "S matrix" (corresponding to a cross-weighting term in a performance index in the control problem or correlated noise in the filtering problem) and an "E matrix" (corresponding to a state equation in the implicit form $Ex_{k+1} = Ax_k + Bu_k$) with no extra difficulty. It should, however, be pointed out that the case of singular E is relatively difficult to handle [26, 28, 27] but the case of nonsingular E, with $E^{-1}A$ being undesirable to form for numerical reasons, is of considerable interest as well. We shall assume nonsingular E in the sequel.

Now, associated with this general class of Riccati equations is the matrix pencil

$$\begin{pmatrix} \hat{A} & 0 \\ -Q+SR^{-1}S^T & E^T \end{pmatrix} - \lambda \begin{pmatrix} E & BR^{-1}B^T \\ 0 & \hat{A}^T \end{pmatrix}.$$

This pencil can be written down directly from the Euler-Lagrange equations of the associated linear-quadratic optimal control problem. The solution of the Riccati equation can then be determined by computing a basis for the deflating subspace (which now plays the role of an invariant subspace) for the generalized eigenvalue problem $Nx = \lambda Lx$ where L and N are defined in the obvious way. The procedure is very similar to before. This time, instead of an "ordered" real Schur form, we compute an ordered generalized real Schur form (using the algorithm in [214]). Specifically, we compute orthogonal matrices Q and Z such that

$$QLZ = \begin{pmatrix} L_{11} & L_{12} \\ 0 & L_{22} \end{pmatrix} \quad , \quad QNZ = \begin{pmatrix} N_{11} & N_{12} \\ 0 & N_{22} \end{pmatrix}$$

where QLZ is upper triangular, QNZ is quasi-upper triangular, and the generalized eigenvalues of the pencil $N_{11} - \lambda L_{11}$ are stable. In this case the columns of $\begin{pmatrix} Z_{11} \\ Z_{21} \end{pmatrix}$ span the stable (deflating) subspace and the Riccati solution is computed from

$$X = Z_{21} Z_{11}^{-1} .$$

If $\hat{A}$ is invertible then L is invertible and it can easily be checked that the matrix $M = L^{-1}N$ is symplectic and has half its eigenvalues inside the unit circle (stable) and half outside (unstable). This symplectic M plays the same role as the Hamiltonian M did previously.

Another significant extension of the generalized eigenvalue formulation follows by considering the extended pencil [213]

$$\begin{pmatrix} A & 0 & B \\ -Q & E^T & -S \\ S^T & 0 & R \end{pmatrix} - \lambda \begin{pmatrix} E & 0 & 0 \\ 0 & A^T & 0 \\ 0 & -B^T & 0 \end{pmatrix}.$$

This extended pencil again follows from the Euler-Lagrange equations except that the "u equation" is not solved explicitly for u and all equations are kept in implicit form. Notice now that even R^{-1} does not appear and this can be important in the so-called deadbeat control problem which can be formulated with $R = 0$ [74, 75]. Again, the Riccati solution can be constructed by computing a basis for the n-dimensional deflating subspace corresponding to the stable eigenvalues of the $(2n + m) \times (2n + m)$ pencil.

The continuous-time problem can also be generalized as above. Consider now the more general equation

$$\hat{A}^T XE + E^T X\hat{A} - E^T XBR^{-1}B^T E + Q - SR^{-1}S^T = 0$$

with $\hat{A}$ as defined above. This time the generalized eigenvalue problem takes the form

$$\begin{pmatrix} \hat{A} & -BR^{-1}B^T \\ -Q+SR^{-1}S^T & -\hat{A}^T \end{pmatrix} - \lambda \begin{pmatrix} E & 0 \\ 0 & E^T \end{pmatrix}.$$

The analog of the extended pencil in this case takes the form

$$\begin{pmatrix} A & 0 & B \\ -Q & -A^T & -S \\ S^T & B^T & R \end{pmatrix} - \lambda \begin{pmatrix} E & 0 & 0 \\ 0 & E^T & 0 \\ 0 & 0 & 0 \end{pmatrix}.$$

The above matrix pencils exhibit the same Hamiltonian and symplectic structure (respectively, λ, $-\lambda$ and λ, $1/\lambda$ symmetry in their spectrum) as do the corresponding Hamiltonian and symplectic matrices. Note that if $\hat{A}$ is singular in the discrete-time case, there will be some (stable) generalized eigenvalues at 0. The symplectic symmetry is nicely preserved by a corresponding number of (unstable) infinite generalized eigenvalues. A fairly complete theory of Hamiltonian and symplectic pencils has been developed in [143, 146].

The deflating subspace approach was an important advance in the numerical solution of Riccati equations because it affords the means to handle directly a broad range of problems (including nonsymmetric) including:

- cross-weighting matrices (correlated noise) (obviating the need to "block diagonalize" the performance index thereby introducing an extra source of roundoff error unnecessarily)
- singular R (particularly important in the discrete-time case; important even when R is nonsingular but nearly singular)
- singular A (crucial in the discrete-time case)
- $E \neq I$.

Unfortunately, it has proven difficult to develop good parallel versions of the ordered generalized eigenvalue approach to solving Riccati equations and other methods have been developed instead. Before turning to these, however, we shall briefly review some other significant developments in Riccati algorithms, especially those based on invariant or deflating subspaces.

7.3 Iterative Refinement, Residuals, and Other Advances

An important recent development in the use of methods based on invariant subspaces has been the use of iterative refinement. As we shall see, this idea is intimately involved with Newton's method for the solution of Riccati equations. Newton's method is an interesting and useful technique in its own right and a rather complete theory for this algorithm has now been developed including computable estimates of the region of convergence. We shall outline this theory here, referring the interested reader to [123] for additional details.

For simplicity, let us consider the "standard" Riccati equation

$$A^T X + XA - XBB^T X + Q = 0 .$$

Suppose we have an approximation S to the true Riccati solution X. Write X in the form

$$X = S + (X - S)$$

and substitute this in the Riccati equation. The result is

$$(A-BB^T S)^T X + X(A-BB^T S) = -SBB^T S - Q + (X-S)BB^T(X-S)$$

and if we assume that the quantity $(X - S)$ is "small" then we can neglect the second-order term $(X - S)BB^T(X - S)$ to get the following linear (Lyapunov) equation:

$$(A - BB^T S)^T S_1 + S_1(A - BB^T S) = -SBB^T S - Q .$$

What we want then are bounds on $\|X - S\|$ to guarantee that S_1 is a better approximation to X than S, i.e.,

$$\|X - S_1\| < \|X - S\|.$$

Newton's method is simply an extension of this idea. Specifically, apply the above idea repeatedly so that, starting from some initial guess $S_0 = S$, we solve the following for $k \geq 0$:

$$(A - BB^T S_k)^T S_{k+1} + S_{k+1}(A - BB^T S_k) = -S_k BB^T S_k - Q .$$

If there is an "E-matrix" in the underlying dynamical system, the above equation is simply replaced with

$$(A - BB^T S_k)^T S_{k+1} E + E^T S_{k+1}(A - BB^T S_k) = -S_k BB^T S_k - Q .$$

It can be shown [125, 194] that if $A - BB^T S$ is asymptotically stable, then for $k \geq 1$:

1. $A - BB^T S_k$ is asymptotically stable
2. $0 \leq X \leq S_{k+1} \leq S_k \leq \cdots \leq S_1$
3. $\lim_{k\to+\infty} S_k = X$.

It follows that $\|X - S_{k+1}\| \leq \|X - S_k\|$ for $k \geq 1$ but what seems to have been overlooked previously is the fact that the error on the first step can increase dramatically. The following example will illustrate. Consider the case $A = 0$ and $B = Q = I$ so that $X = I$. Suppose the initial guess is $S = \epsilon I$. Notice that $A - BB^T S$ is indeed asymptotically stable. Then $S_1 = \frac{1+\epsilon^2}{2\epsilon} I$, and $\|X - S_1\| \approx \frac{1}{2\epsilon}$. Clearly the Newton iteration must then be carried out for a large number of steps before this disastrous initial error is damped out. Thus it is not enough simply to require that $A - BB^T S$ be stable. The following theorem gives conditions which guarantee monotone convergence from the first step on.

Theorem. *If* $\|X-S\| < r := \frac{1}{3\|B\|^2\|\Omega^{-1}\|}$, *then* $\|X-S_1\| \leq \|X-S\|$ *(with equality only when* $S = X$*) where* $(A - BB^T S)^T S_1 + S_1(A - BB^T S) = -SBB^T S - Q$ *and*

$$\Omega(Z) := (A - BB^T X)^T Z + Z(A - BB^T X) .$$

A couple of remarks concerning the parameter r are now in order. First, it is easy to show that $\|X - S\| < r$ implies $A - BB^T S$ stable. Second, it can be established that r is optimal in the scalar case. It is unknown at this time whether r can be improved upon in general.

The principal component in the determination of r is, of course, the computation of $\|\Omega^{-1}\|$. The key result here is based on the work of Hewer and Kenney [103] and can be stated as follows:

(1) Solve the Lyapunov equation $(A - BB^T X)^T H + H(A - BB^T X) = -I$. Then $\|\Omega^{-1}\| = \|H\|$.

Unfortunately, the construction of the Lyapunov equation coefficient matrix requires the Riccati solution X so in some sense the result appears not to offer a useful result for computation. However, it turns out that an approximation to X can be used instead; i.e., one can solve instead the Lyapunov equation

$$(A - BB^T S)^T H + H(A - BB^T S) = -I$$

with S being an approximation to X. The following result from [123] can then be used.

Lemma. *Suppose* $\|X - S\| < \frac{1}{2\|B\|^2\|\Omega^{-1}\|}$. *Then* $A - BB^T S$ *is stable and*

$$\frac{\|\Omega^{-1}\|}{1 + 2\|B\|^2\|X - S\|\|\Omega^{-1}\|} \leq \|\Omega_S^{-1}\| \leq \frac{\|\Omega^{-1}\|}{1 - 2\|B\|^2\|X - S\|\|\Omega^{-1}\|}$$

where $\Omega_S(Z) := (A - BB^T S)^T Z + Z(A - BB^T S)$.

There are other interesting and useful results relating to the quantity $\|\Omega^{-1}\|$ and we list a few here for reference. See [103, 115] for details.

1. Letting A_X denote the closed-loop matrix $(A - BB^T X)$, then

   $$\frac{1}{\|\Omega^{-1}\|} = \frac{1}{\|H\|} = \mathrm{sep}\,(A_X,\ -A_X{}^T)\ ,$$

   where sep is the separation operator [218].
2. If

   $$D := \sup_{\|x_0\|=1} \|x\|_{L_2} = \sup_{\|x_0\|=1} \left[\int_0^{+\infty} \|x(t)\|^2 dt\right]^{1/2}$$

   where x solves $\dot{x} = A_X x$; $x(0) = x_0$, then $\|\Omega^{-1}\| = D^2$.

3. The Riccati condition number is
$\leq \frac{1}{\|X\|}\left[\|H\|\|Q\| + 2\|H\|^{1/2}\|H_2\|^{1/2}\|A\| + \|H_2\|\|B\|^2\right]$
where $H = \Omega^{-1}(-I),\ H_2 = \Omega^{-1}(-X^2)$.

Another important new result related to the above has been derived related to residuals. It is natural to compute residuals when solving Riccati equations; i.e., a solution is computed and then it is "plugged back in" to the Riccati equation to see how well the solution does in the sense of giving nearly zero upon substitution. However, we know from standard examples involving the solution of linear systems of equations that a small residual does not always guarantee small error if the equations are ill-conditioned. The same kind of result can now be proven for Riccati equations. To illustrate, let us again denote by S an approximate solution to the Riccati equation and let us define the RESIDUAL $\mathcal{R}$ by

$$\mathcal{R} := Q + A^T S + SA - SBB^T S .$$

We then have the following theorem [123].

Theorem. *If* $\|X - S\| < r = \frac{1}{3\|B\|^2\|\Omega^{-1}\|}$ *and* $4\|B\|^2\|\Omega^{-1}\|^2\|\mathcal{R}\| < 1$, *then*

$$\|X - S\| \leq \frac{2\|\Omega^{-1}\|\|\mathcal{R}\|}{1 + \sqrt{1 - 4\|B\|^2\|\Omega^{-1}\|^2\|\mathcal{R}\|}} \leq 2\|\Omega^{-1}\|\|\mathcal{R}\|.$$

In other words, what this theorem is saying is that if the residual is small and $\|\Omega^{-1}\|$ is small, then one may conclude that the computed solution is accurate. This is entirely analogous to the situation of the solution of systems of linear equations. In this case, $\|\Omega^{-1}\|$ plays the role of the familiar condition number with respect to inversion of the linear system. If the residual is small but $\|\Omega^{-1}\|$ is large (the latter if the Riccati equation is ill-conditioned), then one cannot necessarily conclude that the computed solution is accurate.

Finally, we mention a recent result related to the optimal scaling of Riccati equations. In a recent article [182], it was claimed that Schur-based methods are numerically unstable. A numerical example was provided wherein inaccurate solutions to an ostensibly well-conditioned Riccati equation were obtained. It turns out that this numerical inaccuracy is actually due mostly to the effects of poor scaling. In [122] it is shown that if the Riccati equation is scaled in the right way by a scalar ρ such that the resulting Riccati solution has 2-norm equal to 1, then the linear equation step at the end is optimally well-conditioned. If this procedure is employed for the example of [182], an accurate solution is indeed computed. Of course, this then raises the question of how the norm of the solution can be estimated accurately *a priori*.

7.4 The Matrix Sign Function

The Schur-based methods, for all their reliability and efficiency, turn out to be relatively difficult to parallelize and vectorize effectively. For that reason, other methods have been re-examined with an eye towards their implementation on various advanced computing architectures. Methods based on the matrix sign function turn out to be especially attractive and, when implemented on parallel computers or large vector "supercomputers," have proven to be particularly suitable for the solution of large-scale Riccati equations. We shall first provide some introductory material regarding the definition and simple properties of the matrix sign function before turning to recent generalizations and other computational advances.

7.4.1 Definition and Properties

Definition. Let $M \in \mathbb{R}^{q \times q}$ with no pure imaginary eigenvalues, i.e., $\Lambda(M) \cap \{\mathrm{Re} z = 0\} = \emptyset$. Let M have a Jordan decomposition $M = T(D+N)T^{-1}$ where $D = diag\{\lambda_1, \ldots, \lambda_q\}$ and N is nilpotent and commutes with D. Then the MATRIX SIGN of M is given by

$$Z = sgn(M) := T \begin{pmatrix} sgn(\mathrm{Re}\lambda_1) & & \bigcirc \\ & \ddots & \\ \bigcirc & & sgn(\mathrm{Re}\lambda_q) \end{pmatrix} T^{-1} .$$

Naturally, the above definition does not generally lend itself to reliable computation of the matrix sign. Before turning to more effective algorithms we will first state some elementary and easily determined properties of $Z = sgn(M)$:

1. Z is diagonalizable with eigenvalues ± 1 and $Z^2 = I$
2. $\begin{aligned} range(Z - I) &= \text{stable invariant subspace of } M \\ &= M\text{−invariant subspace corresponding to} \\ &\quad \{\lambda \in \Lambda(M) : \mathrm{Re}\lambda < 0\} \end{aligned}$
3. for all nonsingular S, $sgn(SMS^{-1}) = S\ sgn(M)S^{-1}$
4. for all scalars c, $sgn(cM) = sgn(c)sgn(M)$
5. $sgn(M^T) = (sgn(M))^T$.

7.4.2 The Classical Matrix Sign Function Iteration

Let $Z_0 = M \in \mathbb{R}^{q \times q}$ be an arbitrary matrix whose sign is desired. For $k \geq 0$, perform the iteration:

$$Z_{k+1} = \frac{1}{2c}\left(Z_k + c^2 Z_k^{-1}\right) \quad ; \quad c = |det\ Z_k|^{1/q}.$$

Then

$$\lim_{k\to+\infty} Z_k = Z = sgn(M) .$$

The constant c is chosen to enhance convergence of this iterative process; [14, 16, 56] may be consulted for details. If $c = 1$ the iteration is usually known as the "classical" sign function iteration and simply amounts to Newton's method for solving $Z^2 - I = 0$. Naturally, it can be shown that the iteration is ultimately quadratically convergent. In addition to Property 2 above, which provides the link to Riccati equations, what makes this iteration so attractive computationally is that it consists of simple matrix components: scaling, matrix addition, and the solution of linear systems of equations, all of which are easily parallelized. We shall examine a particular parallel implementation in the sequel but first we make precise the connection to Riccati equations.

7.4.3 Application to Solving Algebraic Riccati Equations

In this subsection, we shall consider an algebraic Riccati equation of the form

$$A^T X + XA - XGX + Q = 0 \qquad (*)$$

where the matrix $G := BR^{-1}B^T$. As usual, we define the Hamiltonian matrix

$$M = \begin{pmatrix} A & -G \\ -Q & -A^T \end{pmatrix} \in \mathrm{R}^{2n\times 2n}$$

and we shall further assume that (A, B) is stabilizable and $(Q^{1/2}, A)$ is detectable in which case M has no imaginary axis eigenvalues. We then have the following theorem, due to Roberts [191]. The proof of this theorem is quite instructive and will thus be outlined in detail.

Theorem. *Let*

$$Z = \begin{pmatrix} Z_{11} & Z_{12} \\ Z_{21} & Z_{22} \end{pmatrix} = sgn(M)$$

where $Z_{ij} \in \mathrm{R}^{n\times n}$. *Then* X_P, *the unique symmetric nonnegative definite stabilizing solution of* $(*)$, *is a solution of*

$$\begin{pmatrix} Z_{12} \\ Z_{22} + I \end{pmatrix} X_P = -\begin{pmatrix} Z_{11} + I \\ Z_{21} \end{pmatrix} .$$

Proof. Let

$$T = \begin{pmatrix} I & Y \\ 0 & I \end{pmatrix}\begin{pmatrix} I & 0 \\ -X & I \end{pmatrix} = \begin{pmatrix} I - YX & Y \\ -X & I \end{pmatrix}$$

where

$$A^TX + XA - XGX + Q = 0$$

$$(A - GX)Y + Y(A - GX)^T = -G\,.$$

The above is a standard construction. The matrix X is used to construct a block triangularization of M while the matrix Y is then used to block diagonalize the block triangular matrix. It is then trivial to compute an explicit inverse for T:

$$T^{-1} = \begin{pmatrix} I & 0 \\ X & I \end{pmatrix}\begin{pmatrix} I & -Y \\ 0 & I \end{pmatrix} = \begin{pmatrix} I & -Y \\ X & I - XY \end{pmatrix}.$$

We can now compute the complete block diagonalizing similarity:

$$T\begin{pmatrix} A & -G \\ -Q & -A^T \end{pmatrix}T^{-1} = TMT^{-1} = \begin{pmatrix} (A - GX) & 0 \\ 0 & -(A - GX)^T \end{pmatrix}.$$

Thus, using properties of the matrix sign function, we find that

$$\begin{aligned} sgn(M) &= T^{-1} sgn\begin{pmatrix} (A - GX) & 0 \\ 0 & -(A - GX)^T \end{pmatrix}T \\ &= T^{-1}\begin{pmatrix} -I & 0 \\ 0 & I \end{pmatrix}T \\ &= \begin{pmatrix} -I + 2YX & -2Y \\ -2X + 2XYX & I - 2XY \end{pmatrix}. \end{aligned}$$

From

$$sgn(M) + I = \begin{pmatrix} 2YX & -2Y \\ -2X + 2XYX & 2I - 2XY \end{pmatrix}$$

or

$$\begin{pmatrix} Z_{11} + I & Z_{12} \\ Z_{21} & Z_{22} + I \end{pmatrix} = \left(\begin{bmatrix} 2Y \\ 2(XY - I) \end{bmatrix}X\,,\ -\begin{bmatrix} 2Y \\ 2(XY - I) \end{bmatrix}\right)$$

we see by comparing both sides of the above equation that X must satisfy

$$\begin{pmatrix} Z_{12} \\ Z_{22} + I \end{pmatrix}X = -\begin{pmatrix} Z_{11} + I \\ Z_{21} \end{pmatrix}.$$

This completes the proof. □

It is easily seen in the above proof that by working instead with $sgn(M) - I$ we get the dual result

$$X\,(Z_{11} - I\,,\ Z_{12}) = (Z_{21}\,,\ Z_{22} - I)\,.$$

Theorem. *X_N, the unique symmetric nonpositive definite anti-stabilizing solution of $(*)$, is a solution of*

$$\begin{pmatrix} Z_{12} \\ Z_{22} - I \end{pmatrix} X_N = - \begin{pmatrix} Z_{11} - I \\ Z_{21} \end{pmatrix} .$$

Proof. Similar to above except now

$$sgn(A - GX) = +I$$

and

$$sgn(M) - I = \left(\begin{bmatrix} * \\ * \end{bmatrix} X, - \begin{bmatrix} * \\ * \end{bmatrix} \right) . \qquad \square$$

The careful reader will have noticed that the above proofs assume that the (anti-) stabilizing solution exists. We shall now state a theorem which shows that existence of Riccati solutions is actually equivalent to the existence of the $2n \times n$ "matrix sign equations" and that this is true even for indefinite G, Q.

Theorem. *Consider the Riccati equation $(*)$ and associated Hamiltonian matrix M but assume only that G and Q are symmetric. Then:*

1. *there exists a unique symmetric stabilizing solution to $(*)$ if and only if M has no pure imaginary eigenvalues and*

$$\operatorname{rank} \begin{pmatrix} Z_{12} \\ Z_{22} + I \end{pmatrix} = n$$

2. *there exists a unique symmetric anti-stabilizing solution to $(*)$ if and only if M has no pure imaginary eigenvalues and*

$$\operatorname{rank} \begin{pmatrix} Z_{12} \\ Z_{22} - I \end{pmatrix} = n .$$

The proof of this theorem can be found in [121] along with an even more general version for the nonsymmetric Riccati equation

$$A_1 X + X A_2 - XGX + Q = 0 \;\; ; \;\; G, Q \text{ arbitrary} .$$

Sufficient conditions for the eigenvalue hypothesis are, of course, the usual nonnegative definiteness of G and Q together with stabilizability and detectability. Necessary and sufficient conditions for M not to have pure imaginary eigenvalues have been given in [130], namely that there be no pure imaginary uncontrollable or unobservable eigenvalues.

Still further interesting results are available for the dual Riccati equation pair

$$A^T X + XA - XBB^T X + C^T C = 0$$
$$A\Phi + \Phi A^T - \Phi C^T C \Phi + BB^T = 0 .$$

Assume, for example, that (A, B) is controllable and (C, A) is observable. Let

$$Z = \begin{pmatrix} Z_{11} & Z_{12} \\ Z_{21} & Z_{22} \end{pmatrix} = sgn \begin{pmatrix} A & -BB^T \\ -C^TC & -A^T \end{pmatrix} = sgn(M) .$$

Then X_P, X_N solve

$$\begin{pmatrix} Z_{12} \\ Z_{22}+I \end{pmatrix} X_P = - \begin{pmatrix} Z_{11}+I \\ Z_{21} \end{pmatrix} \quad ; \quad \begin{pmatrix} Z_{12} \\ Z_{22}-I \end{pmatrix} X_N = - \begin{pmatrix} Z_{11}-I \\ Z_{21} \end{pmatrix}$$

and $\Phi_P = -X_N^{-1}$, $\Phi_N = -X_P^{-1}$ solve

$$\begin{pmatrix} Z_{21} \\ Z_{11}-I \end{pmatrix} \Phi_P = \begin{pmatrix} Z_{22}-I \\ Z_{12} \end{pmatrix} \quad ; \quad \begin{pmatrix} Z_{21} \\ Z_{11}+I \end{pmatrix} \Phi_N = \begin{pmatrix} Z_{22}+I \\ Z_{12} \end{pmatrix} .$$

The latter follow by using Property 5 of the matrix sign function and by observing further that if M is Hamiltonian then $Z = sgn(M)$ is also Hamiltonian. Thus, four Riccati solutions can all be found efficiently from one matrix sign function calculation. Further details and references to applications such as closed-loop LQG balancing and H^∞ control can be found in [121]. Another efficient algorithmic approach to dual Riccati equations can be found in [227].

7.4.4 Symmetric Version of Matrix Sign Function Iteration

When working with Hamiltonian matrices, a more efficient version of the classical matrix sign function iteration can be derived by working only with symmetric matrices [30, 56] in the following way. Starting from the symmetric matrix

$$Z_0 = JM = \begin{pmatrix} 0 & I \\ -I & 0 \end{pmatrix} \begin{pmatrix} A & -G \\ -Q & -A^T \end{pmatrix} = \begin{pmatrix} -Q & -A^T \\ -A & G \end{pmatrix} ,$$

perform the following iteration for $k \geq 0$:

$$Z_{k+1} = \frac{1}{2c} \left(Z_k + c^2 J Z_k^{-1} J \right) ,$$

where, as before, the parameter c can be chosen to enhance convergence. It is easy to check that each Z_k is symmetric (indefinite) and $\lim_{k \to +\infty} Z_k = J sgn(M)$. That this scheme works is immediate from the following observation:

$$\begin{aligned} J\tilde{Z}_{k+1} &= \frac{1}{2c} \left(J\tilde{Z}_k + c^2 J \left(J\tilde{Z}_k \right)^{-1} J \right) \\ &= \frac{1}{2c} \left(J\tilde{Z}_k + c^2 J \tilde{Z}_k^{-1} \right) . \end{aligned}$$

Since it is necessary to compute only the upper triangle of Z_{k+1}, the iteration can be performed much more efficiently. Algorithms from LINPACK [69] for symmetric indefinite systems can be employed.

For use in the sequel, we shall denote by Y the matrix $Jsgn(M)$ whence

$$sgn(M) = \begin{pmatrix} Z_{11} & Z_{12} \\ Z_{21} & Z_{22} \end{pmatrix} = -JY = \begin{pmatrix} -Y_{21} & -Y_{22} \\ Y_{11} & Y_{12} \end{pmatrix} .$$

Thus the standard matrix sign equations

$$\begin{pmatrix} Z_{12} \\ Z_{22} + I \end{pmatrix} X = - \begin{pmatrix} Z_{11} + I \\ Z_{21} \end{pmatrix}$$

are simply equivalent to the matrix sign equations

$$\begin{pmatrix} Y_{22} \\ Y_{12} + I \end{pmatrix} X = \begin{pmatrix} I - Y_{21} \\ -Y_{11} \end{pmatrix}$$

in the symmetric version of the algorithm.

7.4.5 Generalization of the Matrix Sign Function

We now discuss an extremely important generalization of the matrix sign function to the case of generalized eigenvalue problems. Details may be consulted in [89].

Clearly, the standard matrix sign function iteration

$$Z_0 = M$$

$$Z_{k+1} = \frac{1}{2c}\left(Z_k + c^2 Z_k^{-1}\right) \; ; \; c = |\det \; Z_k|^{1/q}$$

is concerned with the eigenvalue problem $M - \lambda I$. The iteration can be generalized for the generalized eigenvalue problem $M - \lambda E_1 E_2$ in the following way:

$$Z_0 = M$$

$$Z_{k+1} = \frac{1}{2c}\left(Z_k + c^2 E_1 E_2 Z_k^{-1} E_1 E_2\right) \; ; \; c = \left(\frac{|\det \; Z_k|}{|\det \; E_1 E_2|}\right)^{1/q} .$$

In this case we have

$$\lim_{k \to +\infty} Z_k = E_1 \left(sgn \; E_1^{-1} M E_2^{-1}\right) \; E_2$$

where, of course, we are assuming that E_1 and E_2 are nonsingular.

The symmetric version of the above is the following:

$$Z_0 = JM$$

$$Z_{k+1} = \frac{1}{2c}\left(Z_k + c^2 \left(J E_1 E_2\right) Z_k^{-1} \left(J E_1 E_2\right)\right)$$

$$\rightarrow \; J E_1 \; sgn\left(E_1^{-1} M E_2^{-1}\right) E_2 \; .$$

To see how these matrix sign iterations can be useful for generalized Riccati equations consider the continuous-time equation

$$A^T X E + E^T X A - E^T X B R^{-1} B^T X E + Q = 0 \; .$$

Let

$$M = \begin{pmatrix} A & -BR^{-1}B^T \\ -Q & -A^T \end{pmatrix}$$

$$E_1 = \begin{pmatrix} I & 0 \\ 0 & E^T \end{pmatrix}, \quad E_2 = \begin{pmatrix} E & 0 \\ 0 & I \end{pmatrix}.$$

The feedback gain K for the associated linear optimal control problem takes the form $K = -R^{-1}B^T XE$ and it is then easily shown that $\bar{X} := XE$ can be found from the matrix sign equation

$$\begin{pmatrix} Z_{12} \\ Z_{22} + E^T \end{pmatrix} \bar{X} = -\begin{pmatrix} Z_{11} + E \\ Z_{21} \end{pmatrix}$$

or, in the symmetric case,

$$\begin{pmatrix} Y_{22} \\ Y_{12} + E^T \end{pmatrix} \bar{X} = \begin{pmatrix} E - Y_{21} \\ -Y_{11} \end{pmatrix}.$$

The discrete-time Riccati equation

$$A^T XA - E^T XE - A^T XB(R + B^T XB)^{-1}B^T XA + Q = 0$$

can be handled similarly. This time

$$M = \begin{pmatrix} A & 0 \\ -Q & E^T \end{pmatrix}$$

$$E_1 = \begin{pmatrix} I & BR^{-1}B^T \\ 0 & A^T \end{pmatrix}, \quad E_2 = \begin{pmatrix} E & 0 \\ 0 & I \end{pmatrix}.$$

However, in this case it is necessary that the symplectic pencil $M - \lambda E_1 E_2$ be transformed into a pencil with Hamiltonian structure since the matrix sign function discriminates spectra by sign rather than magnitude. The following theorem from [89] is crucial.

Theorem *If the pencil $N - \lambda L$ is symplectic (respectively, Hamiltonian) then the pencil $(N \pm L) - \lambda(N \mp L)$ is Hamiltonian (respectively, symplectic).*

This theorem is clearly a generalization of the classical Cayley-transform-type results (assuming the indicated inverses exist):

$$S \text{ symplectic} \Rightarrow (S+I)^{-1}(S-I) \text{ Hamiltonian}$$
$$H \text{ Hamiltonian} \Rightarrow (H-I)^{-1}(H+I) \text{ symplectic.}$$

Applying the above results, the discrete-time Riccati equation can then be solved by applying the generalized matrix sign function iteration to the pencil

$$\begin{pmatrix} A - E & -BR^{-1}B^T \\ -Q & E^T - A^T \end{pmatrix} - \lambda \begin{pmatrix} A + E & BR^{-1}B^T \\ -Q & E^T + A^T \end{pmatrix}.$$

7.4.6 Parallel Implementation

We shall outline some implementations of the standard matrix sign function iteration for solving algebraic Riccati equations on a 32-processor Intel iPSC/d5 hypercube. Further details can be consulted in [90, 92, 91, 88, 147], as well as [59] which also includes a systolic implementation for the same algorithm.

The basic symmetric iteration which must be implemented takes the form

$$Z_{k+1} = Z_k - F^T Z_k^{-1} F \ .$$

For simplicity, we have removed all the constants here. The equation is written for the symmetric version and with a minus sign rather than a plus sign as before because of the fact that $(JE_1E_2)^T = -(JE_1E_2)$.

There is a good deal of parallelism available in implementing this iteration. The most obvious is data parallelism, and its exploitation is critical in large-scale problems. We used so-called "column-wrapped" distribution of the matrix columns among processors. The symmetric matrices were stored in packed form. This straightforward distribution of the matrix columns results in quite reasonable load balancing among the processors. There is also a coarse-grained functionality whereby a matrix factorization can be overlapped or pipelined with a subsequent matrix multiplication or solve. Finally, additional parallelism can be attained by overlapping the iterations but we did not attempt to use this technique.

The heart of the computation is the symmetric indefinite matrix factorization: $Z = UDU^T$, where D and U are easily invertible. Specifically, we implemented a parallel version of the stable DIAGONAL PIVOTING method of Bunch *et al.* (see Chapter 5 of the *LINPACK Users' Guide* [69] and the references therein), in which case D is block diagonal with 1×1 or 2×2 blocks and U is a product of unit upper triangular and permutation matrices. Thus $F^T Z^{-1} F = (U^{-1}F)^T D^{-1} (U^{-1}F)$ whereupon the next iterate of Z is then obtained easily, at least in theory. Unfortunately, to ensure numerical stability and to preserve symmetry, the diagonal pivoting algorithm requires both row and column operations (pivoting) and hence a lot of interprocessor communication. It turns out that row pivoting can be handled easily but not column pivoting since the latter tends to involve communication in a very irregular pattern.

We tried two other approaches to avoid this difficulty. The first was a TRIDIAGONALIZATION algorithm. This is similar to the diagonal pivoting algorithm but Z is first reduced to tridiagonal form by orthogonal similarity. This method ends up requiring essentially the same amount of communication as diagonal pivoting but approximately four times as many floating-point operations. This was borne out by noticeably poorer performance in timing tests. The second alternative attempted was an UPDATE algorithm suggested by Van Dooren. Here the idea is to use a factorization of Z_k of the form $Z_k = U_k \Sigma_k U_k^T$ to generate directly new factors U_{k+1} and Σ_{k+1} such that $U_{k+1} \Sigma_{k+1} U_{k+1}^T = Z_{k+1}$. Unfortunately, this method again requires column pivoting and also more computation. Various *ad hoc* preconditioning schemes were tried which did help but the existence of such schemes is not guaranteed without pivoting which is what we were trying to avoid in the first place.

Further details on all three algorithms, including complexity analyses and timing results, are available in [92, 87, 91, 88]. The performance of the three algorithms is summarized in Table 1.

Table 1: Comparison of parallel sign function implementations.

n	Diagonal Pivoting (sec)	Tridiagonalization (sec)	Update (sec)	Schur (1 Node) (sec)
20	12.2	31.9	39.9	77.6
40	36.3	117.0	163.0	578.0
60	94.6	266.0	391.0	2000*
80	166.0	518.0	758.0	4600*
100	323.0	977.0	1465.0	9100*
				*cubically extrapolated

The test matrices used were random and all problems were solved using 32 nodes. Note that communication overhead is clearly significant for small n but that computation begins to dominate for larger n. While these are not fast times in absolute terms, it must be remembered that the iPSC/d5 is a very slow and relatively primitive machine. The identical codes run on a later generation hypercube with more nodes will produce some very impressive times.

7.5 A General Theory of Matrix Sign Function Iterations

7.5.1 Motivation

Newton's method applied to the equation $Z^2 - I = 0$ provides generally excellent performance for computing the sign of a matrix. However, it is of interest to derive higher-order formulas with corresponding higher rates of convergence—naturally at the expense of more arithmetic. An obvious formula to try is Halley's method which requires second derivative information (easily available in this case) but offers cubic convergence. For reference, we review the derivation of this important method. First recall a derivation of Newton's method (in the scalar case):

$$\begin{array}{rrcl} & f(x_{n+1}) & = & f(x_n) + (x_{n+1} - x_n) f'(x_n) + \cdots \\ \Rightarrow & 0 & \approx & f(x_n) + (x_{n+1} - x_n) f'(x_n) \\ \Rightarrow & x_{n+1} & = & x_n - \frac{f(x_n)}{f'(x_n)} \quad (\text{Note} : x_{n+1} - x_n \approx -\frac{f(x_n)}{f'(x_n)}). \end{array}$$

Halley's method simply keeps the second-order term in the Taylor expansion and further approximates one of the $x_{n+1} - x_n$ terms with the approximation noted above. Thus we find

$$\begin{array}{rcl} f(x_{n+1}) & = & f(x_n) + (x_{n+1} - x_n)f'(x_n) + \frac{1}{2}(x_{n+1} - x_n)^2 f''(x_n) + \cdots \\ \Rightarrow \qquad 0 & \approx & f(x_n) + (x_{n+1} - x_n)\left[f'(x_n) + \frac{1}{2}(x_{n+1} - x_n)f''(x_n)\right] \\ \Rightarrow \qquad x_{n+1} & = & x_n - \frac{f(x_n)}{f'(x_n) - \frac{1}{2}\frac{f(x_n)f''(x_n)}{f'(x_n)}} . \end{array}$$

We see that Halley's method is essentially Newton's method with a correction factor based on the second derivative. It is this correction which yields cubic convergence. Let us examine an example.

Consider the equation $f(x) = x^2 - 1 = 0$ with $f'(x) = 2x$ and $f''(x) = 2$. Substituting above we get Halley's iteration:

$$\begin{array}{rcl} x_{n+1} & = & x_n - \frac{x_n^2 - 1}{2x_n - \frac{1}{2}\frac{(x_n^2-1)\cdot 2}{2x_n}} \\ & = & \frac{x_n^3 + 3x_n}{3x_n^2 + 1} . \end{array}$$

It then follows easily that the matrix version of Halley's iteration takes the form

$$Z_{k+1} = Z_k\left(3I + Z_k^2\right)\left(I + 3Z_k^2\right)^{-1} .$$

However, this form of the iteration requires 2 matrix "multiplies" and 1 "divide" at each step. The formula can be rewritten in the equivalent form

$$Z_{k+1} = \frac{1}{3}Z_k + \frac{8}{3}Z_k\left(I + 3Z_k^2\right)^{-1}$$

which requires only 1 multiply and 1 divide. Compare this with the 1 divide per step required by Newton's method.

In some situations it may be advantageous to have an iteration with no "divides," i.e., an inverse-free matrix sign function iteration. One approach is to embed an iterative method for computing a matrix inverse within the overall matrix sign function iteration. One such iteration is Schulz's iteration [195] which takes the form

$$U_{k+1} = U_k + (I - U_k X)U_k .$$

It is easy to prove that

$$\lim_{k\to+\infty} U_k = X^{-1} .$$

Thus, in Newton's matrix sign iteration (taking $c = 1$ for convenience), replace Z_k^{-1} by $U_0 + (I - U_0 Z_k)U_0$, i.e., a single step of the Schulz iteration. The only question is then what to take as U_0. The best choice turns out to be to take $U_0 = Z_k$ since $\lim Z_k = sgn(M) = sgn(M)^{-1}$. Thus we have the Newton-Schulz iteration:

$$\begin{aligned} Z_{k+1} &= \frac{3}{2}Z_k - \frac{1}{2}Z_k^3 \\ &= Z_k + \frac{1}{2}Z_k\left(I - Z_k^2\right) . \end{aligned}$$

Similarly, we can develop a Halley-Schulz iteration by replacing $\left(I+3Z_k^2\right)^{-1}$ by the Schulz approximation $U_0 + \left(I - U_0\left(I+3Z_k^2\right)\right)U_0$. This time we choose $U_0 = \frac{1}{4}I$ since, in the limit,

$$sgn\left(I+3M^2\right)^{-1} \rightarrow (4I)^{-1}.$$

The overall iteration becomes:

$$Z_{k+1} = \frac{1}{16}Z_k\left(7I-3Z_k^2\right)\left(3I+Z_k^2\right).$$

The above iterations, both inverse-free and with inverses, are obviously rich in matrix multiplications and hence are ideally suited for implementation on Crays and other supercomputers as well as on certain parallel machines. Unfortunately, the inverse-free algorithms are not globally convergent so that it is necessary to adopt some sort of hybrid scheme. That is, a method such as Newton or Halley is employed until it is determined that the region of convergence of an inverse-free method has been reached whereupon a switch to the latter can be made. This is by analogy with a similar situation in iterative methods for the polar factor of a matrix; see [105, 106] for details. In fact, there are other interesting similarities between polar factor iteration and the matrix sign function iteration and these will be mentioned below.

7.5.2 A General Theory

Recently, a very general theory of matrix sign function iterations has been developed; see [120] for details. The basic procedure is the following:

Step 1. Let $r_{\nu\delta}(x)$ be the $(\nu,\delta)^{th}$ Padé approximation to the hypergeometric function $f(x) = (1-x)^{-\frac{1}{2}}$:

$$r_{\nu\delta}(x) = \frac{p_\nu(x)}{q_\delta(x)} \approx (1-x)^{-\frac{1}{2}}.$$

Here p_ν and q_δ are polynomials of degrees ν and δ, respectively.

Step 2. Beginning at any number x_0 whose sign is sought, define the recursion

$$x_{n+1} = x_n r_{\nu\delta}\left(1-x_n^2\right).$$

The motivation for this at-first-sight strange looking procedure becomes clearer upon noting that

$$sgn(z) = \frac{z}{|z|} = \frac{z}{\sqrt{1-(1-z^2)}}.$$

What makes this family of formulas so powerful is the fact that two very powerful theorems can be proved regarding convergence. Both theorems will be stated in the scalar case but it must be emphasized that the results are easily shown to hold in the matrix case.

Theorem. *The above recursion is accurate to order $\nu + \delta + 1$ and converges to $sgn(x_0)$.*

Theorem. *Convergence is global if $\nu = \delta$ or $\nu = \delta - 1$.*

When convergence is not global, local convergence can be studied by computing Julia sets, i.e., by computing the boundary of the basin of attraction of $\pm$ 1 for

$$h(x) = x r_{\nu\delta}\left(1 - x^2\right) .$$

Specific iteration formulas for $0 \leq \delta \leq 3$ and $0 \leq \nu \leq 3$ are given in Table 2. Notice that the formula for Newton's method is the "inverse" of what we saw previously. This makes no difference for the determination of sign and, in fact, any of the formulas in the table may be inverted. Our previously derived Halley's method appears in the $\nu = 1$, $\delta = 1$ entry. (See Table 2)

In the matrix case, the iteration takes the form (with Z_0 given):

$$Z_{k+1} = Z_k p_\nu\left(I - Z_k^2\right)\left[q_\delta\left(I - Z_k^2\right)\right]^{-1} .$$

The appropriate matrix-valued iteration formulas can then be written directly from Table 2.

7.5.3 A Parallel Implementation

Clearly the formulas derived above are rich in matrix multiplications, making them ideal for vector supercomputers as noted previously. However, it is equally clear that evaluation of higher-order formulas requires considerably more arithmetic. However, by making use of the well-developed theory associated with Padé approximations of hypergeometric functions, this increased computational burden can be greatly minimized by parallelization. To illustrate, let us consider the "fraction part" of the matrix iteration formula above. The zeros of the denominator polynomial q_δ are known explicitly and analytically and thus the rational function can be decomposed into a sum by the method of partial fractions. Now, partial fractions do not normally lend themselves to reliable numerical computing in finite arithmetic because of the difficulty with ill-conditioned zeros. However, in this case those difficulties are not present. Let us illustrate these remarks below for the globally convergent formulas with $\nu = \delta - 1$ and, for simplicity, let us write X_k for the

Table 2: Iteration formulas from Padé approximation.

	$\nu = 0$	$\nu = 1$	$\nu = 2$	$\nu = 3$
$\delta = 0$	x	$\frac{x}{2}(3 - x^2)$	$\frac{x}{8}(15 - 10x^2 + 3x^4)$	$\frac{x}{16}(35 - 35x^2 + 21x^4 - 5x^6)$
$\delta = 1$	$\frac{2x}{1 + x^2}$	$\frac{x(3 + x^2)}{1 + 3x^2}$	$\frac{x}{4}\frac{(15 + 10x^2 - x^4)}{(1 + 5x^2)}$	$\frac{x}{8}\frac{(35 + 35x^2 - 7x^4 + x^6)}{(1 + 7x^2)}$
$\delta = 2$	$\frac{8x}{3 + 6x^2 - x^4}$	$\frac{4x(1 + x^2)}{1 + 6x^2 + x^4}$	$\frac{x(5 + 10x^2 + x^4)}{1 + 10x^2 + 5x^4}$	$\frac{x}{2}\frac{(35 + 105x^2 + 21x^4 - x^6)}{(3 + 42x^2 + 35x^4)}$
$\delta = 3$	$\frac{16x}{5 + 15x^2 - 5x^4 + x^6}$	$\frac{8x(3 + 5x^2)}{5 + 45x^2 + 15x^4 - x^6}$	$\frac{2x(3 + 10x^2 + 3x^4)}{1 + 15x^2 + 15x^4 + x^6}$	$\frac{x(7 + 35x^2 + 21x^4 + x^6)}{1 + 21x^2 + 35x^4 + 7x^6}$

matrix $I - Z_k^2$. We can then write:

$$p_{\delta-1}\left(X_k^2\right) q_\delta^{-1}\left(X_k^2\right) = \sum_{i=1}^{\delta} \frac{1}{2j\alpha_i x_i}\left[(X_k - j\alpha_i I)^{-1} - (X_k + j\alpha_i I)^{-1}\right]$$
$$= \sum_{i=1}^{\delta} \frac{1}{x_i}\left(X_k^2 + \alpha_i^2 I\right)^{-1}$$

where $\alpha_i^2 = \frac{1}{x_i} - 1$ and the x_i's are zeros of (shifted) Chebyshev polynomials and hence are in the interval $(0, 1)$. Details can be consulted in [174].

Table 3 lists execution times based on computations performed on a Cray Y-MP with 4 processors at NASA Ames. The parameter n is the order of the Riccati equation being solved while p is the size of the matrix sign problem ($= 2n$).

Table 3: Cray execution times.

p	n	Newton (sec)	r_{34} formula ($\delta = 4$) (sec)
98	49	1.1	0.3
298	149	3.5	1.3
398	199	8.0	3.1

Finally, we note that further research has been done on scaling for this family of iterations. Effective scaling strategies are available which result in significant enhancement of the rate of convergence. Further details can be consulted in [117]. Scaling can be particularly important in ill-conditioned problems and a thorough study of the sensitivity or condition of the matrix sign function has been done in [119]. It was noted above that iterations for the matrix sign are very similar to iterations for the orthogonal factor of the polar decomposition of a matrix (for which, for example, Newton's iteration looks like $Z_{k+1} = \frac{1}{2}\left(Z_k + Z_k^{-T}\right)$). The seemingly innocuous difference of using Z^{-T} rather than Z^{-1} creates enormous differences. Where the perturbation theory for the polar factor is relatively straightforward and condition estimation fairly inexpensive, the perturbation theory for the matrix sign is difficult and condition estimation relatively expensive. This is perhaps not unexpected in light of the fact that the polar factor is intimately connected with singular values which are numerically rather "pleasant" as compared to eigenvalues which arise in the sign function and are known to be numerically difficult to deal with at times.

7.6 Concluding Remarks

We have given an overview of numerical methods for the solution of Riccati equations based on the intimate connection between Riccati equations and invariant subspaces of certain associated matrices. We have concentrated on direct methods based on Schur vectors and iterative methods based on the matrix sign function. We have made no attempt to discuss other methods of solving Riccati equations.

An interesting phenomenon can be observed over the past ten to fifteen years. Prior to about 1978, solving an algebraic Riccati equation (ARE) was something that was most definitely to be avoided, and with good reason. Existing techniques were often quite unreliable. In some standard textbooks, a method suggested for the solution of ARE's was to solve an associated spectral factorization problem. We now know that most calculations with polynomials and polynomial matrices are replete with numerical pitfalls and, in fact, today it is generally conceded that the best way to solve a spectral factorization problem is via state space methods which involve ultimately the solution of an associated ARE. Solving an ARE has today almost come to be taken for granted.

Numerical Riccati equation methods based on invariant subspaces have proven to be quite dependable for matrix sizes even of the order of many hundred. A great deal is now known about these methods and their performance in the finite arithmetic environment of digital computing machines. However, many important questions remain unanswered and are the subject of continuing research.

What are some of the numerical research problems today related to Riccati equations? We list here a few.

- structure-preserving algorithms: Structure-preserving algorithms may be particularly useful and necessary for solving large-scale problems, especially those with large sparse coefficient matrices. As noted long ago (certainly in 1978 [139, 140], and undoubtedly long before that), the connection between the algebraic Riccati equation and Hamiltonian and symplectic matrices leads naturally to the question of how far one can go using symplectic similarities which preserve this structure. Many important advances have been made in, for example [54, 173] and extensive recent work by Bunse-Gerstner, Byers, and Mehrmann, much of which is nicely summarized in [47, 48].
- nonsymmetric and other "nonstandard" Riccati equations: Nonsymmetric Riccati equations arise in a variety of applications and the theory here is nowhere near as well developed as for the symmetric case. Even for the symmetric case, we are beginning to see a flurry of activity on the indefinite problems which arise in the H^∞ theory. Such problems have previously been studied in the game-theoretic context and this work is likely to play a pivotal role.
- Riccati differential equations: These are likely to come under increasing study as the availability of cheap and reliable computing allows better modeling with time-varying parameters and a concomitant interest in finite-time and transient phenomena. One promising family of algorithms, particularly for stiff Riccati differential equations, has been proposed in [61] and is under continuing study.

- software: This is almost always, of course, the ultimate vehicle of reliable technology transfer. Early attempts at a large comprehensive Fortran-based Riccati package (RICPACK; see [11]) will undoubtedly be superseded by much more easily constructed packages based on software such as MATLAB and its clones and imitators.

Acknowledgements. Much of the research surveyed in this paper represents collaborative efforts with past and present students including Thrasyvoulos Pappas, Kwae Hi Lee, Bo Arnold, Doug Bender, Judy Gardiner, Matt Wette, Chiu Choi, Pascal Gahinet, Pradeep Pandey, Thorkell Gudmundsson, and Ali Ghavimi. It has also been a special privilege to have the opportunity to collaborate with my colleague, Charles Kenney, of the Scientific Computation Laboratory in the Electrical and Computer Engineering Department at UCSB.

We are grateful for research support from the National Science Foundation (and AFOSR) under Grant No. ECS87-18897, the National Science Foundation under Grant No. DMS88-00817, and the Air Force Office of Scientific Research under Contract No. AFOSR-89-0167.

References

1. C. Ahlbrandt. Continued fraction representations of maximal and minimal solutions of a discrete matrix Riccati equation. preprint. (1990).
2. F.A. Aliev, B.A. Bordyug, and V.B. Larin. The spectral method of solving matrix algebraic Riccati equations. *Soviet Math. Dokl.*, 35:121–125, 1987.
3. F.A. Aliev, B.A. Bordyug, and V.B. Larin. Orthogonal projections and the solution of a matrix algebraic Riccati equation. *Soviet Math. Dokl.*, 38:531–534, 1989.
4. B.D.O. Anderson. Solution of quadratic matrix equations. *Electron. Letters*, 2:371–372, 1966.
5. B.D.O. Anderson. Second-order convergent algorithms for the steady-state Riccati equation. *Int. J. Control*, 28:295–306, 1978.
6. B.D.O. Anderson and J.B. Moore. *Linear Optimal Control.* Prentice Hall, Englewood Cliffs, NJ, 1971.
7. B.D.O. Anderson and J.B. Moore. *Optimal Filtering.* Prentice Hall, Englewood Cliffs, NJ, 1979.
8. L.R. Anderson, D.W. Brewer, and A.R. Baykam. Numerical solution of the symmetric Riccati equation through iteration. In *Proc. 1982 Amer. Control Conf.*, pages 1010–1015, Arlington, VA, June 1982.
9. S. Arimoto. Optimal feedback control minimizing the effects of noise distrubances. *Trans. SICE*, 2(1966), pp. 1–7, (in Japanese).
10. W.F. Arnold, III. *On the Numerical Solution of Algebraic Matrix Riccati Equations.* PhD thesis, Univ. of Southern California, Dept. of Elec. Eng.—Systems, Los Angeles, CA, December 1983.
11. W.F. Arnold, III and A.J. Laub. A software package for the solution of generalized algebraic Riccati equations. In *Proc. 22nd IEEE Conf. Decis. Control*, pages 415–417, San Antonio, TX, December 1983.
12. W.F. Arnold, III and A.J. Laub. Generalized eigenproblem algorithms and software for algebraic Riccati equations. *Proc. of IEEE*, 72:1746–1754, 1984.
13. M. Athans, Editor. Special issue on the linear-quadratic-Gaussian estimation and control problem. *IEEE Trans. Auto. Control*, AC-16, December 1971.

14. L. Balzer. Accelerated convergence of the matrix sign function. *Int. J. Control*, 32:1057–1078, 1980.
15. Y. Bar-Ness and G. Langholz. The solution of the matrix equation $XC - BX = D$ as an eigenvalue problem. *Int. J. Syst. Sci.*, 8:385–392, 1977.
16. A.Y. Barraud. Investigation autour de la fonction signe d'une matrice—application à l'équation de Riccati. *R.A.I.R.O. Automatique/Systems Analysis and Control*, 13:335–368, 1979.
17. A.Y. Barraud. Produit étoile et fonction signe de matrice—application à l'équation de Riccati dans le cas discret. *R.A.I.R.O. Automatique/Systems Analysis and Control*, 14:55–85, 1980.
18. R.H. Bartels and G.W. Stewart. Solution of the matrix equation $AX + XB = C$: Algorithm 432. *Commun. ACM*, 15:820–826, 1972.
19. R.W. Bass. Machine solution of high-order matrix Riccati equations. Technical Report 4538, Douglas Aircraft, Missile and Space Systems Division, 1967.
20. A.R. Baykam. Comparison of algebraic Riccati equation solvers. Master's thesis, Virginia Polytechnic and State University, Dept. of Aerospace and Ocean Eng., Blacksburg, VA, 1981.
21. A.N. Beavers and E.D. Denman. A computational method for eigenvalues and eigenvectors of a matrix with real eigenvalues. *Numer. Math.*, 21:389–396, 1973.
22. A.N. Beavers and E.D. Denman. Asymptotic solutions to the matrix Riccati equation. *Mathematical Biosciences*, 20:339–344, 1974.
23. A.N. Beavers and E.D. Denman. A new similarity transformation method for eigenvalues and eigenvectors. *Mathematical Biosciences*, 21:143–169, 1974.
24. A.N. Beavers and E.D. Denman. A new solution method for quadratic matrix equations. *Mathematical Biosciences*, 20:135–143, 1974.
25. D.J. Bender. *Descriptor Systems and Geometric Control Theory*. PhD thesis, Univ. of California, Santa Barbara, ECE Dept., Santa Barbara, CA, September 1985.
26. D.J. Bender and A.J. Laub. The linear-quadratic optimal regulator problem for descriptor systems. In *Proc. 24th IEEE Conf. Decis. Control*, pages 957–962, Ft. Lauderdale, FL, December 1985.
27. D.J. Bender and A.J. Laub. The linear-quadratic optimal regulator for descriptor systems: Discrete-time case. *Automatica*, 23:71–85, 1987.
28. D.J. Bender and A.J. Laub. The linear-quadratic optimal regulator for descriptor systems. *IEEE Trans. Auto. Control*, AC-32:672–688, 1987.
29. G.J. Bierman. Matrix sign function solution to the continuous time algebraic Riccati equation. Technical report, Factorized Estimation Applications, Inc., 7017 Deveron Ridge Road, Canoga Park, CA, 1983.
30. G.J. Bierman. Computational aspects of the matrix sign function solution to the ARE. In *Proc. 23rd IEEE Conf. Decis. Control*, pages 514–519, Las Vegas, NV, December 1984.
31. S.P. Bingulac, M.R. Stojić, and N. Ćuk. Solution of time-invariant Riccati equations. In *Proc. 1971 Joint Automatic Control Conf.*, pages 178–182, St. Louis, MO, 1971.
32. R.R. Bitmead, M. Gevers, I.R. Petersen, and R.J. Kaye. Monotonicity and stabilizability properties of solutions of the Riccati difference equation: Propositions, lemmas, theorems, fallacious conjectures, and counterexamples. *Sys. Control Lett.*, 5:309–315, 1985.
33. S. Bittanti, editor. *Proc. Workshop on the Riccati Equation in Control, Systems, and Signals; Como, Italy*. Pitagora Editrice Bologna, June 1989.
34. T.R. Blackburn. Solution of the algebraic matrix Riccati equation via the Newton-Raphson iteration. In *Proc. Joint Automatic Control Conf.*, pages 940–945, University of Michigan, Ann Arbor, 1968.
35. T.R. Blackburn and J.C. Bidwell. Some numerical aspects of control engineering computations. In *Proc. Joint Automatic Control Conf.*, pages 203–207, University of Michigan, Ann Arbor, 1968.
36. R.D. Brandt. *Smooth Extrapolation and Gradient Methods for Inherently Discrete Applications*. PhD thesis, Univ. of California, Santa Barbara, ECE Dept., Santa Barbara, CA, July 1988.
37. R.D. Brandt and A.J. Laub. Continuous eigenvector decomposition. submitted to *Math. Comp.*, May 1990.

38. A.E. Bryson and W.E. Hall. Optimal control and filter synthesis by eigenvector decomposition. Technical Report No. 436, Stanford University, Dept. of Aeronautics and Astronautics, December 1971.
39. R.S. Bucy. Global theory of the Riccati equation. *J. Comp. Sys. Sci.*, 1:349–361, 1967.
40. R.S. Bucy. Two-point boundary value problems of linear Hamiltonian systems. *SIAM J. Appl. Math.*, 15:1385–1389, 1967.
41. R.S. Bucy. Structural stability for the Riccati equation. *SIAM J. Control Opt.*, 13:749–753, 1975.
42. R.S. Bucy and P.D. Joseph. *Filtering for Stochastic Processes with Applications to Guidance*. Interscience, New York, 1968. (second edition: 1987, Chelsea, New York).
43. J.R. Bunch. The weak and strong stability of algorithms in numerical algebra. *Lin. Alg. Appl.*, 88:49–66, 1987.
44. A. Bunse-Gerstner. Eigenvalue algorithm for matrices with special structure. *Colloquia Math. Soc. Jànos Bolyai 50. Numerical Methods. Miskolc (Hungary)*, pages 141–163, 1986.
45. A. Bunse-Gerstner. Matrix factorization for symplectic QR-like methods. *Lin. Alg. Appl.*, 83:49–77, 1986.
46. A. Bunse-Gerstner. *Symplectic QR-Like Methods*. PhD thesis, Universität Bielefeld, Bielefeld, FRG, 1986. Habilitationsschrift.
47. A. Bunse-Gerstner, R. Byers, and V. Mehrmann. Numerical methods for algebraic Riccati equations. In S. Bittanti, editor, *Proc. Workshop on the Riccati Equation in Control, Systems, and Signals*, pages 107–116, Como, Italy, June 1989.
48. A. Bunse-Gerstner, R. Byers, and V. Mehrmann. A chart of numerical methods for structured eigenvalue problems. *SIAM J. Matrix Anal. Appl.*, 1991. to appear.
49. A. Bunse-Gerstner and V. Mehrmann. A symplectic QR-like algorithm for the solution of the real algebraic Riccati equation. *IEEE Trans. Auto. Control*, AC-31:1104–1113, 1986.
50. A. Bunse-Gerstner and V. Mehrmann. The HHDR-algorithm and its application to optimal control problems. *R.A.I.R.O. Automatique/Productique Informatique Industrielle*, 4:309–329, 1989.
51. N. Burgoyne and R. Cushman. Normal forms for real Hamiltonian matrices. In R. Hermann, editor, *The 1976 Ames Research Center Conference on Geometric Control Theory*, pages 483–529, Brookline, MA, 1976.
52. R. Byers. *Hamiltonian and Symplectic Algorithms for the Algebraic Riccati Equation*. PhD thesis, Cornell University, Dept. Comp. Sci., Ithaca, NY, 1983.
53. R. Byers. Numerical condition of the algebraic Riccati equation. In *Linear Algebra and Its Role in Systems Theory*, volume 47, pages 35–49. Contemp. Math., 1985. AMS.
54. R. Byers. A Hamiltonian QR-algorithm. *SIAM J. Sci. Stat. Comp.*, 7:212–229, 1986.
55. R. Byers. Numerical stability and instability in matrix sign function based algorithms. In C.I. Byrnes and A. Lindquist, editors, *Computational and Combinatorial Methods in Systems Theory*, pages 185–200. Elsevier (North-Holland), New York, 1986.
56. R. Byers. Solving the algebraic Riccati equation with the matrix sign function. *Lin. Alg. Appl.*, 85:267–279, 1987.
57. R. Byers. A Hamiltonian-Jacobi algorithm. *IEEE Trans. Auto. Control*, AC-35:566–570, 1990.
58. R. Byers and V. Mehrmann. Symmetric updating of the solution of the algebraic Riccati equation. In *Proc. 10th Symposium on Operations Research*, volume 54, pages 117–125, Universität München, August 1985.
59. J.P. Charlier and P. Van Dooren. A systolic algorithm for Riccati and Lyapunov equations. *Math. Control, Signals, Sys.*, 2:109–136, 1989.
60. C. Choi and A.J. Laub. Constructing Riccati differential equations with known analytic solutions for numerical experiments. *IEEE Trans. Auto. Control*, AC-35:437–439, 1990.
61. C. Choi and A.J. Laub. Efficient matrix-valued algorithms for solving stiff Riccati differential equations. *IEEE Trans. Auto. Control*, AC-35:770–776, 1990. (see also *Proc. 1989 CDC*, pp. 885-887).
62. C.H. Choi. *Efficient Algorithms for Solving Stiff Matrix-Valued Riccati Differential Equations*. PhD thesis, Univ. of California, Santa Barbara, ECE Dept., Santa Barbara, CA, September 1988.

63. W.A. Coppel. Matrix quadratic equations. *Bull. Austral. Math. Soc.*, 10:377–401, 1974.
64. J. Demmel. On condition numbers and the distance to the nearest ill-posed problem. *Numer. Math.*, 51:251–289, 1987.
65. E.D. Denman and A.N. Beavers. The matrix sign function and computations in systems. *Appl. Math. Comp.*, 2:63–94, 1976.
66. E.D. Denman and J. Leyva-Ramos. Spectral decomposition of a matrix using the generalized sign matrix. *Appl. Math. Comp.*, 8:237–250, 1981.
67. C.E. DeSouza, M.R. Gevers, and G.C. Goodwin. Riccati equations in optimal filtering of nonstabilizable systems having singular state transition matrices. *IEEE Trans. Auto. Control*, AC-31:831–838, 1986.
68. L. Dieci and R.D. Russell. On the computation of invariant subspaces. Technical report, Simon Fraser University, LCCR, Dept. of Math. and Stats., Burnaby, BC, 1987.
69. J.J. Dongarra, J.R. Bunch, C.B. Moler, and G.W. Stewart. LINPACK *Users' Guide*. SIAM, Philadelphia, PA, 1979.
70. J.J. Dongarra, C.B. Moler, and J.H. Wilkinson. Improving the accuracy of computed eigenvalues and eigenvectors. *SIAM J. Numer. Anal.*, 20:46–58, 1983.
71. P. Dorato and A. Levis. Optimal linear regulators: The discrete-time case. *IEEE Trans. Auto. Control*, AC-16:613–620, 1971.
72. J. Doyle, K. Glover, P.P. Khargonekar, and B.A. Francis. State-space solutions to standard H_2 and H_∞ control problems. *IEEE Trans. Auto. Control*, AC-34:831–847, 1989.
73. L. Elsner. Matrix decompositions, symmetries and eigenvalue algorithms. In *Proc. Matrix Theory Conf.*, Auburn, AL, 1986.
74. A. Emami-Naeini and G.F. Franklin. Design of steady-state-quadratic-loss optimal digital controls for systems with a singular system matrix. In *Proc. 13th Asilomar Conf. Circuits, Systems, and Computers*, pages 370–374, November 1979.
75. A. Emami-Naeini and G.F. Franklin. Deadbeat control and tracking of discrete-time systems. *IEEE Trans. Auto. Control*, AC-27:176–181, 1982.
76. F.A. Farrar and R.C. DiPietro. Comparative evaluation of numerical methods for solving the algebraic matrix Riccati equation. Technical Report No. R76-140268-1, United Technologies Res. Center, East Hartford, CT, December 1976.
77. A.F. Fath. Computational aspects of the linear optimal regulator problem. *IEEE Trans. Auto. Control*, AC-14:547–550, 1969.
78. D.S. Flamm and R.A. Walker. Remark on Algorithm 506. *ACM Trans. Math. Software*, 8:219–220, 1982.
79. U. Flaschka and D. Zywietz. Strukturerhaltende zerlegungsalgorithmen zur lösung des allgemeinen symplektischen eigenwertproblems. Master's thesis, Universität Bielefeld, Bielefeld, FRG, 1988. Diplomarbeit.
80. W.C. Freestedt, R.F. Webber, and R.W. Bass. The GASP computer program—an integrated tool for optimal control and filter design. In *Proc. Joint Automatic Control Conf.*, pages 198–202, University of Michigan, Ann Arbor, 1968.
81. P. Gahinet and A.J. Laub. Computable bounds for the sensitivity of the algebraic Riccati equation. *SIAM J. Control Opt.*, 28 1990, pp. 1461-1480.
82. P. Gahinet and A.J. Laub. Estimating the distance to the nearest uncontrollable pair through the algebraic Riccati equation, *SIAM J. Control Opt.*, 1991, to appear.
83. P. Gahinet, A.J. Laub, C. Kenney, and G. Hewer. Sensitivity of the stable discrete-time Lyapunov equation. *IEEE Trans. Auto. Control*, AC-35 (1990), pp. 1209-1217.
84. P. Gahinet, A.J. Laub, M. Sorine, and C. Kenney. Stability margins and Lyapunov equations for linear operators in Hilbert space. *Proc. 29th IEEE Conf. Decis. Control*, pp. 2638-2639, Honolulu, HI, December 1990.
85. P.M. Gahinet. *Perturbational and Topological Aspects of Sensitivity in Control Theory*. PhD thesis, Univ. of California, Santa Barbara, ECE Dept., Santa Barbara, CA, December 1989.
86. B.S. Garbow, J.M. Boyle, J.J. Dongarra, and C.B. Moler. *Matrix Eigensystem Routines—*EISPACK *Guide Extension*, volume 51 of *Lecture Notes in Computer Science*. Springer-Verlag, New York, 1977.

87. J.D. Gardiner. *Iterative and Parallel Algorithms for the Solution of Algebraic Riccati Equations*. PhD thesis, Univ. of California, Santa Barbara, ECE Dept., Santa Barbara, CA, July 1988.
88. J.D. Gardiner and A.J. Laub. Parallel algorithms for algebraic Riccati equations, to appear in *Int. J. Control*, 1991.
89. J.D. Gardiner and A.J. Laub. A generalization of the matrix-sign-function solution for algebraic Riccati equations. *Int. J. Control*, 44:823–832, 1986. (see also *Proc. 1985 CDC*, pp. 1233-1235).
90. J.D. Gardiner and A.J. Laub. Implementation of two control system design algorithms on a message-passing hypercube. In M.T. Heath, editor, *Hypercube Multiprocessors 1987*, pages 512–519. SIAM, Philadelphia, PA, 1987.
91. J.D. Gardiner and A.J. Laub. Matrix sign function implementations on a hypercube multiprocessor. In *Proc. 27th IEEE Conf. Decis. Control*, pages 1466–1471, Austin, TX, December 1988.
92. J.D. Gardiner and A.J. Laub. Solving the algebraic Riccati equation on a hypercube multiprocessor. In G. Fox, editor, *Hypercube Concurrent Computers and Applications, Vol. II*, pages 1562–1568. ACM Press, New York, 1988.
93. J.D. Gardiner, A.J. Laub, J.A. Amato, and C.B. Moler. Solution of the Sylvester Matrix Equation $AXB^T + CXD^T = E$, to appear in *ACM Trans. Math. Software*, 1991.
94. A. Ghavimi, C. Kenney, and A.J. Laub. Convergence analysis of conjugate gradient methods for solving algebraic Riccati equations, to appear in *IEEE Trans. Auto. Control*, 1991.
95. I. Gohberg, P. Lancaster, and L. Rodman. *Invariant Subspaces of Matrices with Applications*. Wiley, New York, 1986.
96. G.H. Golub and C.F. Van Loan. *Matrix Computations*. Johns Hopkins University Press, Baltimore, 1983. (second edition: 1989).
97. G.H. Golub and J.H. Wilkinson. Ill-conditioned eigensystems and the computation of the Jordan canonical form. *SIAM Review*, 18:578–619, 1976.
98. T. Gudmundsson, C. Kenney, and A.J. Laub. Scaling of the discrete-time algebraic Riccati equation to enhance stability of the Schur solution method, to appear in *IEEE Trans. Auto. Control*, 1991.
99. S.J. Hammarling. Newton's method for solving the algebraic Riccati equation. NPL Report DITC 49/81, National Physical Laboratory, 1981.
100. S.J. Hammarling. Numerical solution of the stable, non-negative definite Lyapunov equation. *IMA J. Numer. Anal.*, 2:303–323, 1982.
101. S.J. Hammarling and M.A. Singer. A canonical form for the algebraic Riccati equation. In *Proc. MTNS*, Beer Sheeva, Israel, 1983.
102. G.A. Hewer. An iterative technique for the computation of steady state gains for the discrete optimal regulator. *IEEE Trans. Auto. Control*, AC-16:382–384, 1971.
103. G.A. Hewer and C. Kenney. The sensitivity of the stable Lyapunov equation. *SIAM J. Control Opt.*, 26:321–344, 1988.
104. G.A. Hewer and G. Nazaroff. A survey of numerical methods for the solution of algebraic Riccati equations. Technical report, Michelson Laboratory, Naval Weapons Center, China Lake, CA, 1974.
105. N.J. Higham. Computing the polar decomposition—with applications. *SIAM J. Sci. Stat. Comp.*, 7:1160–1174, 1986.
106. N.J. Higham and R.S. Schreiber. Fast polar decomposition of an arbitrary matrix. *SIAM J. Sci. Stat. Comp.*, 11:648–655, 1990.
107. K.L. Hitz and B.D.O. Anderson. Iterative method of computing the limiting solution of the matrix Riccati differential equation. *Proc. of IEEE*, 60:1402–1406, 1972.
108. W.E. Holley and S.Y. Wei. Improved method for solving the algebraic Riccati equation. *AIAA J. Guid. Control*, 3:190–192, 1980.
109. R.D. Howerton and J.L. Hammond, Jr. A new computational solution of the linear optimal regulator problem. *IEEE Trans. Auto. Control*, AC-16:645–651, 1971.
110. J.L. Howland. The sign matrix and the separation of matrix eigenvalues. *Lin. Alg. Appl.*, 49:221–232, 1983.

111. M. Jamshidi. An overview on the solutions of the algebraic matrix Riccati equation and related problems. *Large Scale Systems*, 1:167–192, 1980.
112. B. Kågström and A. Ruhe. *Matrix Pencils (Proceedings, Pite Havsbad, 1982)*. Springer-Verlag, Berlin, 1983.
113. R.E. Kalman. Contributions to the theory of optimal control. *Boletin Sociedad Matematica Mexicana*, 5:102–119, 1960.
114. R.E. Kalman and T.S. Englar. A user's manual for the automatic synthesis program. RIAS Report CR-475, NASA, June 1966.
115. C. Kenney and G. Hewer. The sensitivity of the algebraic and differential Riccati equations. *SIAM J. Control Opt.*, 28:50–69, 1990.
116. C. Kenney and A.J. Laub. Condition estimates for matrix functions. *SIAM J. Matrix Anal. Appl.*, 10:191–209, 1989.
117. C. Kenney and A.J. Laub. On scaling Newton's method for polar decomposition and the matrix sign function. In *Proc. 1990 Amer. Control Conf.*, pp. 2560–2564, San Diego, CA, May 1990.
118. C. Kenney and A.J. Laub. On scaling Newton's method for polar decomposition and the matrix sign function, *SIAM J. Matrix Anal. Appl.*, 1991, to appear.
119. C. Kenney and A.J. Laub. Polar decomposition and matrix sign function condition estimates. *SIAM J. Sci. Stat. Comp.*, 1991, to appear.
120. C. Kenney and A.J. Laub. Rational iterative methods for the matrix sign function. *SIAM J. Matrix Anal. Appl.*, 1991, to appear.
121. C. Kenney, A.J. Laub, and E.A. Jonckheere. Positive and negative solutions of dual Riccati equations by matrix sign function iteration. *Sys. Control Lett.*, 13:109–116, 1989.
122. C. Kenney, A.J. Laub, and M. Wette. A stability-enhancing scaling procedure for Schur-Riccati solvers. *Sys. Control Lett.*, 12:241–250, 1989.
123. C. Kenney, A.J. Laub, and M. Wette. Error bounds for Newton refinement of solutions to algebraic Riccati equations. *Math. Control, Signals, Sys.*, 3:211–224, 1990.
124. C. Kenney and R.B. Leipnik. Numerical integration of the differential matrix Riccati equation. *IEEE Trans. Auto. Control*, AC-30:962–970, 1985.
125. D.L. Kleinman. On an iterative technique for Riccati equation computations. *IEEE Trans. Auto. Control*, AC-13:114–115, 1968.
126. D.L. Kleinman. An easy way to stabilize a linear constant system. *IEEE Trans. Auto. Control*, AC-15:692, 1970.
127. D.L. Kleinman. Stabilizing a discrete constant linear system, with application to iterative methods for solving the Riccati equation. *IEEE Trans. Auto. Control*, AC-19:479–481, 1974.
128. M.M. Konstantinov, P.Hr. Petkov, and N.D. Christov. Perturbation analysis of the continuous and discrete matrix Riccati equations. In *Proc. 1986 Amer. Control Conf.*, pages 636–639, Seattle, WA, June 1986.
129. V. Kučera. The structure and properties of time optimal discrete linear control. *IEEE Trans. Auto. Control*, AC-16:375–377, 1971.
130. V. Kučera. A contribution to matrix quadratic equations. *IEEE Trans. Auto. Control*, AC-17:344–347, 1972.
131. V. Kučera. On nonnegative definite solutions to matrix quadratic equations. *Automatica*, 8:413–423, 1972.
132. V. Kučera. A review of the matrix Riccati equation. *Kybernetica* (Prague), 9:42–61, 1973.
133. H. Kwakernaak and R. Sivan. *Linear Optimal Control Systems*. Wiley-Interscience, New York, 1972.
134. D.G. Lainiotis. Partitioned Riccati solutions and integration-free doubling algorithms. *IEEE Trans. Auto. Control*, AC-21:677–689, 1976.
135. P. Lancaster, A.C.M. Ran, and L. Rodman. Hermitian solutions of the discrete algebraic Riccati equation. *Int. J. Control*, 44:777–802, 1986.
136. P. Lancaster, A.C.M. Ran, and L. Rodman. An existence and monotonicity theorem for the discrete algebraic Riccati equation. *Lin. Multilin. Alg.*, 20:353–361, 1987.
137. P. Lancaster and L. Rodman. Existence and uniqueness theorems for the algebraic Riccati equation. *Int. J. Control*, 32:285–309, 1980.

138. A.J. Laub. Canonical forms for σ-symplectic matrices. Master's thesis, Univ. of Minnesota, School of Mathematics, Minneapolis, MN, December 1972.
139. A.J. Laub. A Schur method for solving algebraic Riccati equations. LIDS Rept. LIDS-R-859, MIT, Lab. for Info. and Decis. Syst., Cambridge, MA, October 1978. (including software).
140. A.J. Laub. A Schur method for solving algebraic Riccati equations. *IEEE Trans. Auto. Control*, AC-24:913–921, 1979. (see also *Proc. 1978 CDC (Jan. 1979)*, pp. 60-65).
141. A.J. Laub. Further 'Comments on the numerical solution of the discrete-time algebraic Riccati equation'. *IEEE Trans. Auto. Control*, AC-25:1252–1253, 1980.
142. A.J. Laub. Schur techniques for Riccati differential equations. In D. Hinrichsen and A. Isidori, editors, *Feedback Control of Linear and Nonlinear Systems*, pages 165–174. Springer-Verlag, New York, 1982.
143. A.J. Laub. Schur techniques in invariant imbedding methods for solving two-point boundary value problems. In *Proc. 21st IEEE Conf. Decis. Control*, pages 55–61, Orlando, FL, December 1982.
144. A.J. Laub. Numerical aspects of solving algebraic Riccati equations. In *Proc. 22nd IEEE Conf. Decis. Control*, pages 184–186, San Antonio, TX, December 1983.
145. A.J. Laub. Numerical linear algebra aspects of control design computations. *IEEE Trans. Auto. Control*, AC-30:97–108, 1985.
146. A.J. Laub. Algebraic aspects of generalized eigenvalue problems for solving Riccati equations. In C.I. Byrnes and A. Lindquist, editors, *Computational and Combinatorial Methods in Systems Theory*, pages 213–227. Elsevier (North-Holland), 1986.
147. A.J. Laub and J.D. Gardiner. Hypercube implementation of some parallel algorithms in control. In M.J. Denham and A.J. Laub, editors, *Advanced Computing Concepts and Techniques in Control Engineering*, pages 361–390. Springer-Verlag, Berlin, 1988.
148. A.J. Laub and K.R. Meyer. Canonical forms for symplectic and Hamiltonian matrices. *Celestial Mechanics*, 9:213–238, 1974.
149. K.H. Lee. *Generalized Eigenproblem Structures and Solution Methods for Riccati Equations*. PhD thesis, Univ. of Southern California, Dept. of Elec. Eng.—Systems, Los Angeles, CA, January 1983.
150. R.B. Leipnik. Rapidly convergent recursive solution of quadratic operator equations. *Numer. Math.*, 17:1–16, 1971.
151. R.B. Leipnik. A canonical form and solution for the matrix Riccati differential equation. *J. Austral. Math. Soc., Ser. B*, 26:355–361, 1985.
152. J.J. Levin. On the matrix Riccati equation. *Proc. Amer. Math. Soc.*, 10:519–524, 1959.
153. W. Levine and M. Athans. On the optimal error regulation of a string of moving vehicles. *IEEE Trans. Auto. Control*, AC-11:355–361, 1966.
154. W.-W. Lin. A new method for computing the closed loop eigenvalues of a discrete-time algebraic Riccati equation. *Lin. Alg. Appl.*, 6:157–180, 1987.
155. W.-W. Lin. An SDR algorithm for the solution of the generalized algebraic Riccati equation. *IEEE Trans. Auto. Control*, AC-34:875–879, 1989.
156. A.G.J. MacFarlane. An eigenvector solution of the optimal linear regulator problem. *J. Electron. Contr.*, 14:643–654, 1963.
157. M. Maki. Numerical solution of the algebraic Riccati equation for single input systems. *IEEE Trans. Auto. Control*, AC-17:264–265, 1972.
158. K. Mårtensson. On the matrix Riccati equation. *Info. Sci.*, 3:17–49, 1971.
159. K. Mårtensson. Approaches to the numerical solution of optimal control problems. Technical Report No. 7206, Lund. Inst. of Tech., Div. of Automat. Contr., Lund, Sweden, March 1972.
160. V. Mehrmann. Der SR-algorithmus zur berechnung der eigenwerte einer matrix. Master's thesis, Universität Bielefeld, Bielefeld, FRG, 1979. Diplomarbeit.
161. V. Mehrmann. *The Linear Quadratic Control Problem: Theory and Numerical Algorithms*. PhD thesis, Universität Bielefeld, Bielefeld, FRG, 1987. Habilitationsschrift.
162. V. Mehrmann. A symplectic orthogonal method for single input or single output discrete time optimal linear quadratic control problems. *SIAM J. Matrix Anal. Appl.*, 9:221–248, 1988.
163. V. Mehrmann. Uniqueness and stability of solutions to singular, linear-quadratic control problems. *Lin. Alg. Appl.*, 1990. to appear.

164. V. Mehrmann and E. Tan. Defect correction methods for the solution of algebraic Riccati equations. *IEEE Trans. Auto. Control*, AC-33:695–698, 1988.
165. H.B. Meyer. The matrix equation $AZ + B - ZCZ - ZD = 0$. *SIAM J. Appl. Math.*, 30:136–142, 1976.
166. M.L. Michelson. On the eigenvalue, eigenvector method for solution of the stationary discrete matrix Riccati equation. *IEEE Trans. Auto. Control*, AC-24:480–481, 1979.
167. C.B. Moler and G.W. Stewart. An algorithm for generalized matrix eigenvalue problems. *SIAM J. Numer. Anal.*, 10:241–256, 1973.
168. B.P. Molinari. The stabilizing solution of the algebraic Riccati equation. *SIAM J. Control Opt.*, 11:262–271, 1973.
169. B.P. Molinari. The time invariant linear quadratic optimal control problem. *Automatica*, 13:347–357, 1977.
170. J.J. O'Donnell. Asymptotic solution of the matrix Riccati equation of optimal control. In *Proc. Fourth Allerton Conf. Circuit and System Theory*, pages 577–586, University of Illinois, Urbana, IL, October 1966.
171. Y. Oshman and I.Y. Bar-Itzhack. Eigenfactor solution of the matrix Riccati equation—A continuous square root algorithm. *IEEE Trans. Auto. Control*, AC-30:971–978, 1985.
172. M. Pachter and T.E. Bullock. Ordering and stability properties of the Riccati equation. Technical Report No. WISK 264, Nat. Res. Inst. for Math. Sci., Pretoria, June 1977.
173. C.C. Paige and C.F. Van Loan. A Schur decomposition for Hamiltonian matrices. *Lin. Alg. Appl.*, 14:11–32, 1981.
174. P. Pandey, C. Kenney, and A.J. Laub. A parallel algorithm for the matrix sign function. *Int. J. High Speed Computing*, 2 (1990) pp. 181–191.
175. T. Pappas. Solution of discrete-time LQG problems with singular transition matrix. MIT, May 1979. B.S. Thesis, Dept. of Elec. Engr.
176. T. Pappas, A.J. Laub, and N.R. Sandell. On the numerical solution of the discrete-time algebraic Riccati equation. LIDS Rept. LIDS-P-908, MIT, Lab. for Info. and Decis. Syst., Cambridge, MA, May 1979.
177. T. Pappas, A.J. Laub, and N.R. Sandell. On the numerical solution of the discrete-time algebraic Riccati equation. *IEEE Trans. Auto. Control*, AC-25:631–641, 1980.
178. R.V. Patel and M. Toda. On norm bounds for algebraic Riccati and Lyapunov equations. *IEEE Trans. Auto. Control*, AC-23:87–88, 1978.
179. H.J. Payne and L.M. Silverman. On the discrete time algebraic Riccati equation. *IEEE Trans. Auto. Control*, AC-18:226–234, 1973.
180. C.E.M. Pearce. On the solution of a class of algebraic matrix Riccati equation. *IEEE Trans. Auto. Control*, AC-31:252–255, 1986.
181. P.Hr. Petkov, N.D. Christov, and M.M. Konstantinov. Reliability of the algorithms and software for synthesis of linear optimal systems. Dept. of Automatics, Higher Institute of Mechanical and Electrical Engineering, 1986. B1.2; 1156 Sofia, Bulgaria.
182. P.Hr. Petkov, N.D. Christov, and M.M. Konstantinov. On the numerical properties of the Schur approach for solving the matrix Riccati equation. *Sys. Control Lett.*, 9:197–201, 1987.
183. J.E. Potter. Matrix quadratic solutions. *SIAM J. Appl. Math.*, 14:496–501, 1966.
184. M.A. Poubelle, R.R. Bitmead, and M.R. Gevers. Fake algebraic Riccati techniques. *IEEE Trans. Auto. Control*, AC-33:379–381, 1988.
185. M.A. Poubelle, I.R. Petersen, M.R. Gevers, and R.R. Bitmead. A miscellany of results on an equation of Count J.F. Riccati. *IEEE Trans. Auto. Control*, AC-31:651–653, 1986.
186. A.V. Ramesh, S. Utku, and J.A. Garba. Computational complexities and storage requirements of some Riccati equation solvers. *AIAA J. Guid. Control*, 12:469–479, 1989.
187. A.C.M. Ran and R. Vreugdenhil. Existence and comparison theorems for algebraic Riccati equations for continuous- and discrete-time systems. *Lin. Alg. Appl.*, 99:63–83, 1974.
188. W.T. Reid. *Riccati Differential Equations*. Academic Press, New York, 1972.
189. D.W. Repperger. A square root of a matrix approach to obtain the solution to a steady-state matrix Riccati equation. *IEEE Trans. Auto. Control*, AC-21:786–787, 1976.
190. J.R. Rice. A theory of condition. *SIAM J. Numer. Anal.*, 3:287–310, 1966.

191. J.D. Roberts. Linear model reduction and solution of the algebraic Riccati equation by use of the sign function. *Int. J. Control*, 32:677–687, 1980. (reprint of Technical Report No. TR-13, CUED/B-Control, Cambridge University, Engineering Department, 1971).
192. J. Rodriguez-Canabal. The geometry of the Riccati equation. *Stochastics*, 1:129–149, 1973.
193. W.E. Roth. On the matrix equation $X^2 + AX + XB + C = 0$. *Proc. AMS*, 1:586–589, 1950.
194. N.R. Sandell. On Newton's method for Riccati equation solution. *IEEE Trans. Auto. Control*, AC-19:254–255, 1974.
195. G. Schulz. Iterative computation of reciprocal matrices. *Z. Angew. Math. Mech.*, 13:57–59, 1933. (in German).
196. I. Schur. Über die charakteristischen wurzeln einer linearen substitution mit einer anwendung auf die theorie der integralgleichungen (On the characteristic roots of a linear substitution with an application to the theory of integral equations). *Math. Ann.*, 66:488–510, 1909. (in German).
197. M.A. Shayman. Geometry of the algebraic Riccati equation, Part I. *SIAM J. Control Opt.*, 21:375–394, 1983.
198. M.A. Shayman. Geometry of the algebraic Riccati equation, Part II. *SIAM J. Control Opt.*, 21:395–409, 1983.
199. H.A. Shubert. Analytic solution for the algebraic Riccati equation. *IEEE Trans. Auto. Control*, AC-19:255–256, 1974.
200. G.S. Sidhu and G.J. Bierman. Integration-free interval doubling for Riccati equation solutions. *IEEE Trans. Auto. Control*, AC-22:831–834, 1977.
201. L.M. Silverman. Discrete Riccati equations: Alternative algorithms, asymptotic properties, and system theory interpretations. In C.T. Leondes, editor, *Advances in Control Systems, Vol. 12*, pages 313–386. Academic Press, New York, 1976.
202. M.A. Singer and S.J. Hammarling. The algebraic Riccati equation: A summary review of some available results. NPL Report DITC 23/83, National Physical Laboratory, 1983.
203. B.T. Smith, J.M. Boyle, J.J. Dongarra, B.S. Garbow, Y. Ikebe, V.C. Klema, and C.B. Moler. *Matrix Eigensystem Routines*—EISPACK *Guide, Second Edition*, volume 6 of *Lecture Notes in Computer Science*. Springer-Verlag, New York, 1976.
204. G.W. Stewart. Error bounds for approximate invariant subspaces of closed linear operators. *SIAM J. Numer. Anal.*, 8:796–808, 1971.
205. G.W. Stewart. On the sensitivity of the eigenvalue problem $Ax = \lambda Bx$. *SIAM J. Numer. Anal.*, 9:669–686, 1972.
206. G.W. Stewart. Error and perturbation bounds for subspaces associated with certain eigenvalue problems. *SIAM Review*, 15:727–764, 1973.
207. G.W. Stewart. *Introduction to Matrix Computations*. Academic Press, New York, 1973.
208. G.W. Stewart. Gershgorin theory for the generalized eigenvalue problem $Ax = \lambda Bx$. *Math. Comp.*, 29:600–606, 1975.
209. G.W. Stewart. Algorithm 506—HQR3 and EXCHNG: Fortran subroutines for calculating and ordering the eigenvalues of a real upper Hessenberg matrix. *ACM Trans. Math. Software*, 2:275–280, 1976.
210. G.W. Stewart. Perturbation theory for the generalized eigenvalue problem. In C. de Boor and G.H. Golub, editors, *Recent Advances in Numerical Analysis*, pages 193–206. Academic Press, New York, 1978.
211. W.G. Tuel. An improved algorithm for the solution of the discrete regulation problem. *IEEE Trans. Auto. Control*, AC-12:522–528, 1967.
212. P. Van Dooren. The generalized eigenstructure problem in linear system theory. *IEEE Trans. Auto. Control*, AC-26:111–129, 1981.
213. P. Van Dooren. A generalized eigenvalue approach for solving Riccati equations. *SIAM J. Sci. Stat. Comp.*, 2:121–135, 1981.
214. P. Van Dooren. Algorithm 590—DSUBSP and EXCHQZ: Fortran subroutines for computing deflating subspaces with specified spectrum. *ACM Trans. Math. Software*, 8:376–382, 1982.
215. C.F. Van Loan. A symplectic method for approximating all the eigenvalues of a Hamiltonian matrix. *Lin. Alg. Appl.*, 16:233–251, 1984.

216. A.C.M. Van Swieten. *Qualitative Behavior of Dynamical Games with Feedback Strategies.* PhD thesis, University of Groningen, Groningen, The Netherlands, 1977.
217. W.H. Vandevender. On the stability of an invariant imbedding algorithm for the solution of two-point boundary value problems. Technical Report Rept. No. SAND77-1107, Sandia Laboratories, August 1977.
218. J.M. Varah. On the separation of two matrices. *SIAM J. Numer. Anal.*, 16:216–222, 1979.
219. D.R. Vaughan. A negative exponential solution for the matrix Riccati equation. *IEEE Trans. Auto. Control*, AC-14:72–75, 1969.
220. D.R. Vaughan. A nonrecursive algebraic solution for the discrete Riccati equation. *IEEE Trans. Auto. Control*, AC-15:597–599, 1970.
221. G. Von Escherich. Die zweite variation der einfachen integrale. *Wiener Sitzungsberichte*, 107:1191–1250, 1898.
222. R.A. Walker, A. Emami-Naeini, and P. Van Dooren. A general algorithm for solving the algebraic Riccati equation. In *Proc. 21st IEEE Conf. Decis. Control*, pages 68–72, Orlando, FL, December 1982.
223. O.H.D. Walter. Eigenvector scaling in a solution of the matrix Riccati equation. *IEEE Trans. Auto. Control*, AC-15:486–487, 1970.
224. R.C. Ward. The combination shift QZ algorithm. *SIAM J. Numer. Anal.*, 12:835–853, 1975.
225. R.C. Ward. Balancing the generalized eigenvalue problem. *SIAM J. Sci. Stat. Comp.*, 2:141–152, 1981.
226. L.-F. Wei and F.-B. Yez, A modified Schur method for solving algebraic Riccati equations. *Sys. Control Lett.*, 1991, to appear.
227. L.-F. Wei and F.-B. Yeh. On dual algebraic Riccati equations. *IEEE Trans. Auto. Control*, 1991, to appear.
228. M.R. Wette, J.D. Gardiner and A.J. Laub. Algorithm: A Fortran-77 software package for solving the Sylvester matrix equation $AXB^T + CXD^T = E$, to appear in *ACM Trans. Math. Software*, 1991.
229. M. Wette and A.J. Laub. Numerical algorithms and software for spectral factorization problems. In *Proc. 1988 Amer. Control Conf.*, pages 311–316, Atlanta, GA, June 1988.
230. M.R. Wette. *Numerical Algorithms for Spectral Factorization Problems in Linear Filtering and Control.* PhD thesis, Univ. of California, Santa Barbara, ECE Dept., Santa Barbara, CA, July 1988.
231. J.H. Wilkinson. *The Algebraic Eigenvalue Problem.* Oxford University Press, Oxford, 1965.
232. J.C. Willems. Least squares stationary optimal control and the algebraic Riccati equation. *IEEE Trans. Auto. Control*, AC-16:621–634, 1971.
233. H.K. Wimmer. On the algebraic Riccati equation. *Bull. Austral. Math. Soc.*, 14:457–461, 1976.
234. H.K. Wimmer. Monotonicity of maximal solutions of algebraic Riccati equations. *Sys. Control Lett.*, 5:317–319, 1985.
235. M. Womble and J.E. Potter. A prefiltering version of the Kalman filter and new numerical integration formulas for Riccati equations. *IEEE Trans. Auto. Control*, AC-20:378–380, 1975.
236. W.M. Wonham. On a matrix Riccati equation of stochastic control. *SIAM J. Control Opt.*, 6:681–697, 1968.
237. W.M. Wonham. *Linear Multivariable Control: A Geometric Approach.* Springer-Verlag, New York, 2nd edition, 1979.

8 The Dissipation Inequality and the Algebraic Riccati Equation

H.L. Trentelman and Jan C. Willems

8.1 Introduction

Undoubtedly one of the most important concepts in linear systems and control, both from a theoretical as well as from a practical point of view, is the algebraic Riccati equation. Since its introduction in control theory by Kalman [16] the beginning of the sixties, the algebraic Riccati equation has known an impressive range of applications, such as linear quadratic optimal control, stability theory, stochastic filtering and stochastic control, stochastic realization theory, synthesis of linear passive networks, differential games and, most recently, H_∞ optimal control and robust stabilization. The purpose of the present paper is to give an expository survey of the main concepts, results and applications related to the algebraic Riccati equation.

In our opinion, the most appealing framework for studying the Riccati equation is the framework of dissipative systems. In this framework, the Riccati equation emerges in a natural way as a consequense of the dissipation inequality, which expresses the fact that the system under consideration is dissipative. In this framework, real symmetric solutions of the algebraic Riccati correspond to storage functions, i.e. functions on the state space of the system that measure the amount of supply that is stored inside the system.

In this paper we intend to review and slightly extend the existing theory of dissipative systems. We will start with a treatment of dissipativeness for a very general class of systems. Contrary to most of the existing literature on dissipative systems, where the property of dissipativeness is described in terms of the internal (state space) properties of the system, we will consider dissipativeness as a property of the external behavior of the system. The question whether the system is internally dissipative then becomes one of finding a suitable state space representation and a suitable storage function. This expresses the property that, internally, the system can only store a finite amount of supply.

We will show that for linear, time-invariant, finite-dimensional systems with quadratic supply rates, the property of internal dissipativeness leads to solvability of what is called the linear matrix inequality, which in turn leads to solvabilty of the associated algebraic Riccati equation. As mentioned before, real symmetric solutions of the Riccati equation correspond to storage functions of the system. In

Bittanti, Laub, Willems (Eds.)
The Riccati Equation

this vein, it turns out that the Riccati equation has a positive semi-definite solution if and only if there exists a storage function that attains its minimum at the origin. We will explain how the question of existence of positive semi-definite solutions leads to the Kalman-Yakubovich-Popov lemma (sometimes called the Positive Real Lemma) and the Bounded Real Lemma.

A major part of this paper is devoted to a discussion of the main applications of the ideas developed here. Without any attempt to be complete, we will discuss some of the most important problems in the context of linear quadratic optimal control. We will also briefly outline how the concept of storage function can be used to construct Lyapunov functions and how, in this way, the Positive Real Lemma and the Bounded Real Lemma can be used to obtain results on the stability of feedback systems. Next, we will outline how the Positive Real Lemma can be used as an important tool in the synthesis of linear passive networks and in the state space representation theory of stationary Gaussian random processes (the problem of covariance generation). Finally, as a most recent application, we will discuss the application of the Bounded Real Lemma to the problem of H_∞ optimal control.

8.2 Dissipative Systems

In this section we will review the definition of dynamical system as proposed in [43]. For most of the definitions and basic facts we refer to [43], [44]. Furthermore, we will give a definition of the concept of dissipative dynamical system.

Definition 8.2.1. A *dynamical system* Σ is defined as a triple

$$\Sigma = (T, W, \mathcal{B})$$

with $T \subseteq \mathcal{R}$ the *time axis*; W a set, called the *signal alphabet*; and $\mathcal{B} \subseteq W^T$ *the behavior.*

Let $T \subseteq \mathcal{R}$ and let W be a set. If $w_1, w_2 : T \to W$ and if $t \in T$ then we define the *concatenation at* t of w_1 and w_2, $w_1 \underset{t}{\wedge} w_2$, as the mapping from T to W defined by

$$(w_1 \underset{t}{\wedge} w_2)(\tau) = \begin{cases} w_1(\tau) & \text{for } \tau < t \\ w_2(\tau) & \text{for } \tau \geq t \end{cases}$$

In this paper we shall restrict ourselves to continuous time systems with time axis T equal to $\mathcal{R}$. Furthermore, we shall assume throughout that the systems Σ under consideration are *time-invariant*, i.e. $\sigma^t \mathcal{B} = \mathcal{B}$ for all $t \in \mathcal{R}$. Here, σ^t denotes the *t-shift*: $(\sigma^t f)(\tau) := f(\tau + t)$.

Let $\Sigma = (\mathcal{R}, W, \mathcal{B})$ be a time-invariant dynamical system. If $w \in \mathcal{B}$ then we define

$$\mathcal{B}^+(w) := \{\tilde{w} : [0, \infty) \to W \mid w \underset{0}{\wedge} \tilde{w} \in \mathcal{B}\}$$

In other words, $\mathcal{B}^+(w)$ consists of all time functions on the positive real half line with the property that their concatenation with the past of w is an element of the behavior $\mathcal{B}$. Stated differently, $\mathcal{B}^+(w)$ contains all possible future continuations of the past of w. In a similar fashion, for $w \in \mathcal{B}$ we define $\mathcal{B}^-(w)$ to be the set of all possible past continuations of the future of w:

$$\mathcal{B}^-(w) := \{\tilde{w} : (-\infty, 0) \to W \mid \tilde{w} \underset{0}{\wedge} w \in \mathcal{B}\}$$

We shall now define the notion of dissipative dynamical system. Assume $\Sigma = (\mathcal{R}, W, \mathcal{B})$ is a time-invariant dynamical system. Let s be a real valued function defined on W. This function will be called the *supply rate*. We assume that s is such that for all $w \in \mathcal{B}$ the function $s(w(.))$ is locally integrable, i.e., $\int_{t_0}^{t_1} |s(w(t))| dt < \infty$ for all $(t_0, t_1) \in \mathcal{R}_2^+ := \{(\tau_0, \tau_1) \in \mathcal{R}^2 | \tau_0 \leq \tau_1\}$. The notion of dissipativeness requires a condition on periodic signals in $\mathcal{B}$. If $T > 0$ then $w \in \mathcal{B}$ is called *periodic with period T* or *T-periodic* if for all $t \in \mathcal{R}$ we have $w(t) = w(t + T)$.

Definition 8.2.2. The pair (Σ, s) with $\Sigma = (\mathcal{R}, W, \mathcal{B})$ a time-invariant dynamical system and $s : W \to \mathcal{R}$ the supply rate is called *dissipative* if for all for all periodic $w \in \mathcal{B}$ we have

$$\int_0^T s(w(t))dt \geq 0 \tag{8.1}$$

where T is the period of $w \in \mathcal{B}$.

The idea is that the function $s(w(.))$ is the rate at which supply (for example, the power, the rate at which energy is supplied) flows into the system if the system produces the signal $w(.)$. Thus, for $T \geq 0$ $\int_0^T s(w(t))dt$ is equal to the the total supply that flows into the system over the time-interval $[0, T]$. The above definition formalizes the idea that along periodic signals supply flows net *into* the system. The above definition of dissipativeness is taken from [37], where however a somewhat different terminology was used.

In addition to the concept of dissipativeness in terms of periodic signals, we will give the following definition in terms of continuations of signals $w \in \mathcal{B}$.

Definition 8.2.3. The pair (Σ, s) is called *forward dissipative* if for all $w \in \mathcal{B}$ and for all $T \geq 0$ such that $\mathcal{B}^+(w) = \mathcal{B}^+(\sigma^T w)$ we have $\int_0^T s(w(t))dt \geq 0$.

The interpretation of this definition is as follows: the system is called forward dissipative if along any signal with the property that its future continuations at time $t = 0$ and $t = T$ coincide, the supply flow over the time-interval $[0, T]$ is net *into* the system. It is also possible to give a backward version of this concept:

Definition 8.2.4. The pair (Σ, s) is called *backward dissipative* if for all $w \in \mathcal{B}$ and for all $T \geq 0$ such that $\mathcal{B}^-(w) = \mathcal{B}^-(\sigma^T w)$ we have $\int_0^T s(w(t))dt \geq 0$.

It turns out that the properties of forward dissipativeness and backward dissipativeness both imply dissipativeness:

Proposition 8.2.5. *Let Σ be a time-invariant dynamical system and let s be a supply rate. Then we have*

(a) If (Σ, s) is forward dissipative then (Σ, s) is dissipative,
(b) If (Σ, s) is backward dissipative then (Σ, s) is dissipative.

Proof. If $T > 0$ and w is T-periodic then we trivially have $\sigma^T w = w$. This implies $\mathcal{B}^+(w) = \mathcal{B}^+(\sigma^T w)$ and $\mathcal{B}^-(w) = \mathcal{B}^-(\sigma^T w)$. □

We will now show that if the dynamical system is *complete*, then the three notions of dissipativeness introduced above are equivalent. If $(t_0, t_1) \in \mathcal{R}_2^+$ then we define

$$\mathcal{B}\mid_{[t_0,t_1]} := \{w : [t_0, t_1] \to W | w \in \mathcal{B}\}$$

The dynamical system $\Sigma = (\mathcal{R}, W, \mathcal{B})$ is called *complete* if for all $w : \mathcal{R} \to W$ the following implication holds:

$$\{\forall (t_0, t_1) \in \mathcal{R}_2^+ \quad w\mid_{[t_0,t_1]} \in \mathcal{B}\mid_{[t_0,t_1]}\} \Rightarrow \{w \in \mathcal{B}\}.$$

In other words, a system is complete if for any function $w : \mathcal{R} \to W$, in order to check whether it is an element of the behavior of the system, it is sufficient to check whether its restrictions to finite intervals are elements of the behavior restricted to these intervals. For any $w : [0, T) \to W$ we define the *periodic continuation of w to $\mathcal{R}$* as the T-periodic function $\bar{w} : \mathcal{R} \to W$ whose restriction to $[0, T)$ is equal to w. The following lemma states that if Σ is time-invariant and complete, if $T > 0$ and if w is an element in the behavior of Σ with the property that its future continuations at time $t = 0$ and time $t = T$ coincide, then the T-periodic continuation of $w\mid_{[0,T)}$ is also an element of the behavior of Σ:

Lemma 8.2.6. *Let $\Sigma = (\mathcal{R}, W, \mathcal{B})$ be time-invariant and complete. Let $w \in \mathcal{B}$ and $T > 0$ be such that $\mathcal{B}^+(w) = \mathcal{B}^+(\sigma^T w)$. Then the T-periodic continuation of $w\mid_{[0,T)}$ is an element of $\mathcal{B}$.*

Proof. Clearly, $w\mid_{[0,\infty)} \in \mathcal{B}^+(w) = \mathcal{B}^+(\sigma^T w)$. This implies that $\sigma^T w \underset{0}{\wedge} w \in \mathcal{B}$. Define

$$w_1 := \sigma^{-T}(\sigma^T w \underset{0}{\wedge} w).$$

Then by time-invariance $w_1 \in \mathcal{B}$. Furthermore, $w_1(t) = w(t)$ on $[0, T)$ and $w_1(t + T) = w_1(t)$ for all $t \in [0, T)$. Also, $w_1\mid_{(-\infty,0]} = w\mid_{(-\infty,0]}$. The latter implies that $w_1\mid_{[0,\infty)} \in \mathcal{B}^+(w)$. Since $\mathcal{B}^+(w) = \mathcal{B}^+(\sigma^T w)$ we find that $\sigma^T w \underset{0}{\wedge} w_1 \in \mathcal{B}$. Define

$$w_2 := \sigma^{-T}(\sigma^T w \underset{0}{\wedge} w_1).$$

Then $w_2 \in \mathcal{B}$. Also, $w_2(t) = w(t)$ on $[0,T)$, $w_2(t+T) = w_2(t)$ for all $t \in [0,2T)$ and $w_2 \mid_{(-\infty,0]} = w \mid_{(-\infty,0]}$. Carrying on inductively we can for each n construct $w_n \in \mathcal{B}$ with the following properties: (i) $w_n(t) = w(t)$ for all $t \in [0,T)$ and (ii) $w_n(t+T) = w_n(t)$ for all $t \in [0,nT)$. Now define $\bar{w}_n := \sigma^{nT} w_{2n}$. Then we have $\bar{w}_n(t) = w(t)$ for all $t \in [0,T)$ and $\bar{w}_n(t+T) = \bar{w}_n(t)$ for all $t \in [-nT,nT)$. Let $\bar{w}$ be the T-periodic continuation of $w \mid_{[0,T)}$. Let $(t_0,t_1) \in \mathcal{R}_2^+$. For n sufficiently large we have $\bar{w} \mid_{[t_0,t_1]} = \bar{w}_n \mid_{[t_0,t_1]} \in \mathcal{B} \mid_{[t_0,t_1]}$. Completeness then implies that $\bar{w} \in \mathcal{B}$. □

Using the previous lemma it is easy to show that if a time-invariant complete system is dissipative then it is also forward dissipative:

Proposition 8.2.7. *Let $\Sigma = (\mathcal{R}, W, \mathcal{B})$ be time-invariant and complete and let $s : W \to \mathcal{R}$ be a supply rate. Then (Σ, s) is dissipative if and only if (Σ, s) is forward dissipative.*

Proof. Let $T > 0$ and $w \in \mathcal{B}$ be such that $\mathcal{B}^+(w) = \mathcal{B}^+(\sigma^T w)$. Let $\bar{w} \in \mathcal{B}$ be the T-periodic continuation of $w \mid_{[0,T)}$. By dissipativeness we then have $\int_0^T s(w(t))dt = \int_0^T s(\bar{w}(t)dt \geq 0$. □

Completely analogously it can be shown that for time-invariant complete systems dissipativeness implies backward dissipativeness. Thus we arrive at the following:

Theorem 8.2.8. *Let $\Sigma = (\mathcal{R}, W, \mathcal{B})$ be time-invariant and complete and let $s : W \to \mathcal{R}$ be a supply rate. Then the following statements are equivalent:*

(a) (Σ, s) is dissipative,
(b) (Σ, s) is forward dissipative,
(c) (Σ, s) is backward dissipative. □

8.3 Internally Dissipative Systems

The definition of dissipativeness that we gave in the previous section was given completely in terms of the external behavior of the system and is independent of any consideration with respect to state space representations. In the present section we will discuss the notion of state space representation and give a definition of internal dissipativeness of a dynamical system in state space form. We will then study the relations between (external) dissipativeness and internal dissipativeness.

Definition 8.3.1. A *dynamical system in state space form* is defined as a quadruple $\Sigma_s = (T, W, X, \mathcal{B}_s)$ with T and W as in Def. 8.2.1, X the *state space* and where the *state behavior* $\mathcal{B}_s \subseteq (W \times X)^T$ satisfies the *axiom of state*:

$$\begin{aligned}\forall t \in T : &\{(w_1, x_1), (w_2, x_2) \in \mathcal{B}_s \text{ and } x_1(t) = x_2(t)\} \\ &\Rightarrow \{(w_1, x_1) \underset{t}{\wedge} (w_2, x_2) \in \mathcal{B}_s\}\end{aligned}$$

We note that if $(T, W, X, \mathcal{B}_s)$ is a dynamical system in state space form then the triple $(T, W \times X, \mathcal{B}_s)$ defines a dynamical system in the sense of Def 8.2.1. We denote by P_W the projection of $W \times X$ onto W along X, i.e., $P_W(w, x) := w$. Likewise, P_X denotes the projection of $W \times X$ onto X along W. If $(T, W, X, \mathcal{B}_s)$ is a dynamical system in state space form then $P_W \mathcal{B}_s$ will be called the *external behavior* of Σ_s and the triple $(T, W, P_W \mathcal{B}_s)$ will be called the *system induced by* Σ_s. Conversely, we will call $(T, W, X, \mathcal{B}_s)$ a *state space representation* of $(T, W, P_W \mathcal{B}_s)$. Finally, we will call $P_X \mathcal{B}_s$ the *state behavior* of Σ_s. A dynamical system in state space form Σ_s is called time-invariant if the associated dynamical system with signal alphabet $W \times X$ is time-invariant.

Let $\Sigma_s = (\mathcal{R}, W, X, \mathcal{B}_s)$ be a system in state space form. We will say that Σ_s is *observable* if the state trajectory is uniquely determined by the external signal, in the sense that the following implication holds:

$$\{(w, x_1), (w, x_2) \in \mathcal{B}_s\} \Rightarrow \{x_1 = x_2\}$$

Obviously, this condition is equivalent with the existence of a mapping $F : P_W \mathcal{B}_s \to P_X \mathcal{B}_s$ such that $(w, x) \in \mathcal{B}_s$ if and only if $x = F(w)$. It is easy to show that if Σ_s is observable and *time-invariant*, then the mapping F commutes with the shift σ^t for all t.

Assume that Σ_s is time-invariant. We will say that Σ_s is *connected* if for any two elements x_0 and x_1 in X there exists a state trajectory $x \in P_X \mathcal{B}_s$ and $t \geq 0$ such that $x(0) = x_0$ and $x(t) = x_1$, i.e., if any two points in the state space of Σ_s can be connected by means of a suitable state trajectory.

If $\Sigma = (\mathcal{R}, W, \mathcal{B})$ is a dynamical system then there always exists a state space representation of Σ, i.e., a dynamical system in state space form $\Sigma_s = (\mathcal{R}, W, X, \mathcal{B}_s)$ such that $P_W \mathcal{B}_s = \mathcal{B}$. If Σ is time-invariant then Σ_s can also be chosen to time-invariant. Two particular state space representations of a given dynamical system will play an important role in the sequel. Let $\Sigma = (\mathcal{R}, W, \mathcal{B})$ be a dynamical system. Two signals w_1 and w_2 in $\mathcal{B}$ are said to be *past equivalent* if $\mathcal{B}^+(w_1) = \mathcal{B}^+(w_2)$. In this case we will write $w_1 \overset{-}{\sim} w_2$. Clearly, the relation $\overset{-}{\sim}$ defines an equivalence relation on $\mathcal{B}$. Define $X^{past} := \mathcal{B}(mod \overset{-}{\sim})$ and let

$$\mathcal{B}_s^{past} := \{(w, x) : \mathcal{R} \to X^{past} | w \in \mathcal{B}, x(t) = (\sigma^t w)(mod \overset{-}{\sim})\}.$$

It can be shown that $\Sigma_s^{past} := (\mathcal{R}, W, X^{past}, \mathcal{B}_s^{past})$ defines a state space representation of Σ. This state space representation is called *the canonical past induced state space representation of* Σ. Obviously, if Σ is time-invariant then also Σ_s^{past} is time-invariant. In a similar way, two signals w_1 and w_2 in $\mathcal{B}$ are called *future equivalent* if $\mathcal{B}^-(w_1) = \mathcal{B}^-(w_2)$. Again this defines an equivalence relation on $\mathcal{B}$, denoted by $\overset{+}{\sim}$. We define $X^{fut} := \mathcal{B}(mod \overset{+}{\sim})$ and

$$\mathcal{B}_s^{fut} := \{(w,x) : \mathcal{R} \to X^{fut} | w \in \mathcal{B}, x(t) = (\sigma^t w)(mod \overset{+}{\sim})\}.$$

It can be shown fairly easily that also $\Sigma_s^{fut} := (\mathcal{R}, W \times X^{fut}, \mathcal{B}_s^{fut})$ defines a state space representation of Σ (note that this is not trivial, since the state axiom is not really time-symmetric). This particular state space representation of Σ is called *the canonical future induced state space representation of* Σ. Again, Σ_s^{fut} is time-invariant whenever Σ is time-invariant.

Definition 8.3.2. Let $\Sigma_s = (\mathcal{R}, W, X, \mathcal{B}_s)$ be a time-invariant dynamical system in state space form, let $s : W \to \mathcal{R}$ be a supply rate and let $V : X \to \mathcal{R}$ be a function. The triple (Σ_s, s, V) is called *internally dissipative* if for all $(t_0, t_1) \in \mathcal{R}_2^+$ and for all $(w, x) \in \mathcal{B}_s$ we have

$$V(x(t_0)) + \int_{t_0}^{t_1} s(w(t))dt \geq V(x(t_1)). \tag{8.2}$$

Any function $V : X \to \mathcal{R}$ that satisfies the inequality 8.2 is called a *storage function of* (Σ_s, s). Let us single out a special point $x^\star \in X$. If V is a storage function of (Σ_s, s) with the property that $V(x^\star) = 0$, then V is called a *normalized storage function (at $x^\star$).*

The inequality (8.2) is called the *dissipation inequality*. Obviously, if (Σ_s, s, V) is internally dissipative then for each $x^\star \in X$ there exists a normalized storage function $\tilde{V}$, i.e., a storage function with the property that $\tilde{V}(x^\star) = 0$. Indeed for any storage function V and for any real constant C the function $V + C$ is a storage function as well. Thus we can obtain a normalized storage function by defining $\tilde{V}(x) := V(x) - V(x^\star)$.

The idea is that the function $s(w(.))$ is the rate at which supply flows into the system if the system produces the particular external signal $w(.)$. Thus, if $(t_0, t_1) \in \mathcal{R}_2^+$, then $\int_{t_0}^{t_1} s(w(t))dt$ is equal to the amount of supply that flows into the system over the time interval $[t_0, t_1]$. After choosing a reference point $x^\star$, a normalized storage function V is meant to measure the amount of supply that is stored inside the system: if $x_0 \in X$ then $V(x_0)$ is equal to the amount of supply of the system if it is in the particular state x_0. Of course, by normalization the amount of supply in the reference point $x^\star$ has been agreed upon to be equal to zero. For a given point $x_0 \in X$ the statement $V(x_0) < 0$ thus means that the amount of supply of the system in x_0 is less that in $x^\star$, while the statement $V(x_0) > 0$ means that the amount of supply in x_0 is larger than in $x^\star$. The dissipation inequality expresses the property that if a system produces the signal $(w(.), x(.))$ and if $(t_0, t_1) \in \mathcal{R}_2^+$, then the total supply of the system at time t_1 cannot exceed the sum of the supply of the system at time t_0 and the amount that was supplied to the system through the external channels during the time interval $[t_0, t_1]$. Physically, this describes the property that somewhere inside the system supply must have been dissipated, for example in the form of heat.

We will now show that if a given time-invariant observable system in state space form is internally dissipative, then its externally induced system is dissipative:

Proposition 8.3.3. *Let $\Sigma_s = (\mathcal{R}, W, X, \mathcal{B}_s)$ be a time-invariant state space representation of the time-invariant system $\Sigma = (\mathcal{R}, W, \mathcal{B})$. Assume that Σ_s is observable. Let s be a supply rate. If there exists $V : X \to \mathcal{R}$ such that (Σ_s, s, V) is internally dissipative, then (Σ, s) is dissipative.*

Proof. Since Σ_s is observable, there exists a mapping $F : \mathcal{B} \to P_X \mathcal{B}_s$ such that $(w, x) \in \mathcal{B}_s$ if and only if $x = F(w)$. Define $\pi : P_X \mathcal{B}_s \to X$ by $\pi x := x(0)$. Let $w \in \mathcal{B}$ be a T-periodic signal and let $x = F(w)$. Since $\sigma^T w = w$ we have

$$\begin{aligned} x(0) &= \pi x = (\pi \circ F)(w) = (\pi \circ F \circ \sigma^T)(w) \\ &= (\pi \circ \sigma^T \circ F)(w) = (\pi \circ \sigma^T)(x) = x(T). \end{aligned}$$

From this it immediately follows that $V(x(0)) = V(x(T))$ and hence, from the dissipation inequality, that $\int_0^T s(w(t))dt \geq 0$. Since the latter holds for any T-periodic signal $w \in \mathcal{B}$ we conclude that (Σ, s) is dissipative. □

Next we will study the following question: Given a time-invariant dissipative dynamical system, does there exist an internally dissipative state space representation of this system? It will turn out that if the canonical past induced state space representation of the system is connected, then forward dissipativeness of the system is equivalent with internal dissipativeness of the canonical past induced state space representation. A similar result holds for the canonical future induced state space representation (see [37]):

Theorem 8.2.4. *Let Σ be a time-invariant dynamical system and let s be a supply rate. Let Σ_s^{past} and Σ_s^{fut} be the canonical past induced and future induced state space representations of Σ, respectively. Then we have:*

(a) If Σ_s^{past} is connected then (Σ, s) is forward dissipative if and only if there exists $V : X^{past} \to \mathcal{R}$ such that (Σ_s^{past}, s, V) is internally dissipative.
(b) If Σ_s^{fut} is connected then (Σ, s) is backward dissipative if and only if there exists $V : X^{fut} \to \mathcal{R}$ such that (Σ_s^{fut}, s, V) is internally dissipative. □

Recalling that for time-invariant and complete systems the notions of dissipativeness, forward dissipativeness and backward dissipativeness are equivalent, we thus obtain the following result:

Corollary 8.3.5. *Let Σ be a time-invariant complete dynamical system and let s be a supply rate. Let Σ_s^{past} and Σ_s^{fut} be the canonical past induced and future induced state space representations of Σ, respectively. Then we have:*

(a) If Σ_s^{past} is connected then (Σ_s, s) is dissipative if and only if there exists $V : X^{past} \to \mathcal{R}$ such that (Σ_s^{past}, s, V) is internally dissipative.
(b) If Σ_s^{fut} is connected then (Σ_s, s) is dissipative if and only if there exists $V : X^{fut} \to \mathcal{R}$ such that (Σ_s^{fut}, s, V) is internally dissipative. □

Let $\Sigma_s = (\mathcal{R}, W, X, \mathcal{B}_s)$ be a time-invariant system in state space form and let s be a supply rate. Assuming the existence of a function V such that (Σ_s, s, V) is internally dissipative, one would like to obtain general properties of the set of all possible storage functions of (Σ_s, s). As noted before, if V is a storage function then for any real constant C the function $V + C$ is a storage function as well. Thus, instead of making general statements on the set of all possible storage functions, it is more reasonable to choose an arbitrary but fixed element $x^\star \in X$ and to restrict oneself to those storage functions V with the property that $V(x^\star) = 0$, i.e., to the set of normalized storage functions. The set of normalized storage functions associated with the system Σ_s, the supply rate s and the reference point $x^\star$ will be denoted by

$$\mathcal{V}(x^\star) := \{V : X \to \mathcal{R} \,|\, (\Sigma_s, s, V) \text{ is internally dissipative and } V(x^\star) = 0\}.$$

It turns out to be possible to identify a smallest element and a largest element in the set of normalized storage functions associated with a given internally dissipative system and reference point $x^\star$. For a given connected system $\Sigma_s = (\mathcal{R}, W, X, \mathcal{B}_s)$ and supply rate s we define functions $V_a : X \to \mathcal{R} \cup \{\infty\}$ and $V_r : X \to \mathcal{R} \cup \{-\infty\}$ by:

$$V_a(x) := \sup\{-\int_0^{t_1} s(w(t))dt \,|\, t_1 \geq 0, (w, x) \in \mathcal{B}_s, x(0) = x, x(t_1) = x^\star\}$$

and

$$V_r(x) := \inf\{-\int_{t_{-1}}^{0} s(w(t))dt \,|\, t_{-1} \leq 0, (w, x) \in \mathcal{B}_s, x(0) = x, \\ x(t_{-1}) = x^\star\}.$$

The function V_a is called the *available storage* of the dynamical system Σ_s: the quantity $V_a(x)$ is the maximum amount of supply that can be extracted from the system over all state trajectories connecting x to the reference point $x^\star$. The function V_r is called the *required supply* of the system Σ_s. The quantity $V_r(x)$ is equal to the minimum amount of supply that has to be delivered to the system in order to connect the reference point $x^\star$ to the point x. The following important theorem then holds:

Theorem 8.3.6. *Let $\Sigma_s = (\mathcal{R}, W, X, \mathcal{B}_s)$ be a time-invariant system in state space form and let s be a supply rate. Assume that Σ_s is connected. Let $x^\star \in X$. Then the following statements are equivalent:*

(a) There exists $V : X \to \mathcal{R}$ such that (Σ_s, s, V) is internally dissipative,
(b) $V_a(x) < \infty$ for all $x \in X$,
(c) $V_r(x) > -\infty$ for all $x \in X$.

If one of these statements holds then both $V_a \in \mathcal{V}(x^\star)$ as well as $V_r \in \mathcal{V}(x^\star)$. In addition, for all $V \in \mathcal{V}(x^\star)$ we have $V_a \leq V \leq V_r$.

Proof. (a) $\Rightarrow$ (b) Let V be a storage function for (Σ_s, s) such that $V(x^\star) = 0$. Let $x_0 \in X$. By the dissipation inequality, for any $(w, x) \in \mathcal{B}_s$ such that $x(0) = x_0$ and $x(t_1) = x^\star$ we have $V(x_0) + \int_0^{t_1} s(w(t))dt \geq 0$. This yields $-\int_0^{t_1} s(w(t))dt \leq V(x_0)$. By taking the supremum on the left in the latter inequality we obtain $V_a(x_0) \leq V(x_0)$. (a) $\Rightarrow$ (c) Again let V be a storage function such that $V(x^\star) = 0$. Let $x_0 \in X$. For any $(w, x) \in \mathcal{B}_s$ such that $x(t_{-1}) = x^\star$ and $x(0) = x_0$ we have $\int_{t_{-1}}^0 s(w(t))dt \geq V(x_0)$. This yields $V_r(x_0) \geq V(x_0)$. The implications (b) $\Rightarrow$ (a) and (c) $\Rightarrow$ (a) follow from the fact that if V_a and V_r define real (finite) valued function on X then these functions satisfy the dissipation inequality. Consequently, V_a and V_r are storage functions so (Σ_s, s, V_a) and (Σ_s, s, V_r) are internally dissipative. We will now show that $V_a(x^\star) = 0$. The fact that $V_a(x^\star) \geq 0$ follows immediately from the definition (take $t_1 = 0$). To prove the converse inequality, let $(w, x) \in \mathcal{B}_s$ such that $x(0) = x^\star$ and $x(t_1) = x^\star$ Let V be a normalized storage function. It follows from the dissipation inequality that $-\int_0^{t_1} s(w(t))dt \leq 0$. Taking the supremum on the left in this inequality then yields $V_a(x^\star) \leq 0$. A proof that V_r is normalized can be given in a similar way. □

To summarize the above, we have shown that any possible normalized storage function of an internally dissipative system satisfies the *a priori* sharp inequality

$$V_a \leq V \leq V_r.$$

This can be interpreted by saying that an internally dissipative system can never supply to the outside more than what it has stored and can never store more than what has been supplied to it. Of course, not every function that satisfies the above inequality will be a storage function. We will state one more interesting property of the set of all normalized storage functions. It turns out that this set is *convex*:

Theorem 8.3.7. *Let $\Sigma_s = (\mathcal{R}, W, X, \mathcal{B}_s)$ be a time-invariant system in state space form and let s be a supply rate. Assume that there exists a function $V : X \to \mathcal{R}$ such that (Σ_s, s, V) is internally dissipative. Let $x^\star \in X$. Then for any pair of storage functions $V_1, V_2 \in \mathcal{V}(x^\star)$ and any $\alpha \in [0,1]$ we have $\alpha V_1 + (1-\alpha)V_2 \in \mathcal{V}(x^\star)$. Hence, if Σ_s is connected then $\alpha V_a + (1-\alpha)V_r \in \mathcal{V}(x^\star)$ for all $\alpha \in [0,1]$.* □

We shall now turn to the question whether for a given time-invariant system in state space form there exist storage functions that only take *non-negative values*, i.e., storage functions V with the property that $V(x_0) \geq 0$ for all $x_0 \in X$. Let $\Sigma_s = (\mathcal{R}, W, X, \mathcal{B}_s)$ be a time-invariant system in state space form and let s be a supply rate. First observe that there exists a storage function $V \geq 0$ if and only if there exists a storage function $\tilde{V}$ that is bounded from below, i.e., a storage function $\tilde{V}$ for which there exists a real number M such that $\tilde{V}(x_0) \geq M$ for all $x_0 \in X$. Indeed this follows immediately from the fact that if $\tilde{V}$ is a storage function and if $C \in \mathcal{R}$, then also $\tilde{V} + C$ is a storage function. In particular, if there exists a storage function $\tilde{V}$ with the property that there exists $x^\star \in X$ such that $\tilde{V}(x^\star) = \min_{x \in X} \tilde{V}(x)$ (i.e., $x^\star$ is a point of *minimal storage*) then there exists a storage function $V \geq 0$.

An important quantity in connection with the existence of non-negative storage functions is the *free-endpoint available storage*, which is defined by:

$$V_{a,f}(x_0) := \sup\{-\int_0^{t_1} s(w(t))dt \mid t_1 \geq 0, (w,x) \in \mathcal{B}_s, x(0) = x_0\}$$

Note that $V_{a,f}(x_0) \in [0,\infty) \cup \{\infty\}$ for all $x_0 \in X$. It turns out that the free-endpoint available storage is finite if and only if there exists a non-negative storage function. In fact, in this case the free endpoint available storage is the *smallest* non-negative storage function:

Theorem 8.3.8. *Let $\Sigma_s = (\mathcal{R}, W, X, \mathcal{B}_s)$ be a time-invariant system in state space form and let s be a supply rate. There exists a storage function $V \geq 0$ for (Σ_s, s) if and only if $V_{a,f}(x_0) < \infty$ for all $x_0 \in X$. If this is the case then $V_{a,f}$ itself is a storage function and $0 \leq V_{a,f} \leq V$ for every storage function $V \geq 0$.*

Proof. Let $V \geq 0$ be a storage function. For all $(w,x) \in \mathcal{B}_s$ such that $x(0) = x_0$ and for all $t_1 \geq 0$ we have

$$V(x_0) + \int_0^{t_1} s(w(t))dt \geq V(x(t_1)) \geq 0,$$

which implies

$$-\int_0^{t_1} s(w(t))dt \leq V(x_0).$$

Taking the supremum on the left in this inequality yields $V_{a,f}(x_0) \leq V(x_0)$. Conversely, if $V_{a,f}$ is a real valued function then it can easily be shown to satisfy the dissipation inequality. □

If there exists a non-negative storage function and if $x^\star \in X$ then we would like to know whether there also exists a *normalized* non-negative storage function, i.e., a storage function $V \geq 0$ with the property that $V(x^\star) = 0$. The set of normalized non-negative storage functions associated with the system in state space form Σ_s, supply rate s and reference point $x^\star \in X$ will be denoted by $\mathcal{V}_+(x^\star)$. We will show that there exists a normalized non-negative storage function if and only if there exists a (arbitrary) storage function for which $x^\star$ is a point of minimal storage. Furthermore, in that case the free endpoint available storage is the *smallest* normalized non-negative storage function and the required supply is the *largest* normalized non-negative storage function:

Theorem 8.3.9. *Let $\Sigma_s = (\mathcal{R}, W, X, \mathcal{B}_s)$ be a connected time-invariant system in state space form and let s be a supply rate. Let $x^\star \in X$. Then the following statements are equivalent:*

(a) $\mathcal{V}_+(x^\star) \neq \emptyset$,
(b) there exists a storage function V such that $V(x^\star) = \min_{x \in X} V(x)$,
(c) $V_r(x) \geq 0$ *for all* $x \in X$,

(d) $V_{a,f}(x) < \infty$ *for all* $x \in X$ *and* $V_{a,f}(x^\star) = 0$.

If one of the above statements hold then we have $V_{a,f} \in \mathcal{V}_+(x^\star)$ *and* $V_r \in \mathcal{V}_+(x^\star)$. *Furthermore, for any* $V \in \mathcal{V}_+(x^\star)$ *we have:*

$$V_a \leq V_{a,f} \leq V \leq V_r.$$

Here, V_a *and* V_r *denote the available storage and the required supply, respectively, taken with respect to the reference point* $x^\star$.

Proof. The implication (a) ⇒(b) is obvious. (b) ⇒(a) Define a new storage function $\tilde{V} \in \mathcal{V}(x^\star)$ by $\tilde{V}(x) := V(x) - V(x^\star)$. (a) ⇒(d) By th. 2.8 we have $V_{a,f} \leq V$ for any storage function $V \geq 0$. Thus, if there exists such V with the property that $V(x^\star) = 0$ then $V_{a,f}(x^\star) \leq 0$. Since the converse inequality always holds, we conclude that $V_{a,f}$ is normalized. (d) ⇒(a) Is obvious. (a) ⇒(c) By Th. 8.3.6, for every normalized storage function we have $V \leq V_r$. Thus, if there exists a non-negative normalized storage function then we have $V_r \geq 0$. (c) ⇒(a) If $V_r \geq 0$ then it is a non-negative storage function. It follows immediately from the definition that $V_r(x^\star) \leq 0$ and hence $V_r(x^\star) = 0$. □

Example 8.3.10. Consider a simple (nonlinear) series RLC-circuit, containing inductance, capacitance and resistance. The circuit interacts with its environment through the external signal (I, V), where I denotes the current into the network and where V is the voltage across the external terminals. Let V_R, I_R denote the voltage across and the current into the resistor, respectively. The characteristics of the resistor are described by

$$V_R = R(I_R)I_R, \tag{8.3}$$

where $R(.) \geq 0$ is a given smooth function. Likewise, let V_C, I_C be the voltage across and the current into the capacitor, respectively. If Q_C is the charge on the capacitor, then we have $Q_C = C(V_C)$ for some smooth fuction $C(.)$ with $C'(.) > 0$. Hence the characteristics of this element are given by

$$I_C = C'(V_C)\frac{dV_C}{dt}. \tag{8.4}$$

For the inductor, let the voltage and the current be denoted by V_L and I_L. If Φ_L is the flux, then we have $\Phi_L = L(I_L)$ for some smooth function $L(.)$ with $L'(.) > 0$. Hence the characteristics of the inductor are given by

$$V_L = L'(I_L)\frac{dI_L}{dt}. \tag{8.5}$$

Furthermore, the behavior of the network is governed by Kirchoff's laws:

$$I = I_R = I_C = I_L \tag{8.6}$$

$$V = V_R + V_C + V_L \tag{8.7}$$

Thus we see that the electrical network can be modelled as a dynamical system $\Sigma = (T, W, \mathcal{B})$ with $T = \mathcal{R}$, $W = \mathcal{R}^2$ and behavior $\mathcal{B}$ given by

$$\mathcal{B} = ((I, V) : \mathcal{R} \to \mathcal{R}^2 | \exists (I_R, V_R, I_C, V_C, I_L, V_L) : \mathcal{R} \to \mathcal{R}^6 \quad \text{satisfying equations (8.3)–(8.7). } \}.$$

A state space representation of this dynamical system is described as follows. For the state variables we take I_L and V_C and for the state space we take $X = \mathcal{R}^2$. Define a function $f : \mathcal{R}^3 \to \mathcal{R}^2$ by

$$f(I_L, V_C, V) = \left(\frac{-V_C - R(I_L)I_L + V}{L'(I_L)}, \frac{I_L}{C'(V_C)} \right).$$

It is easily verified that $\Sigma_s = (\mathcal{R}, W, X, \mathcal{B}_s)$, with

$$\mathcal{B}_s = \{((I, V), (I_L, V_C)) : \mathcal{R} \to \mathcal{R}^4 | I = I_L \text{ and } (\dot{I_L}, \dot{V_C}) = f(I_L, V_L, V)\}$$

is a state space representation of Σ. The rate at which electrical energy flows into the network is given by the function $I(.)V(.)$. Thus it is reasonable to define a supply rate $s : \mathcal{R}^2 \to \mathcal{R}$ by

$$s(I, V) = IV.$$

The electrical energy in the capacitor in the presence of a voltage V_C is equal to

$$E_C(V_C) = \int_0^{V_C} vC'(v)dv,$$

and the electrical energy in the inductor in the presence of a current I_L is equal to

$$E_L(I_L) = \int_0^{I_L} iL'(i)di.$$

Denote the total amount of electrical energy in the network by

$$E(I_L, V_C) = E_C(V_C) + E_L(I_L).$$

It can easily be shown that (Σ_s, s, E) is internally dissipative, i.e., that $E(.,.)$ is a storage function.

8.4 Linear Systems with Quadratic Supply Rate

In the previous sections we have studied the concept of dissipativeness on a rather high level of generality. Most of our results were concerned with general time-invariant dynamical systems, while sometimes in addition we assumed the system to be complete. In the present section we shall make the additional assumption that the systems under consideration are *linear*. Furthermore, we shall be concerned with supply rates that are *quadratic functions*. We will study time-invariant dynamical systems $\Sigma = (\mathcal{R}, W, \mathcal{B})$ for which the signal alphabet W is equal to $\mathcal{R}^q$, with q a given positive integer. Such a dynamical system is called *linear* if its behavior $\mathcal{B}$ is a linear subspace of the real linear space $(\mathcal{R}^q)^{\mathcal{R}}$. The state space representations of the systems that will be considered in this section will all be finite-dimensional. A system in state space form Σ_s is called *finite-dimensional* if its state space X is equal to $\mathcal{R}^n$ for some positive integer n. A finite-dimensional system in state space form $\Sigma_s = (\mathcal{R}, \mathcal{R}^q, \mathcal{R}^n, \mathcal{B}_s)$ is called linear if $\mathcal{B}_s$ is a linear subspace of $(\mathcal{R}^q \times \mathcal{R}^n)^{\mathcal{R}}$.

An important class of linear, time-invariant and complete dynamical systems is the class of systems for which there exists a state space representation in the form of a *driving variable representation.* A system in state space form $\Sigma_s = (\mathcal{R}, \mathcal{R}^q, \mathcal{R}^n, \mathcal{B}_s)$ is said to have a driving variable representation if there exist a non-negative integer m and matrices $A \in \mathcal{R}^{n\times n}$, $B \in \mathcal{R}^{n\times m}$, $C \in \mathcal{R}^{q\times n}$ and $D \in \mathcal{R}^{q\times m}$ such that the behavior $\mathcal{B}_s$ is equal to

$$\mathcal{B}_{DV}(A,B,C,D) := \{(w,x) : \mathcal{R} \to \mathcal{R}^q \times \mathcal{R}^n \mid \exists\, v \in L_{2,loc}(\mathcal{R}^+) \\ \text{such that } \dot{x} = Ax + Bv, w = Cx + Dv\}$$

It is easily seen that the quadruple $(\mathcal{R}, \mathcal{R}^q, \mathcal{R}^n, \mathcal{B}_{DV}(A,B,C,D))$ indeed defines a linear, finite-dimensional, time-invariant dynamical system in state space form. This system will be denoted by $\Sigma_{DV}(A,B,C,D)$ or sometimes simply by Σ_{DV}. It is easy to see that the externally induced system of Σ_{DV} is linear, time-invariant and complete. In this section we shall restrict ourselves to linear, finite-dimensional and complete dynamical systems $\Sigma = (\mathcal{R}, W, \mathcal{B})$ for which there exists a state space representation $\Sigma_{DV} = (\mathcal{R}, \mathcal{R}^q, \mathcal{R}^n, \mathcal{B}_{DV}(A,B,C,D))$ in driving variable representation. The driving variable state space representation Σ_{DV} of Σ is called *minimal* if both the integer m (i.e., the dimension of the linear space in which the driving variable v takes its values) as well as n (the dimension of the state space) are minimal (over the class of all driving variable representations of the given system Σ). It can be shown that if Σ_{DV} is minimal, then it is observable (in the sense of section 2, i.e., the state trajectory is uniquely determined by the external signal w). Furthermore, it is easily seen that $\Sigma_{DV}(A,B,C,D)$ is connected if and only if the pair (A,B) is controllable.

Throughout this section we will assume that the supply rate s is a quadratic function. More specifically, we will assume that $s : \mathcal{R}^q \to \mathcal{R}$ is given by $s(w) = w^T M w$, where $M \in \mathcal{R}^{q\times q}$ is a given symmetric matrix. It turns out that if the externally induced system of a state space system in driving variable representation is dissipative, then there exists at least one *quadratic* storage function. In fact:

Theorem 8.4.1. *Let $\Sigma_{DV}(A,B,C,D)$ be a state space representation of Σ and assume that Σ_{DV} is connected. Then the following statements are equivalent:*

(a) There exists $V : \mathcal{R}^n \to \mathcal{R}$ such that (Σ_{DV}, s, V) is internally dissipative,
(b) There exists a symmetric matrix $K \in \mathcal{R}^{n\times n}$ such that if we define $V(x) := x^T K x$ then (Σ_{DV}, s, V) is internally dissipative,
(c) For all $(w,x) \in \mathcal{B}_{DV}(A,B,C,D)$ and for all $T \geq 0$ such that $x(0) = x(T) = 0$ we have $\int_0^T s(w(t))dt \geq 0$.

If, in addition, Σ_{DV} is a minimal state space representation of Σ then any of the above statements is equivalent with:

(d) (Σ, s) *is dissipative.*

Proof. (a) $\Rightarrow$(c) Let V be any storage function and let $(w, x) \in \mathcal{B}_{DV}$ be such that $x(0) = x(T) = 0$. It then follows immediately from the dissipation inequality that $\int_0^T s(w(t))dt \geq 0$. (c) $\Rightarrow$(b) Assuming that (c) holds, for $x_0 \in X$ define

$$V(x_0) := \sup\{-\int_0^{t_1} s(w(t))dt \mid t_1 \geq 0, (w, x) \in \mathcal{B}_{DV},$$
$$x(0) = x_0, x(t_1) = 0\}.$$

In a similar way as in the proof of Th. 8.3.6 it can be shown that V defines a storage function for (Σ_{DV}, s). Now, we claim that, in fact, V is a quadratic function of x_0. Indeed, this follows from [23] upon noting that s is quadratic so $V(x_0)$ represents the optimal cost of a linear quadratic optimization problem. (b) $\Rightarrow$(a) Is obvious. (d) $\Rightarrow$(c) If $(w, x) \in \mathcal{B}_{DV}$ is such that $x(0) = x(T)$, then clearly $\mathcal{B}^+(w) = \mathcal{B}^+(\sigma^T w)$. By completeness, the system Σ is forward dissipative. It follows that $\int_0^T s(w(t))dt \geq 0$. Finally, the implication (a) $\Rightarrow$(d) follows from Th. 8.3.3 (use minimality). □

The set of all quadratic storage functions of a given internally dissipative system $\Sigma_{DV}(A, B, C, D)$ can be characterized as the set of real symmetric solutions of a linear matrix inequality involving the system parameters A, B, C and D and the symmetric matrix M defining the supply rate s. Indeed, if $V(x) = x^T K x$ is a quadratic storage function then the dissipation inequality can be reformulated as: for all $(t_0, t_1) \in \mathcal{R}_+^2$ and for all $(w, x) \in \mathcal{B}_{DV}(A, B, C, D)$ we have

$$\int_{t_0}^{t_1} \left(-\frac{d}{dt}(x(t)^T K x(t)) + w(t)^T M w(t)\right) dt \geq 0.$$

Since $(w, x) \in \mathcal{B}_{DV}(A, B, C, D)$ if and only if there exists $v \in L_{2,loc}(\mathcal{R}^+)$ such that $\dot{x} = Ax + Bv$, $w = Cx + Dv$, the latter inequality can be seen to be equivalent to: for all $v \in L_{2,loc}(\mathcal{R}^+)$ and for all x such that $\dot{x} = Ax + Bv$ we have

$$\int_{t_0}^{t_1} (x(t)^T, v(t)^T) \begin{pmatrix} -A^T K - KA + C^T M C & -KB + C^T M D \\ -B^T K + D^T M C & D^T M D \end{pmatrix} \begin{pmatrix} x(t) \\ v(t) \end{pmatrix} dt \geq 0.$$

It is easily seen that the latter is equivalent with the single requirement that the matrix K satisfies the *linear matrix inequality* (LMI):

$$\begin{pmatrix} -A^T K - KA + C^T M C & -KB + C^T M D \\ -B^T K + D^T M C & D^T M D \end{pmatrix} \geq 0.$$

This leads to the following result:

Theorem 8.4.2. *Consider the system $\Sigma_{DV}(A,B,C,D)$, together with the quadratic supply rate $s(w) = w^T M w$. Let $K \in \mathcal{R}^{n\times n}$ be a symmetric matrix. Then the following statements are equivalent:*

(a) $V(x) = x^T K x$ is a storage function for (Σ_{DV}, s),
(b) K is a solution of the linear matrix inequality LMI.

If there exists a function $V : \mathcal{R}^n \to \mathcal{R}$ such that (Σ_{DV}, s, V) is internally dissipative and if Σ_{DV} is connected then there exist symmetric solutions $K^-, K^+ \in \mathcal{R}^{n\times n}$ of the LMI such that for any symmetric solution $K \in \mathcal{R}^{n\times n}$ of the LMI we have

$$K^- \leq K \leq K^+.$$

In fact, for all $x \in \mathcal{R}^n$ we have

$$V_a(x) = x^T K^- x, \; V_r(x) = x^T K^+ x.$$

Here, V_a and V_r denote the available storage and the required supply with respect to the reference point $x^* = 0$.

Proof. The equivalence of statements (a) and (b) was already proven in the above. Consider the available storage V_a and the required supply V_r. Again, being defined as the optimal costs corresponding to linear quadratic optimization problems, these functions are quadratic. Let K^- and K^+ be real symmetric matrices such that $V_a(x) = x^T K^- x$ and $V_r(x) = x^T K^+ x$ for all $x \in \mathcal{R}^n$. By the above, K^- and K^+ are solutions to the LMI. The claim that these solutions are extremal then follows from Th. 8.3.6 (with $x^* = 0$). □

In addition to the necessary and sufficient conditions that were derived in Th. 8.4.1, we will now derive a frequency domain condition for internal dissipativeness. In order to be able to explain the idea, assume for the moment that we allow all signals to be complex valued. Consider the system $\Sigma_{DV}(A,B,C,D)$ and define a real rational matrix $G(s)$ by $G(s) := C(Is - A)^{-1}B + D$. We contend that if the complex number $i\omega$ is not an eigenvalue of the matrix A, then for each $v_0 \in \mathcal{C}^m$ the signal

$$(w,x) = (e^{i\omega t}G(i\omega)v_0, e^{i\omega t}(Ii\omega - A)^{-1}Bv_0)$$

is an element of (the complexification of) $\mathcal{B}_{DV}(A,B,C,D)$. Indeed, as driving variable take $v(t) := e^{i\omega t}v_0$ and as initial condition take $x_0 := (Ii\omega - A)^{-1}Bv_0$. It is a matter of straightforward calculation to verify that the signal (w,x) given above indeed satisfies the equations $\dot{x} = Ax + Bv$, $w = Cx + Dv$. Now, let the symbol $*$ denote conjugate transpose. We obviously have

$$\begin{aligned} v_0^* G(-i\omega)^T M G(i\omega) v_0 &= v(t)^* G(-i\omega)^T M G(i\omega) v(t) \\ &= w(t)^* M w(t) = s(w(t)). \end{aligned}$$

Assume now that Σ_{DV} is internally dissipative and let V be a storage function. Since x is periodic (with period $2\pi/\omega$), by the dissipation inequality we obtain

$\int_0^{2\pi/\omega} s(w(t))dt \geq 0$. This immediately implies that $v_0^* G(-i\omega)^T M G(i\omega) v_0 \geq 0$. Thus we find that the hermitian matrix $G(-i\omega)^T M G(i\omega)$ is positive semi-definite. In this way we are lead to the following result:

Theorem 8.4.3. *Assume that the system $\Sigma_{DV}(A, B, C, D)$ is connected. Then the following statements are equivalent:*

(a) (Σ_{DV}, s) is internally dissipative,
(b) $G(-i\omega)^T M G(i\omega) \geq 0$ for all $\omega \in \mathcal{R}$, $i\omega \notin \sigma(A)$.

Proof. A proof of the implication (a) $\Rightarrow$(b) can be given using the above ideas. We will prove the converse implication (b) $\Rightarrow$(a). Let $(w, x) \in \mathcal{B}_{DV}$ be such that $x(0) = 0$ and $x(T) = 0$. Define $(\bar{w}, \bar{x})$ as the concatenation $(0,0) \underset{0}{\wedge} (w, x) \underset{T}{\wedge} (0,0)$. Then clearly $(\bar{w}, \bar{x}) \in \mathcal{B}_{DV}$. Let v be a corresponding driving variable. We can take $v(t) = 0$ for $t \notin [0, T]$. Let $x(i\omega)$, $v(i\omega)$ and $w(i\omega)$ be the Fourier-transforms of $\bar{x}$, v and $\bar{w}$, respectively. Then for all $i\omega \notin \sigma(A)$ we have $x(i\omega) = (Ii\omega - A)^{-1} Bv(i\omega)$ so $w(i\omega) = G(i\omega)v(i\omega)$. Using Parseval's theorem this yields $\int_0^T s(w(t))dt \geq 0$. The claim then follows from Th. 8.4.1 □

Summarizing, we obtain the following result on the existence of solutions of the linear matrix inequality LMI:

Corollary 8.4.4. *Let $A \in \mathcal{R}^{n\times n}$, $B \in \mathcal{R}^{n\times m}$, $C \in \mathcal{R}^{q\times n}$ and $D \in \mathcal{R}^{q\times m}$. Assume that (A, B) is a controllable pair. Let $M \in \mathcal{R}^{q\times q}$ be symmetric. Define $G(s) := C(Is - A)^{-1}B + D$. Then the following two statements are equivalent:*

(a) the LMI has a symmetric solution $K \in \mathcal{R}^{n\times n}$,
(b) $G(-i\omega)^T M G(i\omega) \geq 0$ for all $\omega \in \mathcal{R}$, $i\omega \notin \sigma(A)$.

In that case there exist symmetric solutions $K^-, K^+ \in \mathcal{R}^{n\times n}$ such that for any symmetric solution $K \in \mathcal{R}^{n\times n}$ we have $K^- \leq K \leq K^+$. □

We will now study the existence of non-negative storage functions in the context of linear systems with quadratic supply rate. As a first observation, note that for a given linear system $\Sigma_{DV}(A, B, C, D)$ there is a one-to-one correspondence between the set of all non-negative *quadratic* storage functions and the set of real *positive semi-definite* solutions of the linear matrix inequality LMI. We will show that if there exists a non-negative storage function, then there always exists a non-negative *quadratic* storage function and hence a positive semi-definite solution to the LMI. This result will imply that if the linear matrix inequality has a real positive semi-definite solution, then it always has a *smallest* real positive semi-definite solution and a *largest* real positive semi-definite solution.

Theorem 8.4.5. *Consider the system $\Sigma_{DV}(A, B, C, D)$, together with the quadratic supply rate $s(w) = w^T M w$. Assume that Σ_{DV} is connected. Then the following statements are equivalent:*

(a) There exists a storage function $V \geq 0$ for (Σ_{DV}, s),

(b) There exists $K \in \mathcal{R}^{n\times n}$, $K \geq 0$ such that $V(x) := x^T K x$ is a storage function for (Σ_{DV}, s),
(c) The LMI has a solution $K \geq 0$,
(d) For all $(w, x) \in \mathcal{B}_{DV}$ such that $x(0) = 0$ and $T \geq 0$ we have $\int_0^T s(w(t))dt \geq 0$,
(e) $V_r(x) \geq 0$ for all $x \in \mathcal{R}^n$,
(f) $V_{a,f}(x) < \infty$ for all $x \in \mathcal{R}^n$.

If one of these statements holds then we have $K^+ \geq 0$ and there exists a real solution $K_f \geq 0$ of the LMI such that for any real solution $K \geq 0$ of the LMI we have

$$K_f \leq K \leq K^+.$$

In fact, for all $x \in \mathcal{R}^n$ we have $V_{a,f}(x) = x^T K_f x$ and $V_r(x) = x^T K^+ x$. Here, V_r denotes the required supply with respect to the reference point $x^\star = 0$.

Proof. The equivalences (a) $\Leftrightarrow$ (f) and (b) $\Leftrightarrow$ (c) follow immediately from Th. 8.3.9 and Th. 8.4.2, respectively. The implications (e) $\Rightarrow$(a) and (b) $\Rightarrow$(e) follow from Th. 8.3.9. The claim (f) $\Rightarrow$(b) follows by noting that $V_{a,f}(x_0)$ is the optimal cost of a linear quadratic optimization problem. Thus, there exists a real matrix $K_f \geq 0$ such that $V(x_0) = x_0^T K_f x_0$. The implication (b) $\Rightarrow$(d) follows by writing out the dissipation inequality for the quadratic storage function $V(x) = x^T K x$. A proof of the implication (d) $\Rightarrow$(f) can be given completely analogously to the proof of Th. 8.3.4. The remaining statements follow from Th. 8.3.9 and Th. 8.4.2. □

The condition appearing in statement (d) of the above theorem is often used as a definition of *passivity* of the system Σ_{DV}: the system Σ_{DV} is called passive if for any signal $(w, x) \in \mathcal{B}_{DV}$ with $x(0) = 0$ we have $\int_0^T s(w(t))dt \geq 0$ for all $T \geq 0$. Thus we see that a connected system is passive if and only if there exists a non-negative storage function.

To conclude this section we shall derive frequency domain conditions for the existence of non-negative storage functions for the linear system $\Sigma_{DV}(A, B, C, D)$. Again, define $G(s) := C(Is - A)^{-1}B + D$ and for the moment, assume that we allow all signals to be complex valued. Let s be a complex number that is not an eigenvalue of A. Then for each $v_0 \in \mathcal{C}^m$ the signal

$$(w, x) = (e^{st}G(s)v_0, e^{st}(Is - A)^{-1}Bv_0)$$

is an element of the complexification of $\mathcal{B}_{DV}(A, B, C, D)$. Indeed, if we take as driving variable $v(t) := e^{st}v_0$ then the equations $\dot{x} = Ax + Bv$, $w = Cx + Dv$ are satisfied. Let $\bar{s}$ denote the complex conjugate of s. It is easily verified that

$$e^{2t\Re e\, s}v_0^* G(\bar{s})^T M G(s) v_0 = v(t)^* G(\bar{s})^T M G(s) v(t) = w(t)^* M w(t) = s(w(t)).$$

Now, assume that there exists a non-negative storage function, say, $V \geq 0$. Then the free endpoint available storage $V_{a,f}(x_0)$ is finite and for all $T \geq 0$ we must have

$$-\int_0^T s(w(t))dt \leq V_{a,f}(x_0)$$

By the above, this implies that for all $s \notin \sigma(A)$, for all $T \geq 0$ we have

$$\int_0^T e^{2t\Re e\, s} dt \,.\, v_0^* G(\bar{s})^T M G(s) v_0 \geq -V_{a,f}(x_0).$$

This can of course only be true if for all $s \in \mathcal{C}$ with $s \notin \sigma(A)$ and $\Re e\ s \geq 0$, we have $v_0^* G(\bar{s})^T M G(s) v_0 \geq 0$. This argument holds for all $v_0 \in \mathcal{C}^m$ and thus we find that the hermitian matrix $G(\bar{s})^T M G(s)$ must be positive semi-definite for all such s. This leads to the following theorem:

Theorem 8.4.9. *Assume that the system $\Sigma_{DV}(A, B, C, D)$ is connected. Then the following statements are equivalent:*

(a) There exists a non-negative storage function for (Σ_{DV}, s)
(b) $G(\bar{s})^T M G(s) \geq 0$ for all $s \in \mathcal{C}$ with $s \notin \sigma(A)$, $\Re e\ s \geq 0$. □

Summarizing, we obtain the following results on the existence of positive semi-definite solutions of the LMI:

Corollary 8.4.7. *Let $A \in \mathcal{R}^{n\times n}$, $B \in \mathcal{R}^{n\times m}$, $C \in \mathcal{R}^{q\times n}$ and $D \in \mathcal{R}^{q\times m}$. Assume that (A, B) is a controllable pair. Let $M \in \mathcal{R}^{q\times q}$ be symmetric. Define $G(s) := C(Is - A)^{-1}B + D$. Then the following statements are equivalent:*

(a) the LMI has a real symmetric solution $K \geq 0$,
(b) $G(\bar{s})^T M G(s) \geq 0$ for all $s \in \mathcal{C}$ with $s \notin \sigma(A)$, $\Re e\ s \geq 0$.

In that case the LMI has real symmetric solutions K^-, K_f and K^+ with $K_f \geq 0$ and $K^+ \geq 0$ with the following properties: for any real symmetric solution K of the LMI we have $K^- \leq K \leq K^+$ and for any real symmetric solution $K \geq 0$ we have $K_f \leq K \leq K^+$. □

8.5 Positive Real and Bounded Real Functions

A special case of the class of systems considered in the previous section is the class of linear, finite-dimensional, time-invariant systems in input/state/output form. Such a system is defined as a time-invariant linear system in state space form $\Sigma_s = (\mathcal{R}, W, X, \mathcal{B}_s)$ for which the signal alphabet W is equal to the cartesian product $U \times Y$, with $U = \mathcal{R}^m$ and $Y = \mathcal{R}^p$, the state space X is equal to $\mathcal{R}^n$ and $\mathcal{B}_s$ is equal to

$$\mathcal{B}_{i/s/o}(A, B, C_0, D_0) = \{(\binom{u}{y}, x) : \mathcal{R} \to \mathcal{R}^m \times \mathcal{R}^p \times \mathcal{R}^n \mid u \in L_{2,loc}(\mathcal{R}^+) \text{ and } \dot{x} = Ax + Bu,\ y = C_0 x + D_0 u\}.$$

Here, A, B, C_0 and D_0 are matrices in $\mathcal{R}^{n\times n}$, $\mathcal{R}^{n\times m}$, $\mathcal{R}^{p\times n}$ and $\mathcal{R}^{p\times m}$, respectively. Thus, the external signal w is equal to col(u, y). The function u is called the *input*, the function y is called the *output*. Likewise, $U = \mathcal{R}^m$ is called the *input alphabet*, while $Y = \mathcal{R}^p$ is called the *output alphabet*. The system $(\mathcal{R}, \mathcal{R}^m \times \mathcal{R}^p, \mathcal{R}^n, \mathcal{B}_{i/s/o}(A, B, C_0, D_0))$ will be denoted by $\Sigma_{i/s/o}(A, B, C_0, D_0)$ or simply by $\Sigma_{i/s/o}$. The *transfer matrix* of $\Sigma_{i/s/o}$ is defined as the real rational matrix $G_0(s) = C_0(Is - A)^{-1}B + D_0$. It is easy to see that any system $\Sigma_{i/s/o}$ can be considered as a system in state space form in driving variable representation Σ_{DV}, with the driving variable equal to the input. Indeed, we always have

$$\Sigma_{i/s/o}(A, B, C_0, D_0) = \Sigma_{DV}(A, B, C, D),$$

with

$$C := \begin{pmatrix} 0 \\ C_0 \end{pmatrix}, \; D := \begin{pmatrix} I \\ D_0 \end{pmatrix}.$$

Now, let $\Sigma_{i/s/o}(A, B, C_0, D_0)$ be given. Assume that $U = Y$ so, in particular, that $p = m$. By taking the supply rate s defined by $s(u, y) = u^T y$ or, equivalently, $s(w) = w^T M w$ with M given by

$$M := \frac{1}{2}\begin{pmatrix} 0 & I \\ I & 0 \end{pmatrix}$$

we obtain the following special case of Cor. 8.4.7:

Corollary 8.5.1. *Assume that the system $\Sigma_{i/s/o}(A, B, C_0, D_0)$ is connected (equivalently: the pair (A, B) is controllable). Let $G_0(s) = C_0(Is - A)^{-1}B + D_0$ be the transfer matrix of $\Sigma_{i/s/o}$. Define a supply rate by $s(u, y) = u^T y$. Then the following statements are equivalent:*

(a) There exists a function $V \geq 0$ such that $(\Sigma_{i/s/o}, s, V)$ is internally dissipative,
(b) $G_0(\bar{s})^T + G_0(s) \geq 0$ for all $s \in \mathbb{C}$ with $s \notin \sigma(A)$ and $\Re e\, s \geq 0$,
(c) the linear matrix inequality

$$\begin{pmatrix} -A^T K - KA & -KB + C_0^T \\ -B^T K + C_0 & D_0^T + D_0 \end{pmatrix} \geq 0 \tag{8.8}$$

has a real symmetric solution $K \geq 0$.

In that case there exist real symmetric solutions $K_f \geq 0$ and $K^+ \geq 0$ such that for any solution $K \geq 0$ we have $K_f \leq K \leq K^+$. □

A transfer function $G_0(s)$ that satisfies the condition in statement (b) of the above corollary is called *positive real*. The result on the equivalence of statements (b) and (c) is known as the *positive real lemma* or the *Kalman-Yakubovich-Popov lemma* and plays an important role in stability theory of control systems (see [46], [47], [17], [1], [27], see also section 8.10. It also plays an important role in the synthesis theory of passive networks (see [4], [18], see also section 8.10) and in the covariance generation problem (see section 8.11). It can in fact be shown that if $G_0(s)$ is positive real, then *all* real symmetric solutions of the linear matrix inequality 8.8

satisfy the inequality $K_f \leq K \leq K^+$. In particular this implies that if the LMI has a real positive semi-definite solution, then all real symmetric solutions are positive semi-definite and the smallest (overall) real symmetric solution coincides with the smallest positive semi-definite solution, i.e., $K^- = K_f$.

Let us consider one more example.

Again consider the system

$$\Sigma_{i/s/o}(A, B, C_0, D_0).$$

Define a supply rate s by $s(u, y) = \|u\|^2 - \|y\|^2$ or, equivalently, by $s(w) = w^T M w$ with M the symmetric matrix given by $M := \begin{pmatrix} I_m & 0 \\ 0 & -I_p \end{pmatrix}$. Here, I_m and I_p denote the $m \times m$ and $p \times p$ identity matrices, respectively. As a special case of Cor. 8.4.4 we then obtain:

Corolloary 8.5.2. *Assume that the system $\Sigma_{i/s/o}(A, B, C_0, D_0)$ is connected (equivalently: the pair (A, B) is controllable). Let $G_0(s) = C_0(Is - A)^{-1}B + D_0$ be the transfer matrix of $\Sigma_{i/s/o}$. Define a supply rate by $s(u, y) = \|u\|^2 - \|y\|^2$. Then the following statements are equivalent:*

(a) There exists a function $V \geq 0$ such that $(\Sigma_{i/s/o}, s, V)$ is internally dissipative,
(b) $G_0(\bar{s})^T G(s) \leq I$ for all $s \in C$ with $s \notin \sigma(A)$ and $\Re e\ s \geq 0$,
(c) the linear matrix inequality

$$\begin{pmatrix} -A^T K - KA - C_0^T C_0 & KB - C_0^T D_0 \\ B^T K - D_0^T C_0 & I - D_0^T D_0 \end{pmatrix} \geq 0 \tag{8.9}$$

has a real symmetric solution $K \geq 0$.

In that case there exist real symmetric solutions $K_f \geq 0$ and $K^+ \geq 0$ such that for any solution $K \geq 0$ we have $K_f \leq K \leq K^+$. □

A transfer matrix $G_0(s)$ that satisfies the condition in statement (b) of the above corollary is called *bounded real.* The result on the equivalence between (b) and (c) is known as the *bounded real lemma* (see [4]).

8.6 The Dissipation Rate

Let $\Sigma_{DV}(A, B, C, D)$ be a linear finite-dimensional system in driving variable representation, let $s(w) = w^T M w$ be a quadratic supply rate and let $V(x) = x^T K x$ be a quadratic storage function. If the system produces the signal (w, x), then the amount of supply that is dissipated in the system during the time-interval $[t_0, t]$ is equal to

$$V(x(t_0)) - V(x(t)) + \int_{t_0}^{t} s(w(t))dt.$$

The rate at which the supply is dissipated is equal to the derivative of this function and is equal to

$$\frac{d}{dt}\left(-x(t)^T K x(t)\right) + w(t)^T M w(t)$$

If v is a driving variable associated with the signal (w, x) (i.e., (x, w, v) satisfies $\dot{x} = Ax + Bv, w = Cx + Dv$) then the latter can be seen to be equal to

$$(x(t)^T, v(t)^T)\, F(K) \begin{pmatrix} x(t) \\ v(t) \end{pmatrix}.$$

where

$$F(K) := \begin{pmatrix} -A^T K - KA + C^T M C & -KB + C^T M D \\ -B^T K + D^T M C & D^T M D \end{pmatrix}.$$

Since $V(x) = x^T K x$ is a storage function, we know that the real symmetric matrix K satisfies the linear matrix inequality $F(K) \geq 0$. Thus we can factorize

$$F(K) = \begin{pmatrix} M_K^T \\ N_K^T \end{pmatrix} (\, M_K \quad N_K \,), \tag{8.10}$$

with $(M_K \;\; N_K)$ a suitable real matrix with $n + m$ columns and, say, r rows. Therefore the rate at which the supply is dissipated 'along' a signal (w, x) with driving variable v is given by

$$\|M_K x(t) + N_K v(t)\|^2.$$

The quadratic function $d(x, v) = \|M_K x + N_K v\|^2$ is called a *dissipation rate* associated with the quadratic storage function $V(x) = x^T K x$. We note that there is no a priori upper bound to the number of rows r of the matrix $(M_K \;\; N_K)$. Of course, the number of rows of $(M_K \;\; N_K)$ is equal to the rank of $F(K)$ and, of course, $r = \text{rank} F(K)$ if and only if the matrix $(M_K \;\; N_K)$ is of full row rank. There is a close connection between the factorization (8.10) of $F(K)$ and certain factorizations of the matrices $G(-s)^T M G(s)$ and, more general, $G(z)^T M G(s)$. Indeed, if we factorize $F(K)$ and define a real rational matrix $W_K(s)$ by

$$W_K(s) := N_K + M_K (Is - A)^{-1} B \tag{8.11}$$

then it can be shown by straightforward calculation that for all $s, z \in \mathbb{C}$ with $s, z \notin \sigma(A)$ we have

$$G(z)^T M G(s) = W_K(z)^T W_K(s) + (s + z) B^T (Iz - A^T)^{-1} K (Is - A)^{-1} B.$$

From this, it immediately follows that

$$G(-s)^T M G(s) = W_K(-s)^T W_K(s) \tag{8.12}$$

(where the latter should be interpreted as an equality between real rational matrices). A factorization 8.12 of the real rational matrix $G(-s)^T M G(s)$ is often called a *spectral factorization* (see [48], [1], [2], [3], [4]). The real rational matrix $W_K(s)$ is called the spectral factor corresponding to the solution K of the linear matrix inequality. The spectral factorization equation (8.12) plays an important role in stochastic realization theory and filtering [10], [11].

The above can be used to obtain an a priori lower bound to the number r of rows of the matrix $(M_K\ N_K)$ in the factorization of $F(K)$. Let $r^\star$ denote the rank of the real rational matrix $G(-s)^T MG(s)$ (i.e., the rank of the latter matrix considered as a matrix with entries in the field of real rational functions) Furthermore, let $r(K)$ denote the rank of the real matrix $F(K)$. The following result states that for every real symmetric solution of the linear matrix inequality we have $r(K) \geq r^\star$, while this lower bound is attained for the matrix K if and only if the corresponding rational matrix $W_K(s)$ is right-invertible:

Theorem 8.6.1. *Let $A \in \mathcal{R}^{n\times n}$, $B \in \mathcal{R}^{n\times m}$, $C \in \mathcal{R}^{q\times n}$ and $D \in \mathcal{R}^{q\times n}$ and let $M \in \mathcal{R}^{q\times q}$ be symmetric. Assume that the LMI has a symmetric solution, i.e., there exists $K \in \mathcal{R}^{n\times n}$ such that $F(K) \geq 0$. Then for each symmetric solution $K \in \mathcal{R}^{n\times n}$ of the LMI we have $r(K) \geq r^\star$. Furthermore, if $K \in \mathcal{R}^{n\times n}$ is a symmetric solution of the LMI and if we factorize $F(K)$ as in (8.10) with $(M_K\ N_K)$ of full row rank then $r(K) = r^\star$ if and only if the real rational matrix $W_K(s)$ has full row rank (considered as a matrix with entries in the field of real rational functions).*

Proof. Factorize the matrix $F(K)$ as in (8.10), such that $(M_K\ N_K)$ is of full row rank. The number of rows of $(M_K\ ,N_K)$ is then equal to $r(K)$. Let $W_K(s)$ be given by (8.11). Since the number of rows of $W_K(s)$ is equal to the number of rows of $(M_K\ N_K)$, the number of rows of $W_K(s)$ is equal to $r(K)$. On the other hand, it follows from (8.12) that the number of rows of $W_K(s)$ is larger that or equal to $r^\star$, with equality if and only if $W_K(s)$ is a full row rank real rational matrix. This completes the proof. □

At this point of course the question arises whether there always *exists* a symmetric solution to the LMI such that the lower bound $r^\star$ is actually attained or, equivalently, such that the real rational matrix $W_K(s)$ is right-invertible. It can indeed be shown that if the pair (A, B) is controllable and if the LMI has at least one real symmetric solution, then for the largest solution K^+ as well as for the smallest solution K^- the lower bound $r^\star$ is attained, i.e., $r(K^+) = r(K^-) = r^\star$. Also, if the LMI has least one positive semi-definite solution then the lower bound is attained by the smallest positive semi-definite solution K_f, i.e., $r(K_f) = r^\star$ (see [30], [13]). In the following section we shall give a proof of these facts for the special case that the matrix $D^T MD$ is positive definite.

8.7 The Algebraic Riccati Equation

Consider again the driving variable system $\Sigma_{DV}(A, B, C, D)$ studied in section 8.4 and let $s(w) = w^T Mw$ be a quadratic supply rate. In the present section we will assume that the following regularity assumption holds:

$$D^T MD > 0.$$

If this assumption holds then a real symmetric matrix K satisfies the linear matrix inequality $F(K) \geq 0$ if and only if it satisfies the inequality $R(K) \geq 0$, where $R(K)$ is defined by

$$R(K) := -A^T K - KA + C^T MC - \\ (-KB + C^T MD)(D^T MD)^{-1}(-B^T K + D^T MC).$$

(This follows by noting that $R(K)$ is equal to the Schur-complement of $-A^T K - KA + C^T MC$ in the matrix $F(K)$). The equation

$$R(K) = 0$$

is called the *algebraic Riccati equation* (ARE) associated with $\Sigma_{DV}(A, B, C, D)$ and the quadratic supply rate $s(w) = w^T Mw$. Of course, if K is a real symmetric solution of the ARE then it also satifies the linear matrix inequality LMI and hence $V(x) = x^T Kx$ defines a quadratic storage function for (Σ_{DV}, s). Thus, if the ARE has a real symmetric solution and if, in addition, the pair (A, B) is controllable (equivalently: the system $\Sigma_{DV}(A, B, C, D)$ is connected), then we have $G(-i\omega)^T MG(i\omega) \geq 0$ for all $i\omega \notin \sigma(A)$. In this section it will be shown that also the converse holds: if the inequality $G(-i\omega)^T MG(i\omega) \geq 0$ holds for all $i\omega \notin \sigma(A)$, and if the pair (A, B) is controllable, then the ARE has a real symmetric solution. Furthermore, it will turn out that in this case the algebraic Riccati equation has a *smallest* real symmetric solution and a *largest* real symmetric solution which exactly coincide with the smallest real symmetric solution K^- and the largest real symmetric solution K^+, respectively, of the linear matrix inequality LMI. In order to prove these claims, first note that under the regularity assumption $D^T MD > 0$ we have the following explicit expression for the rank of the matrix $F(K)$:

$$r(K) = m + \text{rank } R(K). \tag{8.13}$$

From this we immediately see that for every real symmetric matrix K we have $r(K) \geq m$, with equality if and only if K satisfies the algebraic Riccati equation $R(K) = 0$. Now, let $\Sigma_{DV}(A, B, C, D)$ be connected, assume that $G(-i\omega)^T MG(i\omega) \geq 0$ for all $i\omega \notin \sigma(A)$ and let K^- and K^+ denote the smallest and the largest real symmetric solution of the linear matrix inequality, respectively. We claim that both K^- as well as K^+ are solutions to the algebraic Riccati equation. By the above, this is equivalent to saying that $r(K^-) = m$ and $r(K^+) = m$. Following the notation introduced in the previous section, let $r^\star$ denote the rank of the real rational matrix $G(-s)^T MG(s)$. It is immediately clear that under the regularity condition $D^T MD > 0$ we have $r^\star = m$. Next, recall from the previous section that if K is a real symmetric solution of the LMI and if we factorize $F(K)$ as in (8.10) with $(M_K \; N_K)$ of full row rank, then $r(K) = r^\star$ if and only if the real rational matrix $W_K(s)$ defined by (8.11) has full row rank. Thus, we conclude that K^- and K^+ are solutions to the ARE if and only if $W_{K^-}(s)$ and $W_{K^+}(s)$

are full row rank real rational matrices. This indeed follows from the following lemma:

Lemma 8.7.1 *Let $\Sigma_{DV}(A,B,C,D)$ be a connected system in driving variable representation and assume that D has full column rank. Define a supply rate s by $s(w) := \|w\|^2$ (i.e,. take $M = I$). Let*

$$V_a(x_0) = \sup\{-\int_0^{t_1} \|w(t)\|^2 dt \mid (w,x) \in \mathcal{B}_{DV}(A,B,C,D), t_1 \geq 0, \quad x(0) = 0 \text{ and } x(t_1) = 0\}$$

be the available storage. Then we have: if $V_a(x_0) = 0$ for all x_0 then the real rational matrix $G(s) := C(Is - A)^{-1}B + D$ has full row rank.

Proof. Let $v \in L_{2,loc}(\mathcal{R}^+)$ denote the driving variable of $\mathcal{B}_{DV}(A,B,C,D)$. For all $x_0 \in \mathcal{R}^n$ we have

$$J^*(x_0) := \inf\{\int_0^{t_1} \|w(t)\|^2 dt \mid v \in L_{2,loc}(\mathcal{R}^+), t_1 \geq 0 \text{ such that } x(0) = x_0, x(t_1) = 0\} = 0.$$

For $T \geq 0$ define

$$J_T^\star(x_0) = \inf\{\int_0^{t_1} \|w(t)\|^2 dt \mid v \in L_{2,loc}(\mathcal{R}^+), t_1 \geq T \text{ such that } x(0) = x_0, x(t_1) = 0\}.$$

It is easily shown that for all $T \geq 0$ we have $J_T^\star(x_0) = J^\star(x_0)$ and hence that $J_T^\star(x_0) = 0$. In the following, take a fixed $T > 0$. Obviously we have

$$\inf\{\int_0^{T} \|w(t)\|^2 dt \mid v \in L_{2,loc}(\mathcal{R}^+),\ x(0) = x_0\} \leq J_T^\star(x_0) \ = \ 0.$$

Substitute $v = -(D^TD)^{-1}D^TCx + u$ and denote $\tilde{A} = A - B(D^TD)^{-1}D^TC$ and $\tilde{C} = C - D(D^TD)^{-1}D^TC$. Then our system equations become $\dot{x} = \tilde{A}x + Bu, w = \tilde{C}x + Du$ and we find that

$$\inf\{\int_0^{T} \|\tilde{C}x(t)\|^2 + \|Du(t)\|^2 dt \mid u \in L_{2,loc}(\mathcal{R}^+),\ x(0) = x_0\} \ = \ 0$$

Now, let $\{u_n\}$ be an infimizing sequence with x_n the corresponding state trajectories with $x_n(0) = x_0$. Then, since D has full column rank, we have $u_n \to 0$ and $\tilde{C}x_n \to 0$ as $n \to \infty$ in $L_2[0,T]$. Denote $x(t) := e^{\tilde{A}t}x_0$. Since $u_n \to 0$ in $L_2[0,T]$, we also have $x_n \to x$ in $L_2[0,T]$ and hence $\tilde{C}x_n \to \tilde{C}x$. This implies that $\tilde{C}x$ is identically equal to zero on $[0,T]$. In particular we have $\tilde{C}x_0 = 0$. Since the latter holds for all x_0, we conclude that $\tilde{C} = 0$. Now define a real rational matrix $R(s)$ by

$$R(s) = -(D^TD)^{-1}D^TC(Is - A + B(D^TD)^{-1}D^TC)^{-1}B.$$

It is a easily seen that $R(s)$ is a right inverse of $G(s)$. □

Using the previous lemma we will prove that W_{K^+} is has full row rank. Let $t_1 \geq 0$, let $(w, x) \in \mathcal{B}_{DV}$ be such that $x(0) = x_0$ and $x(t_1) = 0$ and let v be a driving variable associated with (w, x). As in the previous section it can be seen that

$$-\int_0^{t_1} s(w(t))dt = x_0^T K^+ x_0 - \int_0^{t_1} \|M_{K^+}x(t) + N_{K^+}v(t)\|^2 dt.$$

By taking suprema with respect to t_1 and $(w, x) \in \mathcal{B}_{DV}$ such that $x(0) = x_0$ and $x(t_1) = 0$ on both sides in this equality we find that

$$\sup\{-\int_0^{t_1} \|M_{K^+}x(t) + N_{K^+}v(t)\|^2 dt \mid t_1 \geq 0, v \in L_{2,loc}(\mathcal{R}^+),\ x(0) = x_0$$
$$\text{and } x(t_1) = 0\} \ = \ 0.$$

The latter statement can be reformulated as: the available storage of the auxiliary system $\Sigma_{DV}(A, B, M_{K^+}, N_{K^+})$ with supply rate given by $M = I$ is equal to zero for every initial condition. According to the previous lemma this immediately implies that $W_{K^+}(s)$ has full row rank (note that N_{K^+} has full column rank). A similar argument can be given to prove that the real rational matrix $W_{K^-}(s)$ has full row rank. Summarizing, we have proven the following theorem:

Theorem 8.7.2. *Let $A \in \mathcal{R}^{n\times n}$, $B \in \mathcal{R}^{n\times m}$, $C \in \mathcal{R}^{q\times n}$ and $D \in \mathcal{R}^{q\times n}$, let $M \in \mathcal{R}^{q\times q}$ be symmetric and assume that $D^T M D > 0$. Furthermore, assume that the pair (A, B) is controllable. Then the following statements are equivalent:*

(a) $G(-i\omega)^T M G(i\omega) \geq 0$ for all $\omega \in \mathcal{R}$, $i\omega \notin \sigma(A)$,
(a) there exists a symmetric solution $K \in \mathcal{R}^{n\times n}$ of the algebraic Riccati equation ARE.

Assume that one of these statements holds. Let K^- and K^+ be the smallest and the largest real symmetric solution, respectively, of the linear matrix inequality LMI. Then K^- and K^+ are solutions of the ARE and for any real symmetric solution $K \in \mathcal{R}^{n\times n}$ of the ARE we have $K^- \leq K \leq K^+$ □

Summarizing, we conclude that the set of real symmetric solutions of the ARE is a subset of the set of real symmetric solutions of the LMI. We see that the LMI has a real symmetric solution if and only if the ARE has a real symmetric solution. Furthermore, in that case both for the LMI as well as for the ARE there exist a smallest and a largest real symmetric solution. In addition, these smallest real symmetric solutions of the LMI and the ARE, respectively, coincide and the same holds for the largest real symmetric solution of the LMI and the ARE. Of course, in contrast with the solution set of the LMI, the solution set of the ARE will in general *not be convex* (in fact, generically it is a finite set). In the following we will give a complete characterization of the set of all real symmetric solutions of the algebraic Riccati equation. We introduce the following notation:

$$A_+ := A - B(D^TMD)^{-1}(-B^TK^+ + D^TMC),$$
$$A_- := A - B(D^TMD)^{-1}(-B^TK^- + D^TMC),$$
$$\Delta := K^+ - K^-.$$

Note that we always have $\Delta \geq 0$. For a given matrix M, $\sigma(M)$ will denote its set of eigenvalues. We will also denote $\mathcal{C}^- := \{s \in \mathcal{C} \mid \Re e\ s < 0\}$. Likewise, we will denote $\mathcal{C}^0 := \{s \in \mathcal{C} \mid \Re e\ s = 0\}$ and $\mathcal{C}^+ := \{s \in \mathcal{C} \mid \Re e\ s > 0\}$. Now, we claim that

$$\sigma(A^-) \subset \mathcal{C}^- \cup \mathcal{C}^0, \tag{8.14}$$

$$\sigma(A^+) \subset \mathcal{C}^+ \cup \mathcal{C}^0. \tag{8.15}$$

Indeed, since K^- and K^+ satisfy the ARE it is easily seen that

$$(A^+)^T\Delta + \Delta A^+ = \Delta B(D^TMD)^{-1}B^T\Delta$$

and

$$(A^-)^T\Delta + \Delta A^- = -\Delta B(D^TMD)^{-1}B^T\Delta.$$

Assume now that λ is an eigenvalue of A^+ with corresponding eigenvector x. By pre- and post-multiplying the first of the above two equations with x^* and x, respectively, we obtain

$$2(\Re e\ \lambda)\, x^*\Delta x = x^*\Delta B(D^TMD)^{-1}B^T\Delta x.$$

From this we immediately see that $\Re e\ \lambda \geq 0$. In the same way, the second of the above two equations can be used to show that $\sigma(A^-) \subset \mathcal{C}^- \cup \mathcal{C}^0$. In the sequel, if K is a solution of the ARE, we shall denote

$$A_K := A - B(D^TMD)^{-1}(-B^TK + D^TMC),$$

It was shown in [40] that the extremal solutions of the ARE are in fact uniquely determined by the conditions (8.14) and (8.15):

Theorem 8.7.3. *Let $A \in \mathcal{R}^{n\times n}$, $B \in \mathcal{R}^{n\times m}$, $C \in \mathcal{R}^{q\times n}$ and $D \in \mathcal{R}^{q\times n}$, let $M \in \mathcal{R}^{q\times q}$ be symmetric and assume that $D^TMD > 0$. Furthermore, assume that the pair (A, B) is controllable. Assume that the ARE has a real symmetric solution. Let K^- and K^+ be the smallest and the largest real symmetric solution of the ARE, respectively. Then we have $\sigma(A^-) \subset \mathcal{C}^- \cup \mathcal{C}^0$ and $\sigma(A^+) \subset \mathcal{C}^+ \cup \mathcal{C}^0$. Furthermore, if K is a solution of the ARE with the property that $\sigma(A_K) \subset \mathcal{C}^- \cup \mathcal{C}^0$ then we have $K = K^-$. If K is a solution of the ARE such that $\sigma(A_K) \subset \mathcal{C}^+ \cup \mathcal{C}^0$ then we have $K = K^+$.* □

In order to be able to give a characterization of all real symmetric solutions of the ARE, we need the concept of *modal subspace* of a given matrix. If M is a real $n \times n$ matrix, then we will denote by $\mathcal{X}^-(M)$ ($\mathcal{X}^0(M)$, $\mathcal{X}^+(M)$) the largest M-invariant subspace $\mathcal{V}$ of $\mathcal{R}^n$ such that $\sigma(M|\mathcal{V}) \subset \mathcal{C}^-$ ($\sigma(M|\mathcal{V}) \subset \mathcal{C}^0$, $\sigma(M|\mathcal{V}) \subset \mathcal{C}^+$).

The subspace $\mathcal{X}^-(M)$ is called the modal subspace of M corresponding to $\mathcal{C}^-$. Likewise, $\mathcal{X}^0(M)$ and $\mathcal{X}^+(M)$ are called the modal subspaces of M corresponding to $\mathcal{C}^0$ and $\mathcal{C}^+$, respectively. An important role will be played by the modal subspace $\mathcal{X}^+(A^+)$. Let Ω denote the set of all A^+-invariant subspaces of $\mathcal{X}^+(A^+)$. Let Γ denote the set of all real symmetric solutions of the ARE. It turns out that there is a one-to-one correspondence between the set Γ and the set Ω:

Theorem 8.7.4. *Let $A \in \mathcal{R}^{n\times n}$, $B \in \mathcal{R}^{n\times m}$, $C \in \mathcal{R}^{q\times n}$ and $D \in \mathcal{R}^{q\times n}$, let $M \in \mathcal{R}^{q\times q}$ be symmetric and assume that $D^TMD > 0$. Furthermore, assume that the pair (A,B) is controllable. Assume that the ARE has a real symmetric solution, i.e., assume that $\Gamma \neq \emptyset$. Let K^- and K^+ be the smallest and the largest real symmetric solution of the ARE, respectively. Then the following holds: if $\mathcal{V}$ is an A^+-invariant subspace of $\mathcal{X}^+(A^+)$ (that is, if $\mathcal{V} \in \Omega$) then $\mathcal{R}^n = \mathcal{V} \oplus \Delta^{-1}\mathcal{V}^\perp$. There exists a bijection $\gamma : \Omega \to \Gamma$ defined by*

$$\gamma(\mathcal{V}) := K^+P_{\mathcal{V}} + K^-(I - P_{\mathcal{V}}),$$

where $P_{\mathcal{V}}$ is the projection onto $\mathcal{V}$ along $\Delta^{-1}\mathcal{V}^\perp$. If $K = \gamma(\mathcal{V})$ then

$$\mathcal{X}^+(A_K) = \mathcal{V},$$
$$\mathcal{X}^0(A_K) = \ker \Delta,$$
$$\mathcal{X}^-(A_K) = \mathcal{X}^-(A^-) \cap \Delta^{-1}\mathcal{V}^\perp. \qquad \square$$

We want to stress that $\Delta^{-1}\mathcal{V}^\perp$ denotes the inverse image of $\mathcal{V}^\perp$ under Δ, i.e. the subspace $\{x \in \mathcal{R}^n \mid \Delta x \in \mathcal{V}^\perp\}$. For a proof of this theorem we refer to [40], [8] (see also [29], [22]). As noted before, the above result states that there exists a one-to-one correpondence between the set of all real symmetric solutions of the ARE and the set of all A^+-invariant subspaces of $\mathcal{X}^+(A^+)$. If K is a real symmetric solution of the ARE then the corresponding subspace $\gamma^{-1}(K)$ is given by $\gamma^{-1}(K) = \mathcal{X}^+(A_K)$. If $K = \gamma(\mathcal{V})$ then we say that the solution K is supported by the subspace $\mathcal{V}$.

We want to conclude this section with a result on the existence of positive semi-definite solutions of the algebraic Riccati equation. A proof of this theorem can be given analogous to theorem 8.7.2

Theorem 8.7.5. *Let $A \in \mathcal{R}^{n\times n}$, $B \in \mathcal{R}^{n\times m}$, $C \in \mathcal{R}^{q\times n}$ and $D \in \mathcal{R}^{q\times n}$, let $M \in \mathcal{R}^{q\times q}$ be symmetric and assume that $D^TMD > 0$. Furthermore, assume that the pair (A,B) is controllable. Then the following statements are equivalent:*

(a) $G(\bar{s})^TMG(s) \geq 0$ for all $s \in \mathcal{C}$, $s \notin \sigma(A)$, $\Re e\, s \geq 0$,
(b) there exists a positive semi-definite solution $K \in \mathcal{R}^{n\times n}$ of the algebraic Riccati equation ARE.

Assume that one of these statements holds. Let K^+ be the largest real symmetric solution of the ARE. Furthermore, let K_f be the smallest positive semi-definite solution of the LMI. Then K_f is a solution of the ARE and for any positive semi-definite solution $K \in \mathcal{R}^{n\times n}$ of the ARE we have $K_f \leq K \leq K^+$. $\square$

8.8 Linear Quadratic Problems

Undoubtedly the best-known application of the linear matrix inequality and the algebraic Riccati equation is linear quadratic optimal control theory. Consider the finite-dimensional linear time-invariant system

$$\dot{x} = Ax + Bu, \; x(0) = x_0. \tag{8.16}$$

where, as usual, x and u are assumed to take their values in $\mathcal{R}^n$ and $\mathcal{R}^m$, respectively. We will be dealing with optimization problems of the type

$$\inf \int_0^\infty \omega(x,u)dt, \tag{8.17}$$

where ω is a general real quadratic form on $\mathcal{R}^n \times \mathcal{R}^m$ defined by

$$\omega(x,u) := u^T Ru + 2u^T Sx + x^T Qx.$$

Here, R, S and Q are assumed to be real matrices such that $R = R^T$ and $Q = Q^T$. The infimization in (8.17) should be interpreted as follows. For a given control function $u \in L_{2,loc}(\mathcal{R}^+)$, let $x(x_0, u)$ denote the state trajectory of (8.16) and if $T \geq 0$ let

$$J_T(x_0,u) := \int_0^T \omega(x(x_0,u)(t), u(t))dt. \tag{8.18}$$

Define the following classes of control functions:

$$U(x_0) := \{u \in L_{2,loc}(\mathcal{R}^+) \mid \lim_{T\to\infty} J_T(x_0,u) \text{ exists in } \mathcal{R} \cup \{-\infty,+\infty\}\},$$

$$U_s(x_0) := \{u \in U(x_0) \mid \lim_{t\to\infty} x(x_0,u)(t) = 0\}.$$

Note that if (A, B) is controllable, then $U(x_0) \neq \emptyset$ and $U_s(x_0) \neq \emptyset$ for all $x_0 \in \mathcal{R}^n$. For $u \in U(x_0)$ we now define the cost associated with u by

$$J(x_0,u) := \lim_{T\to\infty} J_T(x_0,u).$$

Note that $J(x_0,u) \in \mathcal{R} \cup \{-\infty,+\infty\}$. We are now ready to define the linear quadratic problems that we want to consider. The first problem that we shall introduce is the *zero-endpoint linear quadratic problem* and consist of the minimization of the functional $J(x_0,u)$ over the class of input functions $U_s(x_0)$. The optimal cost associated with this problem is

$$J^+(x_0) := \inf\{J(x_0,u) \mid u \in U_s(x_0)\}.$$

In addition to the zero-endpoint problem we have the *free-endpoint problem*, which consists of minimizing the functional $J(x_0,u)$ over the (larger) class $U(x_0)$ of input functions. The optimal cost for the latter problem is

$$J_f^+(x_0) := \inf\{J(x_0,u) \mid u \in U_s(x_0)\}.$$

If in any of the the above linear quadratic optimization problems the matrix R appearing in the quadratic form $\omega(x,u)$ is positive definite, then the corresponding optimization problem is called *regular*. If the quadratic form ω itself is positive semi-definite then the corresponding linear quadratic problem is called *positive-semi-definite*.

In most of the existing literature on the linear quadratic problem it is assumed that the problem under consideration is both regular as well as positive semi-definite. In fact, under these assumptions the linear quadratic problem has become quite standard and is treated in many basic textbooks in systems and control [5], [6], [21], [45]. For a treatment of the more general problem in which the quadratic form ω is arbitrary, we refer to [27], [40], [35], [31]. If the problem is not regular, then it is called *singular* (see [7], [15], [42], [12]).

It turns out that it is quite easy to characterize the optimal cost J^+ for the zero-endpoint problem in its most general formulation, while for the free-endpoint optimal cost J_f^+ this is, up to now, only possible for the important special cases that the optimization problem is either regular or positive semi-definite. We will first discuss the zero-endpoint problem. In order to provide some intuition we would like to consider this problem in a context of dissipative systems. To this end we consider the system in driving variable form $\mathcal{B}_{DV}(A,B,C,D)$, with

$$C := \begin{pmatrix} I \\ 0 \end{pmatrix}, \; D := \begin{pmatrix} 0 \\ I \end{pmatrix}. \tag{8.19}$$

For this system the signal alphabet W is equal to $X \times U$ so we have $w = (x,u)$. In connection with this system we consider the quadratic supply rate $s(x,u) = -\omega(x,u)$, i.e., the matrix M defining the supply rate is equal to

$$M := -\begin{pmatrix} Q & S^T \\ S & R \end{pmatrix}.$$

Let $V_a(x_0)$ be the associated available storage (with respect to the reference point $x^* = 0$). Under the assumption that (A,B) is controllable, it is quite easy to verify that, in fact, for all x_0 we have $J^+(x_0) = -V_a(x_0)$. Thus we see that the optimal cost $J^+(x_0)$ is finite for every x_0 if and only if the system $\mathcal{B}_{DV}(A,B,C,D)$ with supply rate $-\omega(x,u)$ is internally dissipative. For this particular choice of system we have

$$G(s) := C(Is-A)^{-1} + D = \begin{pmatrix} (Is-A)^{-1} \\ I \end{pmatrix}$$

and hence $G(z)^T M G(s) = H(z,s)$, where

$$H(z,s) :=$$
$$B^T(Iz-A^T)^{-1}Q(Is-A)^{-1}B + S(Is-A)^{-1}B + B^T(Iz-A^T)^{-1}S^T + R$$

Thus we obtain the following:

Theorem 8.8.1. *Let (A,B) be controllable. Then the following statements are equivalent:*

(a) $J^+(x_0)$ *is finite for all* $x_0 \in \mathcal{R}^n$,
(b) $H(-i\omega, i\omega) \geq 0$ *for all* $i\omega \notin \sigma(A)$,
(c) there exists a real symmetric solution K *to the linear matrix inequality*

$$\begin{pmatrix} -A^TK - KA + Q & -KB + S^T \\ -B^TK + S & R \end{pmatrix} \geq 0. \tag{8.20}$$

If one of these statements holds then we have $J^+(x_0) = x_0^T(-K^-)x_0$, *where* K^- *is the smallest real symmetric solution of the linear matrix inequality 8.20.* □

Of course, a real symmetric matrix K is a solution to the linear matrix inequality 8.20 if and only if the matrix $-K$ is a solution to the following linear matrix inequality:

$$\begin{pmatrix} A^TK + KA + Q & KB + S^T \\ B^TK + S & R \end{pmatrix} \geq 0. \tag{8.21}$$

If K^- is the smallest real symmetric solution of (8.20), then $-K^-$ is the largest real symmetric solution of (8.21). Thus we see that the optimal cost $J^+(x_0)$ is equal to $x_0^T\bar{K}^+x_0$, where $\bar{K}^+$ is the *largest* real symmetric solution of the linear matrix inequality (8.21).

We shall now discuss the issue of optimal controls. For a given x_0, an input u^* is called optimal for the zero-endpoint problem if $u^* \in U_s(x_0)$ and if $J(x_0, u^*) = J^+(x_0)$, i.e. the control input attains the optimal value of the cost functional. It is well known that the question of the existence of optimal controls is closely connected to the question whether the optimization problem is regular, i.e., whether the weighting matrix R is *positive definite*. In fact, if this is not the case then optimal controls in general will not exist unless we extend the class of inputs $U_s(x_0)$ to include *distributions*. In this paper we do not want to go into the intricacies of distribution theory and therefore we will assume that $R > 0$. As before, the role of the linear matrix inequality is then taken over by an algebraic Riccati equation, which, in this particular case, is given by

$$-A^TK - KA + Q - (-KB + S^T)R^{-1}(-B^TK + S) = 0. \tag{8.22}$$

Obviously, if K^- is the smallest real symmetric solution of this algebraic Riccati equation, then $\bar{K}^+ := -K^-$ is the largest real symmetric solution of the following algebraic Riccati equation:

$$A^TK + KA + Q - (KB + S^T)R^{-1}(B^TK + S) = 0. \tag{8.23}$$

In the following, let $\bar{\Gamma}$ denote the set of all real symmetric solutions of the algebraic Riccati equation (8.23). If $\bar{\Gamma} \neq \emptyset$, let $\bar{K}^-$ denote the smallest element of $\bar{\Gamma}$. Furthermore, let $\bar{\Delta}$ denote the difference $\bar{K}^+ - \bar{K}^-$ between the largest and the smallest real symmetric solution of the latter ARE. The following result was proven in [40]:

Theorem 8.8.2. *Let* (A, B) *be controllable and assume that* $R > 0$. *Then*

(a) the following statements are equivalent:
(i) $J^+(x_0)$ *is finite for all* $x_0 \in \mathcal{R}^n$,
(ii) $H(-i\omega, i\omega) \geq 0$ *for all* $i\omega \notin \sigma(A)$,
(iii) there exists a real symmetric solution K *to the algebraic Riccati equation (8.23), i.e.* $\bar{\Gamma} \neq \emptyset$.

Assume that one of the above statements hold. Then:

(b) for all $x_0 \in \mathcal{R}^n$ *we have* $J^+(x_0) = x_0^T \bar{K}^+ x_0$,
(c) for all $x_0 \in \mathcal{R}^n$ *there exists an optimal input* u^* *if and only if* $\bar{\Delta} > 0$,
(d) if $\bar{\Delta} > 0$ *then for each* x_0 *there exists exactly one optimal input* u^* *and, moreover, this input* u^* *is given by the feedback control law* $u^* = -R^{-1}(B^T\bar{K}^+ + S)x$. □

We will now turn to the second of the linear quadratic optimization problems we shall consider in this section, the free-endpoint problem. In our discussion of this problem, we will restrict ourselves to two important special cases, the case that the optimization problem is positive semi-definite and the case that the problem is regular. First we shall treat the positive semi-definite case, i.e., the case that the quadratic form $\omega(x, u)$ is positive semi-definite. In that case there exist matrices C_0 and D_0 of appropriate dimensions such that

$$\begin{pmatrix} Q & S^T \\ S & R \end{pmatrix} = (C_0 \; D_0)^T (C_0 \; D_0), \tag{8.24}$$

so the cost functional J is given by

$$J(x_0, u) = \int_0^\infty \|C_0 x(t) + D_0 u(t)\|^2 dt,$$

and it is clear that, since the latter is a priori bounded from below by zero, finiteness of the optimal cost is no issue in this case. The linear matrix inequality 8.20 in this case reads

$$\begin{pmatrix} -A^TK - KA + C_0^TC_0 & -KB + C_0^TD_0 \\ -B^TK + D_0^TC_0 & D_0^TD_0 \end{pmatrix} \geq 0 \tag{8.25}$$

and clearly this inequality always has a solution (take for example $K = 0$). In order to characterize the optimal cost for the free-endpoint problem it turns out that we have to look at a particular subset of the set of all negative semi-definite solutions of the linear matrix inequality, more specifically, the subset of all negative semi-definite *rank-minimizing* solutions of the linear matrix inequality. Note that the above linear matrix inequality is exactly the linear matrix inequality associated with the system in input/state/output form $\mathcal{B}_{i/s/o}(A, B, C_0, D_0)$ and supply rate given by $s(u, y) := \|y\|^2$. Thus it can be easily seen that for this special case we have

$$G(-s)^T M G(s) = G_0(-s)^T G_0(s),$$

where $G_0(s) = C_0(Is - A)^{-1}B + D_0$. By applying theorem 8.6.1 we thus find that for each real symmetric solution K of the linear matrix inequality (8.25) we have

$r(K) \geq r^\star$, where $r^\star$ is equal to the rank of the rational matrix $G(-s)^T M G(s)$ which, by the above, is equal to the rank of the rational matrix $G_0(s)$. Now, instead of working with the linear matrix inequality (8.25), we will work with the following linear matrix inequality:

$$\begin{pmatrix} A^T K + KA + C_0^T C_0 & KB + C_0^T D_0 \\ B^T K + D_0^T C_0 & D_0^T D_0 \end{pmatrix} \geq 0, \tag{8.26}$$

Obviously, also for any real symmetric solution K of the latter linear matrix inequality we have $r(K) \geq r^\star$. Furthermore, (8.26) always has at least one positive semi-definite solution. Following [12] we introduce the set

$$\bar{\Gamma}^+_{min} := \{K \in \mathcal{R}^{n \times n} \mid K \text{ satisfies the LMI 8.26, } K \geq 0 \text{ and } r(K) = r^\star\}$$

of positive semi-definite rank-minimizing solutions of the linear matrix inequality given by (8.26). It was shown in [12] that if (A, B) is controllable then $\bar{\Gamma}^+_{min}$ contains a smallest element, say $\bar{K}_f$, characterized by the following properties:

- $\bar{K}_f \in \bar{\Gamma}^+_{min}$,
- $K \in \bar{\Gamma}^+_{min} \Rightarrow \bar{K}_f \leq K$.

It turns out that the optimal cost for the positive semi-definite free-endpoint problem is determined by the latter solution of the LMI:

Theorem 8.8.3. *Consider the free-endpoint linear quadratic problem associated with the system (A, B) and the quadratic form $\omega(x, u) := \|C_0 x + D_0 u\|^2$. Assume that (A, B) is controllable. For all $x_0 \in \mathcal{R}^n$ we have $J_f^+(x_0) = x_0^T \bar{K}_f x_0$, where $\bar{K}_f$ is the smallest element of $\bar{\Gamma}^+_{min}$.* □

It can be shown that if, in addition, we assume that the problem is regular then the optimal cost for the free-endpoint problem is in fact determined by the smallest positive semi-definite solution of the algebraic Riccati equation associated with (8.26) (see [24], [42]).

As a last subject of this section we shall consider the *regular* free-endpoint problem. Thus, in the remainder we will drop the assumption that the quadratic form ω is positive semi-definite. Instead, we will assume that the matrix R is positive definite. Again let $\bar{\Gamma}$ be the set of all real symmetric solutions of the algebraic Riccati equation (8.23). If $\bar{\Gamma} \neq \emptyset$, let $\bar{K}^-$ and $\bar{K}^+$ denote the smallest and the largest real symmetric solution of the ARE (8.23), respectively. Denote

$$\bar{A}^+ := A - BR^{-1}(B^T \bar{K}^+ + S),$$
$$\bar{A}^- := A - BR^{-1}(B^T \bar{K}^- + S).$$

Let $\bar{\Omega}$ be the set of all $\bar{A}^-$-invariant subspaces of the subspace $\mathcal{X}^+(\bar{A}^-)$. By applying theorem 6.4 to the system in driving variable form $\mathcal{B}_{DV}(A, B, C, D)$ with C and D given by (8.19), we find that there is a one to one correspondence between the sets $\bar{\Omega}$ and $\bar{\Gamma}$:

Theorem 8.8.4. *Let $A \in \mathcal{R}^{n\times n}$, $B \in \mathcal{R}^{n\times m}$, let $Q \in \mathcal{R}^{n\times n}$ be symmetric, let $S \in \mathcal{R}^{n\times m}$ and let $R \in \mathcal{R}^{m\times m}$ be positive definite. Furthermore, assume that the pair (A, B) is controllable. Assume that the ARE given by (8.23) has a real symmetric solution, i.e., assume that $\bar{\Gamma} \neq \emptyset$. Let $\bar{K}^-$ and $\bar{K}^+$ be the smallest and the largest real symmetric solution of the ARE, respectively. Let $\bar{\Delta} := \bar{K}^+ - \bar{K}^-$. Then the following holds: if $\mathcal{V}$ is an $\bar{A}^-$-invariant subspace of $\mathcal{X}^+(\bar{A}^-)$ (that is, if $\mathcal{V} \in \bar{\Omega}$) then $\mathcal{R}^n = \mathcal{V} \oplus \Delta^{-1}\mathcal{V}^\perp$. There exists a bijection $\bar{\gamma} : \bar{\Omega} \to \bar{\Gamma}$ defined by*

$$\bar{\gamma}(\mathcal{V}) := \bar{K}^- P_{\mathcal{V}} + \bar{K}^+(I - P_{\mathcal{V}}),$$

where $P_{\mathcal{V}}$ is the projection onto $\mathcal{V}$ along $\Delta^{-1}\mathcal{V}^\perp$. □

In the characterization of the free-endpoint optimal cost J_f^+ the following subspace of the state space plays a central role:

$$\mathcal{N} :=< \ker\bar{K}^- \mid \bar{A}^- > \cap\, \mathcal{X}^+(\bar{A}^-) \tag{8.27}$$

Here, for an arbitrary subspace $\mathcal{L}$ and an arbitrary linear map M, $< \mathcal{L} \mid M >$ denotes the smallest M-invariant subspace contained in $\mathcal{L}$. Note that $\mathcal{L}$ is an $\bar{A}^-$-invariant subspace contained in $\mathcal{X}^+(\bar{A}^-)$ and hence an element of $\bar{\Omega}$. Let $\bar{K}_f^+$ be the real symmetric solution of the ARE corresponding to the subspace $\mathcal{N}$, i.e., let $\bar{K}_f^+ = \bar{\gamma}(\mathcal{N})$. It was shown in [35] that the free-endpoint optimal cost is determined by this particular solution of the ARE. For a given x_0, an input u^* is called optimal for the free-endpoint problem if $u^* \in U(x_0)$ and if $J(x_0, u^*) = J_f^+(x_0)$. Let $\bar{\Gamma}_-$ be the set of all negative semi-definite solutions of the ARE (8.23). Then we have:

Theorem 8.8.5. *Let (A, B) be controllable and assume that $R > 0$. Then*

(a) the following statements are equivalent:
   (i) $H(\bar{s}, s) \geq 0$ for all $s \in \mathbb{C}$, $s \notin \sigma(A)$, $\Re e\, s \geq 0$,
   (ii) there exists a negative semi-definite solution K to the algebraic Riccati equation 8.23, i.e. $\bar{\Gamma}_- \neq \emptyset$.

Assume that one of the above statements hold. Then:

(b) $J_f^+(x_0)$ is finite for all $x_0 \in \mathcal{R}^n$,
(c) for all $x_0 \in \mathcal{R}^n$ we have $J_f^+(x_0) = x_0^T \bar{K}_f^+ x_0$,
(d) for all $x_0 \in \mathcal{R}^n$ there exists an optimal input u^* if and only if $\ker \bar{\Delta} \subset \ker \bar{K}^-$,
(e) if $\ker \bar{\Delta} \subset \ker \bar{K}^-$ then for each x_0 there exists exactly one optimal input u^* and, moreover, this input u^* is given by the feedback control law $u^* = -R^{-1}(B^T\bar{K}_f^+ + S)x$. □

8.9 Stability Theory

In this section we will discuss the application of the ideas in this paper to stability theory. We start by giving the main underlying idea in the context of dissipative systems. Let $\Sigma_s^1 = (\mathcal{R}, W, X_1, \mathcal{B}_s^1)$ and $\Sigma_s^2 = (\mathcal{R}, W, X_2, \mathcal{B}_s^2)$ be two dynamical systems in state space form. Then their *interconnection* $\Sigma_s = \Sigma_s^1 \times \Sigma_s^2$ is defined as $\Sigma_s = (\mathcal{R}, W, X_1 \times X_2, \mathcal{B}_s)$ with

$$\mathcal{B}_s := \{(w,(x_1,x_2)) \mid (w,x_1) \in \mathcal{B}_s^1 \text{ and } (w,x_2) \in \mathcal{B}_s^2\}.$$

Now assumne that (Σ_s^1, s_1, V_1) and (Σ_s^2, s_2, V_2) are both dissipative. It follows immediately that $(\Sigma_s, s_1+s_2, V_1+V_2)$ is also dissipative. Now, in many applications $s_1 + s_2 = 0$ which shows that $V_1 + V_2$ will be a Lyapunov function for Σ_s in the sense that $V_1(x_1(t)) + V_2(x_2(t))$ will be non-increasing along elements of $\mathcal{B}_s$.

We will now apply this in order to derive stability conditions for the system described by the differential equation

$$\Sigma: \quad \dot{x} = Ax - Bf(Cx,t) \tag{8.28}$$

with $A \in \mathcal{R}^{n\times n}$, $B \in \mathcal{R}^{n\times m}$, $C \in \mathcal{R}^{p\times n}$, and $f : \mathcal{R}^p \times \mathcal{R} \to \mathcal{R}^m$. We will derive conditions on (A,B,C) and f such that the solutions of (8.28) are either bounded on $[0,\infty)$ or converge to zero as $t \to \infty$. We will do this by viewing (8.28) as the (feedback) interconnection of the following two systems:

$$\Sigma_1: \quad \dot{x} = Ax + Bu, \; y = Cx \tag{8.29}$$

$$\Sigma_2: \quad u = -f(y,t) \tag{8.30}$$

Now use as supply rates $s_1(u,y) = u^T y$ and $s_2(u,y) = -u^T y$ to obtain

Theorem 8.9.1. *Assume in (8.28) that (A,B,C) is minimal and $m = p$. Let $G(s) := C(Is - A)^{-1}B$. There exists an $M < \infty$ such that for every solution $x : \mathcal{R} \to \mathcal{R}^n$ and $t \geq 0$ there holds*

$$\|x(t)\| \leq M\|x(0)\|$$

if $G(\bar{s})^T + G(s) \geq 0$ for all $s \in \mathbb{C}$ with $s \notin \sigma(A)$ and $\Re e\, s \geq 0$ and if $y^T f(y,t) \geq 0$ for all $y \in \mathcal{R}^p$.

Proof. According to Corollary 8.5.1 there exists a positive semi-definite solution K to the linear matrix inequality (8.8) with $C_0 = C$ and $D = 0$, or, equivalently, to

$$-A^T K - KA \geq 0, \quad KB = C^T.$$

Using the assumption that (C,A) is observable it is easily shown that, in fact, $K > 0$. Now compute the derivative $\dot{V}_\Sigma$ of $V(x) = x^T Kx$ along solutions of (8.28). It follows that

$$\dot{V}_\Sigma(x(t)) \leq -2(Cx(t))^T f(Cx(t),t).$$

From this it follows that

$$x^T(t)Kx(t) \leq x^T(0)Kx(0)$$

for all $t \geq 0$. This yields the result. □

The above result can be generalized in many different directions:

(1) If we use the supply rates $\|u\|^2 - \|y\|^2$ for (8.28) and $\|y\|^2 - \|u\|^2$ for (8.30) we see that $\|x(t)\| \leq M\|x(0)\|$ can be concluded by assuming $G(\bar{s})^T G(s) \leq I$ for all $s \in C$ with $s \notin \sigma(A)$ and $\Re e\ s \geq 0$ and $\|f(y,t)\| \leq \|y\|$ for all $y \in \mathcal{R}^p$ and $t \in \mathcal{R}$.
(2) If we assume certain 'observability' properties of the dissipation rate, it can be proven that in addition to $\|x(t)\| \leq M\|x(0)\|$ we will have $\lim_{t\to\infty} x(t) = 0$.
(3) By assuming $G(-i\omega)^T + G(i\omega) \geq 0$ for all $\omega \in \mathcal{R}$ with $i\omega \notin \sigma(A)$, but $G(\bar{s})^T + G(s) \ngeq 0$ for some $s \in C$ with $\Re e\ s \geq 0$, in addition to $y^T f(y,t) \geq 0$, one can prove that there are solutions such that $\lim_{t\to\infty} x(t) \neq 0$. By assuming also certain 'observability' properties of the dissipation rate one can prove that in this case there are solutions which are unbounded on $[0,\infty)$.

The constuction of the quadratic Lyapunov function $x^T K x$ which lies at the basis of these stability results is identical to the analysis of the linear matrix inequality as in (8.8) and (8.9) (see also [27], [46], [47], [17], [41]).

8.10 Electrical Network Synthesis

A formal definition of a synthesis question may be given as follows: given certain ideal elements which may be interconnected according to certain interconnection laws, what systems can be realized this way and, for a given system, what elements should be used and how should they be interconnected? In linear passive electrical network synthesis, these elements are taken to be linear resistors ($V = RI,\ R > 0$), linear capacitors ($I = C\frac{dV}{dt},\ C > 0$), linear inductors ($V = L\frac{dI}{dt},\ L > 0$), transformers ($V_2 = nV_1,\ I_1 = -nI_2$), and gyrators ($I_1 = gV_2,\ I_2 = -gV_1$). Here (V,I) denote the port variables of a one-port, while $((V_1,I_1),(V_2,I_2))$ denote the port variables of a two-port:

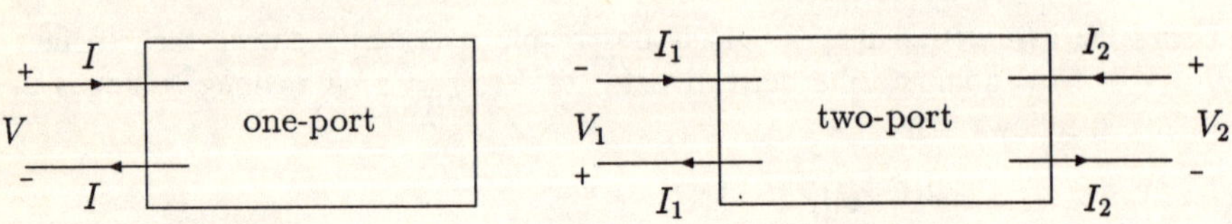

The R's, L's, C's, T's and G's are assumed to be interconnected by the usual electrical interconnections obeying Kirchoff's current and voltage laws. Assume that such an interconnection is set up yielding an N-port:

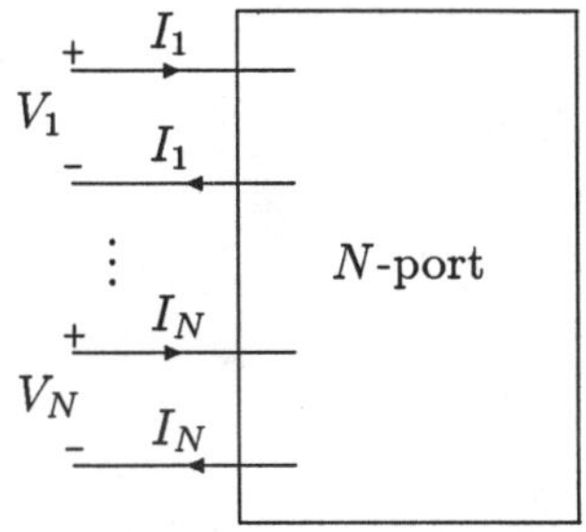

it can be shown that if the N-port contains only linear passive R's, L's, C's, T's and G's, the resulting N-port will allow a hybrid description meaning that there will exist a componentwise partition of the vectors $I = \mathrm{col}(I_1, I_2, \ldots, I_N)$ and $V = \mathrm{col}(V_1, V_2, \ldots, V_N)$ into $I = \mathrm{col}(I^1, I^2)$ and $V = \mathrm{col}(V^1, V^2)$ such that the network is described by a proper transfer function $G(s)$:

$$\begin{pmatrix} V^1 \\ I^2 \end{pmatrix} = G(s) \begin{pmatrix} I^1 \\ V^2 \end{pmatrix}. \tag{8.31}$$

A representation as (8.31) is called a hybrid description. If $V^1 = V$, then we speak about an impedance description and if $V^2 = V$, we speak about an admittance description.

The question arises what properties on $G(s)$ follow from the assumption that the N-port contains only (linear passive) R, L, C, T and G's. In network synthesis we ask for necessary and sufficient conditions, in the sense that we look for conditions on $G(s)$, and a blue-print for synthesizing the network in case these conditions are satisfied. In this section we will describe in broad lines how this problem is solved. We will see how the positive real lemma (Corollary 8.5.1) enters this procedure in an essential way.

It can be shown that the memoryless multiport

$$\begin{pmatrix} V^1 \\ I^2 \end{pmatrix} = R \begin{pmatrix} I^1 \\ V^2 \end{pmatrix} \tag{8.32}$$

can be synthesized using R's, T's and G's (memoryless elements) if and only if the matrix R satisfies $R + R^T \geq 0$. We will take this result to be our starting point for the synthesis of a dynamic N-port with transfer matrix $G(s)$.

Let (A, B, C, D) be a minimal realization of $G(s)$. Hence $G(s) = D + C(Is - A)^{-1}B$ and (A, B) is controllable and (C, A) is observable. Now assume that $G(s)$ is positive real, that is, assume that $G(\bar{s})^T + G(s) \geq 0$ for all $s \in C$ with $s \notin \sigma(A)$ and $\Re e\ s \geq 0$. Then there exists a matrix $K = K^T > 0$ such that

$$\begin{pmatrix} A^T K + KA & KB - C^T \\ B^T K - C & -D - D^T \end{pmatrix} \leq 0 \tag{8.33}$$

Now, by using a suitable basis transformation S such that $S^T S = K$ and considering the new realization $(\bar{A}, \bar{B}, \bar{C}, \bar{D})$ of $G(s)$ with $\bar{A} = SAS^{-1}$, $\bar{B} = SB$,

$\bar{C} = CS^{-1}$, and $\bar{D} = D$, we obtain a realization with

$$\begin{pmatrix} \bar{A}^T + \bar{A} & \bar{B} - \bar{C}^T \\ \bar{B}^T - \bar{C} & -\bar{D} - \bar{D}^T \end{pmatrix} \leq 0. \tag{8.34}$$

Now realize the memoryless multiport

$$\begin{pmatrix} I \\ V^1 \\ I^2 \end{pmatrix} = \begin{pmatrix} -\bar{A} & -\bar{B} \\ \bar{C} & \bar{D} \end{pmatrix} \begin{pmatrix} V \\ I^1 \\ V^2 \end{pmatrix} \tag{8.35}$$

using R's, T's and G's. Next, terminate the first n ports of this multiport using unit capacitors. This imposes the condition

$$\dot{V} = -I,$$

yielding the behavioral equations

$$\dot{I} = \bar{A}I + \bar{B}\begin{pmatrix} I^1 \\ V^2 \end{pmatrix}, \quad \begin{pmatrix} V^1 \\ I^2 \end{pmatrix} = (\bar{C} \quad \bar{D}) \begin{pmatrix} I \\ I^1 \\ V^2 \end{pmatrix}. \tag{8.36}$$

The transfer matrix of (8.36) is $G(s)$, as desired. We summarize this result in:

Theorem 8.10.1. *The N-port hybrid description (8.31) can be realized using linear passive R's, L's, C's, T's and G's if and only if $G(s)$ is positive real: $G(\bar{s})^T + G(s) \geq 0$ for all $s \in \mathcal{C}$ with $s \notin \sigma(A)$ and $\Re e\ s \geq 0$.* □

Similar synthesis procedures can be obtained for networks without gyrators, without resistors, without L's and G's, without C's and G's, or without R's and G's. For this we refer to [18], [36], [4].

8.11 Covariance Generation

Let $y(t)$ $(t \in \mathcal{R}, y(t) \in \mathcal{R}^p)$ be a zero mean stationary Gaussian stochastic vector process defined on a probability space $(\Omega, \mathcal{A}, P)$. Define its autocorrelation function by

$$R(t) = \mathcal{E}\{y(t' + t)y^T(t')\} \tag{8.37}$$

Obviously $R(t) = R^T(-t)$ and, as is easily derived from a direct calculation, $R(.)$ is non-negative definite in the sense that

$$\int_{-\infty}^{\infty}\int_{-\infty}^{\infty} v^T(t')R(t' - t'')v(t'')dt'dt'' \geq 0 \tag{8.38}$$

for all $v(.) : \mathcal{R} \to \mathcal{R}^p$ for which the above double integral exists.

The stochastic realization problem consists of finding a *Markov representation* of y. By this, we mean a zero-mean, stationary Gauss-Markov process $x(t)$, $(t \in \mathcal{R}, x(t) \in \mathcal{R}^n)$ and a matrix $C \in \mathcal{R}^{p \times n}$ such that $Cx(.)$ has the same autocorrelation function as $y(.)$. Equivalently, such that $Cx(.)$ and $y(.)$ are stochastically equivalent. Two processes $y_1(t), y_2(t)$ $(t \in \mathcal{R},\ y_1(t), y_2(t) \in \mathcal{R}^p)$ are said to be stochastically equivalent if for all $t_1, t_2, \ldots, t_k \in \mathcal{R}$ the vector random variables $(y_1(t_1), y_1(t_2), \ldots, y_1(t_k))$ and $(y_2(t_1), y_2(t_2), \ldots, y_2(t_k))$ have the same distributution.

The Markov process $x(t)$ $(t \in \mathcal{R}, x(t) \in \mathcal{R}^n)$ and the matrix C are called a Markov representation of y. A Markov representation (x, C) is said to be *minimal* if n, the dimension of x, is as small as possible. It can be shown that if (x, C) is minimal, then x is mean-square continuous. We proceed by classifying the mean-square continuous Markov processes.

It is easy to see that a mean square continuous zero-mean stationary Markov process $x(t)$ $(t \in \mathcal{R}, x(t) \in \mathcal{R}^n)$ is, up to stochastic equivalence, completely specified by the matrices $Q \in \mathcal{R}^{n \times n}$ and $A \in \mathcal{R}^{n \times n}$ by

$$\mathcal{E}\{x(t)x^T(0)\} = e^{At}Q \quad t \geq 0. \tag{8.39}$$

Thus $Q = \mathcal{E}\{x(t)x^T(t)\}$ while A is such that $\mathcal{E}\{x(t)|x(0)\} = e^{At}x(0)$, $t \geq 0$. Clearly, $Q = Q^T \geq 0$. The question arises what matrices $Q, A \in \mathcal{R}^{n \times n}$ can arise in this way. In fact:

Lemma 8.11.1. *Let $Q, A \in \mathcal{R}^{n \times n}$. Then there exist a mean square continuous zero- mean stationary Gauss-Markov process $x(t)$ $(t \in \mathcal{R}, x(t) \in \mathcal{R}^n)$ such that 8.39 holds if and only if*

$$Q = Q^T \geq 0 \text{ and } AQ + QA^T \leq 0 \tag{8.40}$$

Proof. ($\Rightarrow$) The first condition in (8.40) is obvious. In order to prove the second, calculate $0 \leq \mathcal{E}\{(x(t) - x(0))(x(t) - x(0))^T\}$. This yields $e^{At}Q + Qe^{A^T t} \leq 2Q$ for $t \geq 0$. Now take the limit as $t \downarrow 0$ to obtain $AQ + QA^T \leq 0$.($\Leftarrow$) Factor $AQ + QA^T$ as $-BB^T$ and consider the solution of the stochastic differential equation $dx = Axdt + Bdw$ with $x(0)$ a zero mean Gaussian random vector with $\mathcal{E}\{x(0)x^T(0)\} = Q$ and $w(t), t \in \mathcal{R}$ a Wiener process independent of $x(0)$. □

We now return to the stochastic realization problem. This question has now been reduced to that of finding an n and matrices $A, Q \in \mathcal{R}^{n \times n}$ and $C \in \mathcal{R}^{p \times n}$ such that 8.40 holds and such that $Ce^{At}QC^T = R(t)$ for $t \geq 0$.

Theorem 8.11.2. *Let $y(t)$ $(t \in \mathcal{R}, y(t) \in \mathcal{R}^p)$ be a zero mean stationary Gaussian process. Let its autocorrelation be $R(t) = \mathcal{E}\{y(t)y^T(t)\}$. The following are equivalent:*

(a) There exists a zero-mean stationary finite-dimensional Gauss-Markov process $x(t)$ $(t \in \mathcal{R}, x(t) \in \mathcal{R}^n)$ and a matrix $C \in \mathcal{R}^{p \times n}$ such that (x, C) is a Markov representation of y,

(b) There exists $n < \infty$ and matrices $A, Q \in \mathcal{R}^{n\times n}$, $C \in \mathcal{R}^{p\times n}$ such that (8.40) holds and such that $Ce^{At}QC^T = R(t)$ for $t \geq 0$,

(c) $R(t)$ is a Bohl function, that is, there exist matrices F, G, H such that $R(t) = He^{At}G$ for $t \geq 0$,

(d) The spectral density matrix $S(\omega)$ of $R(t)$, that is, the Fourier transform of R:

$$S(\omega) = \int_{-\infty}^{\infty} R(t)e^{-i\omega t}\,dt$$

is rational.

Proof. The equivalence of (a) and (b) follows from the pre-amble to the theorem. (b) ⇒(c) is immediate and (c) ⇒(d) can be seen from

$$S(\omega) = H(I i\omega - F)^{-1}G + G^T(-Ii\omega - F^T)H^T.$$

We will now prove that (c) ⇒(b). Let (F, G, H) be a triple such that $R(t) = He^{Ft}G$. Now observe that (8.38) implies by Corollary 8.5.1 that there exists a matrix $K = K^T \geq 0$ such that $FK + KF^T \leq 0$ and $KH^T = G$. Now take $A = F$, $Q = K$, and $C = H$. Then $Q = Q^T \geq 0$, $A^TQ + QA \leq 0$, and $Ce^{At}QC^T = He^{At}G = R(t)$ for $t \geq 0$. □

The proof of the above theorem contains an algorithm for computing all minimal Markov representations for y. Let (F, G, H) be a minimal triple such that $R(t) = He^{At}G$ for $t \geq 0$. Then all minimal Markov representations may be obtained by choosing any solution $K = K^T$ such that $FK + KF^T \leq 0$ and $KH^T = G$ (every solution $K = K^T$ will be positive definite and there exist solutions K^- and K^+ such that $K^- \leq K \leq K^+$). Now take $A = F$, $Q = K$ and $C = H$. The only remaining freedom is now a basis choice: $A \to SAS^{-1}$, $Q \to SQS^T$ and $C \to CS^{-1}$.

What we have described above can be called the *covariance generation problem.* What we have done is describe a way of matching by (x, C), the given autocorrelation $R(t)$. Note that the Markov process x can be easily simulated by the stochastic differential equation $dx = Axdt + Bdw$ with B such that $AQ + QA^T = -BB^T$. A related problem is the strong stochastic realization problem in which it is required that x is supported by the given probability space $(\Omega, \mathcal{A}, P)$ which supports y. This version of the stochastic realization problem leads to the question what minimal Markov representations (x, C) are such that $x(t)$ is $y(.)$-measurable. It turns out that this is the case if and only if the solution K to $FK + KF^T \leq 0$, $KH^T = G$ is such that $FK + KF^T$ is of minimal rank. In particular, K^- (in which case $x(t)$ is a function of the past of y) and K^+ (in which case $x(t)$ is a function of the future of y) yield solutions of the strong stochastic realization problem. For details and further references, we refer to [2], [10], [11], [26].

8.12 The H_∞ Control Problem

As one of the most recent applications of the ideas developed in this paper, in this section we will discuss the application to the problem of H_∞ optimal control. We will consider the following linear time-invariant system:

$$\dot{x} = Ax + Bu + Ed, \quad z = Cx + Du. \tag{8.41}$$

In these equations, as usual, x and u are assumed to take their values in $\mathcal{R}^n$ and $\mathcal{R}^m$, respectively. The variable d represents an unknown disturbance, which is assumed to take its values in $\mathcal{R}^l$. Finally, z represents the output to be controlled, which is assumed to take its values in $\mathcal{R}^q$. A, B, E, C and D are real matrices of appropriate dimensions. We will restrict ourselves here to the H_∞ control problem with static state feedback. If F is a real $m \times n$ matrix then the closed system resulting from the state feedback control law $u = Fx$ is given by

$$\dot{x} = (A + BF)x + Ed, \quad z = (C + DF)x. \tag{8.42}$$

The transfer matrix of this system is called the closed loop transfer matrix and is equal to

$$G_F(s) = (C + DF)(Is - A - BF)^{-1}E. \tag{8.43}$$

Obviously, if we put $x(0) = 0$ then the closed loop system defines an operator mapping disturbances d to outputs z. If we restrict ourselves to disturbances $d \in L_2(\mathcal{R}^+)$ and if the closed loop system is asymptotically stable, i.e., if $\sigma(A+BF) \subset \mathcal{C}^-$ then this convolution operator is a bounded operator from $L_2(\mathcal{R}^+)$ to $L_2(\mathcal{R}^+)$. The influence of a disturbance d on the output z can then be measured by the induced norm of this operator. It is well known that this norm is equal to the H_∞ norm of the closed loop transfer matrix, which is denoted by

$$\|G_F\|_\infty := \sup_{\omega \in \mathcal{R}} \rho[G_F(i\omega)].$$

Here, $\rho[M]$ denotes the largest singular value of the complex matrix M. Now, the problem that we shall consider in this section is the following: given a positive real number γ, find a real $m \times n$ matrix F such that $\sigma(A + BF) \subset \mathcal{C}^-$ and such that $\|G_F\|_\infty < \gamma$. It will turn out that if the matrix D has full column rank then the existence of such matrix F is equivalent to the existence of a given solution of a certain algebraic Riccati equation.

Before embarking on the details, first note that under the assumption that D has full column rank, we can assume without loss of generality that $D^TD = I$ and that $D^TC = 0$. Indeed, if these conditions do not hold then we may apply a preliminary feedback of the form $u = -(D^TD)^{-1}D^TCx + (D^TD)^{-1/2}v$ in order to obtain a system for which the conditions do hold. Now, assume that for our system there exists an F such that $\sigma(A + BF) \subset \mathcal{C}^-$ and $\|G_F\|_\infty \le \gamma$. Then clearly $G_F(-i\omega)^T G_F(i\omega) - \gamma^2 I \le 0$ for all $\omega \in \mathcal{R}$. By applying Th. 8.7.2 to the system in driving variable representation

$$\Sigma\left(A+BF,\ E,\ \begin{pmatrix}0\\ C+DF\end{pmatrix},\ \begin{pmatrix}I\\ 0\end{pmatrix}\right)$$

with $w=(d,z)$ and supply rate $s(w)=w^T Mw$ with M given by

$$M:=\begin{pmatrix}\gamma^2 I & 0\\ 0 & -I\end{pmatrix},$$

we find that there exists a real symmetric solution Z to the following algebraic Riccati equation:

$$-(A+BF)^T Z-Z(A+BF)-(C+DF)^T(C+DF)-\gamma^{-2}ZEE^T Z=0. \quad (8.44)$$

Note that strictly speaking one needs controllabilty of the pair $(A+BF,E)$ in order to be able to apply theorem 8.7.2. However, using the fact that the matrix $A+BF$ is asymptotically stable it can be shown that also without this controllability assumption equation (8.44) has a real symmetric solution. By interpreting the above Riccati equation as a Lyapunov equation it can be shown that, in fact, $Z \geq 0$. Writing out (8.44) we find

$$A^T Z+ZA+C^T C+\gamma^{-2}ZEE^T Z+(B^T Z+F)^T(B^T Z+F)-ZBB^T Z=0 \quad (8.45)$$

Now, temporarily let us assume that the pair (C,A) is observable. Using (8.45) it can be shown that Z is positive definite: assume $Zx=0$. Then we find $\|Cx\|^2+\|(B^T Z+F)x\|^2=0$. Obviously, this implies $Cx=0$ and $Fx=0$. Thus, again using (8.45), we find that $ZAx=0$, which contradicts the assumption that (C,A) is observable. Next, define $P:=Z^{-1}$. Then from (8.45) we find that P satisfies the inequality

$$-PA^T-AP+BB^T-\gamma^{-2}EE^T-PC^T CP\geq 0$$

or, equivalently,

$$\begin{pmatrix}-PA^T-AP+BB^T-\gamma^{-2}EE^T & PC^T\\ CP & I\end{pmatrix}\geq 0. \quad (8.46)$$

Consider the system in driving variable representation Σ_{DV} given by the equations $\dot{x}=A^T x-C^T v$, $w=(x,v)$. In connection with this system consider the supply rate s given by $s(w):=w^T Mw$, with M given by

$$\begin{pmatrix}\gamma^{-2}EE^T-BB^T & 0\\ 0 & I\end{pmatrix}.$$

The inequality (8.46) then expresses the fact that the pair (Σ_{DV},s) is internally dissipative. Consequently, by theorem 8.7.2, also the associated algebraic Riccati equation

$$-YA^T-AY+BB^T-\gamma^{-2}EE^T-YC^T CY=0$$

has a real symmetric solution Y. We claim that, in fact, there exists a positive definite solution Y^+. Indeed, let Y^+ be the largest real symmetric solution of the latter algebraic Riccati equation. Since Y^+ coincides with the largest real symmetric solution of (8.46) and since $P > 0$ is a solution of (8.46) this proves our claim. Finally, define $K := (Y^+)^{-1}$. Then K satisfies

$$A^T K + KA - K(BB^T + \gamma^{-2}EE^T)K + C^T C = 0. \tag{8.47}$$

Thus, if there exists F such that $A + BF$ is asymptotically stable and the H_∞ norm of G_F is less than or equal to γ, then there exists a positive definite solution K of the algebraic Riccati equation (8.47). If we assume that the *strict* inequality $\|G_F\|_\infty < \gamma$ holds then it can be shown that, in fact, such K can be found with the property that

$$\sigma(A - BB^T K + \gamma^{-2}EE^T K) \subset \mathcal{C}^-. \tag{8.48}$$

It turns out that also the converse holds: if the Riccati equation (8.47) has a positive semi-definite solution K such that condition (8.48) holds, then there exists F with the desired properties. In that case one such F is given by $F = -B^T K$. It can be shown that the condition that (C, A) should be observable can in fact be replaced by the weaker condition that (C, A) should have no unobservable eigenvalues on the imaginary axis, i.e. by the condition

$$\operatorname{rank}\begin{pmatrix} Ii\omega - A \\ C \end{pmatrix} = n \quad \forall \omega \in \mathcal{R}.$$

In the general situation that we do not necessarily have $D^T D = 0$ and $D^T C = 0$, the condition that there should be no unobservable eigenvalues on the imaginary axis should be replaced by the condition that the quadruple (A, B, C, D) should have no *invariant zeros* on the imaginary axis. An invariant zero of (A, B, C, D) is any complex number s_0 with the property that

$$\operatorname{rank}\begin{pmatrix} Is_0 - A & -B \\ C & D \end{pmatrix} < \operatorname{normrank}\begin{pmatrix} Is - A & -B \\ C & D \end{pmatrix}.$$

Here, for a given real rational matrix $R(s)$, normrank R denotes the rank of R considered as a matrix with entries in the field of real rational functions. We can thus obtain the following theorem:

Theorem 8.12.1. *Consider the system (8.41). Let $\gamma > 0$. Assume that the matrix D has full column rank, and that the quadruple (A, B, C, D) has no invariant zeros on the imaginary axis. Then the following two statements are equivalent:*

(a) There exists a static state feedback law $u = Fx$ such that $\sigma(A + BF) \subset \mathcal{C}^-$ and $\|G_F\|_\infty < \gamma$,

(b) There exists a positive semi-definite solution K of the algebraic Riccati equation

$$A^T K + KA + \gamma^{-2}KEE^T K + C^T C -$$
$$(KB + C^T D)(D^T D)^{-1}(B^T K + D^T C) = 0$$

such that

$$\sigma(A+\gamma^{-2}EE^TK - B(D^TD)^{-1}(B^TK+D^TC)) \subset \mathcal{C}^-$$

If one of the above condition hold then a suitable F is given by

$$F = -(D^TD)^{-1}(D^TC + B^TK). \qquad \square$$

The first contributions on the application of Riccati equations in the context of H_∞ optimal control theory were given in [19], [20], [25], [49]. The first references in which the result of Th. 8.12.1 appears in the form as given above are [9], [34] (see also [28]).

Of course, if the matrix D is *not of full column rank* then the above algebraic Riccati equation does not exist. It was shown in [32] that in the more general case that D is not necessarily of full column rank, the role of the algebraic Riccati equation is taken over by a *quadratic matrix inequality*. For any real number $\gamma > 0$ and matrix $K \in \mathcal{R}^{n\times n}$ we define a matrix $F_\gamma(K) \in \mathcal{R}^{(n+m)\times(n+m)}$ by

$$F_\gamma(K) := \begin{pmatrix} A^TK + KA + \gamma^{-2}KEE^TK + C^TC & KB + C^TD \\ B^TK + D^TC & D^TD \end{pmatrix}. \tag{8.49}$$

If a real symmetric matrix K satisfies the inequality $F_\gamma(K) \geq 0$ then it is said to satisfy the quadratic matrix inequality. In addition to (8.49), for any $\gamma > 0$ and $K \in \mathcal{R}^{n\times n}$ we define a $n \times (n+m)$ polynomial matrix $L_\gamma(K,s)$ by

$$L_\gamma(K,s) := (Is - A - \gamma^{-2}EE^TK \quad -B)$$

Finally, let $G(s) = D + C(Is-A)^{-1}B$ be the open loop transfer matrix from u to z. The previous theorem can then in fact be shown to be a corrolary of the following more general result:

Theorem 8.12.2. *Consider the system (8.41). Let $\gamma > 0$. Assume that the quadruple (A,B,C,D) has no invariant zeros on the imaginary axis. Then the following two statements are equivalent:*

(a) There exists a static state feedback law $u = Fx$ such that $\sigma(A+BF) \subset \mathcal{C}^-$ and $\|G_F\|_\infty < \gamma$,

(b) There exists a positive semi-definite solution K to the quadratic matrix inequality $F_\gamma(K) \geq 0$ such that

$$\text{rank } L_\gamma(K,s) = \text{normrank } G$$

and

$$\text{rank}\begin{pmatrix} L_\gamma(K,s) \\ F_\gamma(K) \end{pmatrix} = n + \text{normrank } G \quad \forall s \in \mathcal{C}^0 \cup \mathcal{C}^+. \qquad \square$$

A more realistic problem formulation is obtained if, instead of static state feedback, we require the control law to be given by *dynamic measurement feedback*. Conditions for the existence of a dynamic compensator that makes the closed loop system internally stable and which makes the H_∞ norm of the closed loop transfer matrix smaller than an a priori given positive real number can be given in terms of *a pair* of algebraic Riccati equations [9], [14]. Again, if due to the singularity of certain system parameters these algebraic Riccati equations do not exist, then these conditions can be reformulated in terms of a pair of quadratic matrix inequalities [33].

Acknowledgement. Research supported by EEC Science Program Contract No. SCI-0433-C(A).

References

1. B.D.O. Anderson, "A system theory criterion for positive real matrices" *SIAM J. Contr. & Opt.*, Vol.5, No.2, 1967, pp.171-182.
2. B.D.O. Anderson, "The inverse problem of stationary covariance generation", *J. Statistical Physics*, Vol. 1, 1969, pp. 133-147.
3. B.D.O. Anderson, "Algebraic properties of minimal degree spectral factors" *Automatica*, Vol. 9, 1973, pp. 491-500.
4. B.D.O. Anderson and S. Vongpanitlerd, *Network Analysis and Synthesis* Prentice Hall, Englewood Cliffs, New Jersey, 1973.
5. B.D.O. Anderson and J.B. Moore, *Linear Optimal Control*, Prentice Hall, Englewood Cliffs, New Jersey, 1971
6. R.W. Brockett, *Finite Dimensional Linear Systems*, J. Wiley & Sons, 1970.
7. D.J. Clements and B.D.O. Anderson, *Singular Optimal Control: the Linear Quadratic Problem*, Springer Lecture Notes Contr. & Inf Sciences, No. 5, Springer, Berlin, 1978.
8. W.A. Coppel, "Matrix quadratic equations", *Bull. Austral. Math. Soc.*, Vol. 10, 1974, pp. 377-401.
9. J. Doyle, K. Glover, P.P. Khargonekar, B.A. Francis, "State space solutions to standard H_2 and H_∞ control problems", *IEEE Trans. Aut. Contr.*, Vol. 34, No. 8, 1989, pp. 831-847.
10. P.L. Faurre, M. Clerget, F. Germain, *Opérateurs rationnels positifs*, Dunod, Paris, 1979.
11. P.L. Faurre, "Stochastic realization algorithms", *System Identification: Advances and Case Studies*, R.K. Mehra, D.G. Lainiotis (eds.), Academic Press, New York 1976, pp. 1-25.
12. T. Geerts, "All optimal controls for the singular linear-quadratic problem without stability; a new interpretation of the optimal cost", *Lin. Alg. Appl.*, Vol. 122, 1989, pp. 65-104.
13. T. Geerts, *Structure of Linear-Quadratic Control*, Doctoral dissertation, Eindhoven University of Technology, 1989.
14. K. Glover, J. Doyle, "State-space formulaes for all stabilizing controllers that satisfy an H_∞ norm bound and relations to risk sensitivity", *Syst. & Contr. Letters* Vol. 11, 1988, pp. 167-172.
15. M.L.J. Hautus and L.M. Silverman, "System structure and singular control", *Lin. Alg. Appl.*, Vol. 50, 1983, pp. 369-402.
16. R.E. Kalman, "Contributions to the theory of optimal control", *Bol. Soc. Mat. Mex.,*, Vol. 5, 1960, pp. 102-199.
17. R.E. Kalman, "Lyapunov functions for the problem of Lur'e in Automatic Control", *Proceedings of the National Academy of Sciences*, Vol.49, No.2, 1963, pp. 201-205.
18. R.E. Kalman, "On a new characterization of linear passive systems", *Proceedings of the First Allerton Conference on Circuit and System Theory*, University of Illinois, 1963, pp. 456-470.
19. P.P. Khargonekar, I.R. Petersen and K. Zhou, "Robust stabilization of uncertain systems and H_∞ optimal control", *Proc. Allerton Conf. on Comm., Contr. & Comp.*, Monticello, Iliinois, 1987, pp. 88-95.
20. P.P. Khargonekar, I.R. Petersen, M.A. Rotea, "H_∞ optimal control with state feedback", *IEEE Trans. Aut. Contr.*, Vol. 33, 1988, pp. 786-788.
21. H.Kwakernaak and R. Sivan, *Linear Optimal Control Systems*, Wiley Interscience, New York, 1972.

22. K. Mårtensson, "On the matrix Riccati equation", *Inform. Sci.*, Vol. 3, 1971, pp. 17-49.
23. B.P. Molinari, "Nonnegativity of a quadratic functional", *SIAM J. Contr. & Opt.*, Vol. 13, 1975, pp. 792-806.
24. B.P. Molinari, "The time-invariant linear quadratic optimal control problem", *Automatica*, Vol. 13, 1977, pp. 347-357.
25. I.R. Petersen, "Disturbance attenuation and H_∞ optimization: a design method based on the algebraic Riccati equation", *IEEE Trans. Aut. Contr.*, Vol. 32, No. 5, 1987, pp. 427-429.
26. G. Picci, "Stochastic realization for stochastic processes", *IEEE Proc.*, Vol. 64, No 1, 1976, pp. 112-122.
27. V.M. Popov, "Hyperstability and optimality of automatic systems with several control functions", *Rev. Roumaine Sci. Tech.- Elektrotechn. et Energ.*, Vol. 9, No. 4, 1964, pp. 629-690.
28. C. Scherer, "H_∞ control by state feedback: An iterative algorithm and characterization of high gain occurence", *Syst. & Contr. Letters*, Vol. 12, 1989, pp. 383-391.
29. M. Shayman, "Geometry of the algebraic Riccati equation - part 1", *SIAM J. Contr. & Opt.*, Vol. 21, 1983, pp. 375-393.
30. J.M. Schumacher, "The role of the dissipation matrix in singular optimal control", *Syst. & Contr. Letters*, Vol. 2, 1983, pp. 262-266.
31. J.M. Soethoudt and H.L. Trentelman, "The regular indefinite linear quadratic problem with linear endpoint constraints", *Systems & Control Letters*, Vol. 12, 1989, pp. 23-31.
32. A.A. Stoorvogel, H.L. Trentelman, "The quadratic matrix inequality in singular H_∞ control with state feedback", *SIAM J. Contr. & Opt.*, Vol. 28, No. 5, 1990, pp. 1190-1208.
33. A.A. Stoorvogel, "The singular H_∞ control problem with dynamic measurement feedback", To appear in *SIAM J. Contr. & Opt.*.
34. G. Tadmor, "H_∞ in the time domain: the standard four blocks problem", To appear in *Math. Contr. Sign. & Syst.*.
35. H.L. Trentelman, "The regular free-endpoint linear quadratic problem with indefinite cost", *SIAM J. Contr. & Opt.*, Vol. 27, No. 1, 1989, pp. 27-42.
36. S. Vongpanitlerd and B.D.O. Anderson, "Scattering matrix synthesis via reactance extraction", *IEEE Trans. on Circuit Theory*, Vol. CT-17, 1970, pp. 511-517.
37. S. Weiland and J.C. Willems, "Dissipative dynamical systems in a behavioural context", preprint, Febr. 1990, submitted for publication.
38. J.C. Willems, "Dissipative dynamical systems - Part I: General theory", *Archive for Rational Mechanics and Analysis*, Vol. 45, 1972, pp. 321-351.
39. J.C. Willems, "Dissipative dynamical systems - Part II: Linear systems with quadratic supply rates", *Archive for Rational Mechanics and Analysis*, Vol. 45, 1972, pp. 352-393.
40. J.C. Willems, "Least squares stationary optimal control and the algebraic Riccati equation", *IEEE Trans. Aut. Contr.*, Vol. 16, 1971, pp. 621-634.
41. J.C. Willems, "Mechanisms for the stability and instability in feedback systems", *IEEE Proc.*, Vol.64, No 1, 1976.
42. J.C. Willems, A. Kitapçi, L.M. Silverman, "Singular optimal control: a geometric approach", *SIAM J. Contr. & Opt.*, Vol. 24, 1986, pp. 323-337.
43. J.C. Willems, "From time series to linear system - part 1, finite dimensional linear time-invariant systems", *Automatica*, Vol. 22, No.6, pp. 675-694, 1986.
44. J.C. Willems, "Models for dynamics", *Dynamics Reported*, Vol. 2, 1989, pp. 171-269,
45. W.M. Wonham, *Linear Multivariable Control: A Geometric Approach*, 2nd ed., Springer Verlag, New York, 1979.
46. V.A. Yakubovich, "The solution of certain matrix inequalities in automatic control theory", *Doklady Akademii Nauk, SSSR*, Vol. 143, 1962, pp. 1304-1307.
47. V.A. Yakubovich, "Absolute stability of nonlinear control in critical cases, I and II", *Automation and Remote Control*, Vol. 24, pp.273-282 (1963) and pp.655-668 (1964).
48. D.C. Youla, "On the factorization of rational matrices", *IEEE Trans. Inf. Theory*, Vol. 7, 1961, pp. 172-189.
49. K. Zhou, P.P. Khargonekar, "An algebraic Riccati equation approach to H_∞ optimization", *Syst. & Contr. Letters*, Vol. 11, 1988, pp. 85-91.

9 The Infinite Horizon and the Receding Horizon LQ-Problems with Partial Stabilization Constraints

Jacques L. Willems and Frank M. Callier

9.1 Introduction

One of the main reasons why the Riccati equation and its generalizations has become very important in the theory of control, systems, and signals, is that it shows up in a very straightforward way in the analysis of two benchmark problems in control system design and signal filtering.

A fundamental problem in optimal control theory is the design of a regulator for a linear plant which minimizes a quadratic cost function, characterizing the control effort and the deviations of the plant from the ideal operation. On the other hand, in signal analysis, a basic problem is the optimal estimation, in the least squares sense, of a useful signal from observations corrupted by additive noise. The solutions of both problems, respectively referred to as the LQ-optimal control problem and the least squares estimation problem, depend in a crucial way on the solution of Riccati-like differential equations.

In the present chapter the solutions of both problems are first reviewed. It is shown that they are related to the so-called Matrix Riccati Differential Equation and the Algebraic Riccati Equation, depending on the time interval involved. If the time interval in the cost of the control problem or the observation interval in the filtering problem is finite, then the Riccati differential equation (RDE) is relevant. For infinite time intervals, on the other hand, the algebraic Riccati equation (ARE) is to be considered.

The central topic discussed in the present chapter is the behavior of the solutions of the Riccati differential equation for increasing time interval. This is interesting mainly for the following two reasons:

1. A plausible procedure for solving the infinite horizon LQ-optimal control problem is to consider the LQ-problem for a receding horizon and to approximate the desired optimal control strategy by the control strategy which would be obtained for a very large finite time horizon. Similar remarks hold for the consideration of the least squares estimation problem with infinite observation interval as the limiting case of the least squares estimation problem with finite observation interval.

Bittanti, Laub, Willems (Eds.)
The Riccati Equation

2. The solutions of the ARE are the equilibrium solutions of the RDE. A reasonable question is whether the solutions of the ARE can be obtained by integrating the RDE until the solution reaches a steady state.

These observations motivate the development of criteria for convergence of the solution of the matrix Riccati differential equation and of algorithms for the identification of the equilibrium the solution converges to. Since the equations are non-linear, the convergence properties and the steady state may indeed depend on the initial condition. This initial condition corresponds to the final state penalty matrix in the finite horizon optimal control problem and to the initial error covariance matrix for the estimation problem.

In the present contribution the convergence properties of the solutions of the Riccati differential equation are analyzed as well as their interpretation with respect to the relation between the infinite horizon and the receding horizon LQ-optimal control problems. The paper mainly deals with the results and their interpretation; for a detailed exposition of the proofs the reader is referred to the papers mentioned in the bibliography. The discussion is mainly restricted to LQ-optimal control problems. The interpretation of the results with respect the the least squares estimation problem is very similar.

9.2 The Riccati Equation for the Solution of LQ-Problems

In this section it is shown how the Riccati equation is related to the solution of LQ-optimal control or least-squares estimation problems. Consider the linear time-invariant control system

$$\dot{x}(t) = Ax(t) + Bu(t) \tag{9.1}$$

where x denotes the state vector and u the control input vector, with n and r components respectively. A and B are constant matrices of appropriate dimension. The notations used in this chapter and the main definitions from control theory needed are briefly summarized in Appendix 9.5.1.

The aim is to design the optimal control strategy [15], [18] for system (9.1) minimizing the quadratic cost function

$$J(t_1) = I(t_1) + x^T(t_1)Sx(t_1) \tag{9.2}$$

with

$$I(t_1) = \int_{t_0}^{t_1} [x^T(t)Qx(t) + u^T(t)u(t)]\mathrm{d}t \,. \tag{9.3}$$

The matrices $Q = C^TC$ and $S = F^TF$ are constant, symmetric and positive semi-definite. C and F have full rank. The input weighting matrix is assumed to be the identity matrix, instead of a general positive definite matrix; it is well known that this entails no loss of generality for regular control problems, i.e., problems where the cost function is positive definite with respect to the control input. To avoid misunderstanding, it should be emphasized that some properties of the optimally controlled system, e.g. the stability margins, depend on the input weighting matrix [28]; however, for the questions discussed in this chapter, a problem with an input weighting matrix different from the identity, can always be transformed to an equivalent problem where the input weighting matrix is the identity matrix [18].

The above problem statement is a mathematical translation of the fact that one wants to design a control to drive the state or part of the state of the system to zero, without using excessive input energy. Moreover, at the final time t_1 the deviation of the state from the zero state has particular importance; it is therefore penalized separately. In many applications the penalization of the state indeed corresponds to the following objectives:

1. Some state variables should be kept small during the transient period; they are penalized by $x^T(t)Qx(t)$.
2. Some state variables are not important during the transient period, but should be controlled towards zero; they are penalized by $x^T(t_1)Sx(t_1)$, but not by $x^T(t)Qx(t)$ in the integral term I of J.
3. It may also occur that some state variables are not to be controlled; for example, for a machine speed control system it is not reasonable to include the angular position in the cost function. Such state variables are neither penalized in $x^T(t)Qx(t)$, nor in the final state penalization $x^T(t_1)Sx(t_1)$.

It has been shown that the optimal control [15], [18] is given by

$$u(t) = -B^T P(t)x(t) \tag{9.4}$$

for $t_0 \leq t \leq t_1$, where $P(t)$ is the solution of the matrix Riccati differential equation (RDE)

$$\dot{P}(t) = -A^T P(t) - PA(t) + P(t)BB^T P(t) - Q \tag{9.5}$$

with terminal condition

$$P(t_1) = S\,.$$

The minimum cost equals

$$J^*(t_1) = x^T(t_0)P(t_0)x(t_0) \tag{9.6}$$

This optimal control problem is called the "finite horizon control problem". The matrix differential equation is called a Riccati equation because it reduces to such equation in the scalar case.

The length of the time interval $[t_0, t_1]$ is rather artificial. To avoid this problem an infinite time interval is considered. In this way the "infinite horizon control problem" is obtained, which corresponds to minimizing the cost function

$$J = \lim_{t_1 \to \infty} J(t_1)\,. \tag{9.7}$$

If some technical conditions are satisfied, as discussed in Section 4, the optimal control strategy for the infinite horizon control problem is given by [15], [18]

$$u(t) = -B^T P x(t) \tag{9.8}$$

where P satisfies the equation

$$A^T P + PA - PBB^T P + Q = 0\,. \tag{9.9}$$

This equation is called the algebraic Riccati equation (ARE), because it is also the equation for the stationary solutions of the Riccati differential equation (RDE). This equation has multiple solutions. Which one should be used in (9.8) is discussed in the sequel of this chapter. The situation is rather simple in an important special case, the so-called standard regulator problem [15], [18], where the pair (C, A) is detectable and the pair (A, B) stabilizable. The definitions of these properties are summarized in Appendix 9.5.1. Then the matrix to be used [9.18, theorem 3.7] in (9.8) is the unique symmetric positive semi-definite solution of (9.9). It is also the only solution of (9.9) with the property that $A - BB^T P$ has only eigenvalues with negative real parts; such a matrix is called a Hurwitz matrix. Such solution of (ARE) is said to be a stabilizing solution. Moreover it is well known that for any positive semi-definite matrix S, the solution $P(t_0)$ of (9.5) tends to the stabilizing solution of (ARE) as the time interval $t_1 - t_0$ tends to infinity.

The situation becomes much more complicated if (C, A) is not detectable; this is relevant in practical applications, since it may not always be necessary to require that all state variables remain small. It is hence quite reasonable to assume that some state variables are not assumed in the cost function. This may lead to cases where the pair (C, A) is not observable. The non-penalized state variables may correspond to modes which are unstable; this yields problems where (C, A) is undetectable. In the latter case (ARE) has several positive semi-definite solutions [4], [5], [7]. The main features of the solutions of (ARE), relevant for the analysis of the present contribution, are summarized in Appendix 9.5.2. It is obvious that all the positive semi-definite solutions of (9.9) are equilibrium solutions of (9.5). For suitable choices of the positive semi-definite terminal penalty matrix S they may hence be limiting solutions of (RDE) as the time interval tends to infinity. It can also be shown that, for some positive semi-definite matrices S, the solution of (RDE) may even fail to converge to a constant matrix as the time interval tends to infinity, but it may converge to a periodic or an almost periodic matrix function. This shows that the convergence properties of the solution of (RDE) strongly depend on the final state penalty matrix S. This problem is discussed in the next section.

The same equations and questions show up in the least-squares estimation problem [4], [8], [18], [22]. Consider indeed a system driven by a control input and a (stochastic) disturbance. Let the disturbance be modeled by a white noise stochastic process. The output measurements are corrupted by a stochastic process which is also assumed to be white. It can be proved that the assumption that the system and measurement noises are white entails no loss of generality. The plant and the measurement are hence described by the stochastic equations

$$\mathrm{d}x(t) = Ax(t)\mathrm{d}t + \mathrm{d}w(t)$$

$$\mathrm{d}y(t) = Cx(t)\mathrm{d}t + \mathrm{d}v(t)$$

where for simplicity and without loss of generality no control input has been included. The stochastic processes $w(t)$ and $v(t)$ are assumed to be independent zero-mean Wiener processes with

$$E[\mathrm{d}w(t)\mathrm{d}w(t)^T] = W\mathrm{d}t$$

$$E[\mathrm{d}v(t)\mathrm{d}v(t)^T] = V\mathrm{d}t \, .$$

The matrices W and V are positive semi-definite. To avoid singular filtering problems, the matrix V is assumed to be positive definite. A less rigorous, but more familiar model is obtained by replacing $\mathrm{d}w(t)/\mathrm{d}t$ and $\mathrm{d}v(t)/\mathrm{d}t$ by white noise processes, representing the system equation as a differential equation and considering $\mathrm{d}y(t)/\mathrm{d}t$ as the output measurement.

The formulation of the optimal least squares state estimation problem is as follows: design a (linear) estimation algorithm to reconstruct the state of the system from the output measurements in an optimal manner. Optimality means that the mean square state estimation error is as small as possible. It turns out that the optimal estimation is realized by the Kalman-Bucy filter. If the outputs are available from the initial time t_0, this filter is described by the following equation [18, section 4.3]

$$\mathrm{d}x(t) = Ax(t)\mathrm{d}t + H(t)[\mathrm{d}y(t) - Cx(t)\mathrm{d}t]$$

with initial condition

$$x(t_0) = \mu_0$$

and

$$H(t) = P(t)C^T V^{-1} \, .$$

The matrix $P(t)$ is the solution of the matrix differential equation

$$\dot{P}(t) = AP(t) + P(t)A^T - P(t)C^T V^{-1} CP(t) + W$$

with initial condition

$$P(t_0) = Q_0 \, .$$

In these expressions μ_0 and Q_0 denote the a priori estimate of $x(t_0)$ and its estimation error variance matrix. If the observation interval is infinite, i.e., if all past output measurements on $]-\infty, t]$ are available for the estimate of the state $x(t)$, then the filter equations are stationary:

$$\mathrm{d}x(t) = Ax(t) + H[\mathrm{d}y(t) - Cx(t)\mathrm{d}t]$$

with

$$H = PC^T V^{-1}$$

and P a solution of

$$AP + PA^T - PC^TV^{-1}CP + W = 0$$

where further analysis is necessary to find the particular solution one should use. One immediately sees that the same equations show up as in the optimal control problem. In particular the matrices W and C in the filtering problem correspond to respectively Q and B^T in the control problem. The matrix V corresponds to the input weighting matrix; without loss of generality it may be assumed to be the identity matrix. In the filtering problem the Riccati differential equation should be solved forwards in time instead of backwards. However this entails no essential difference. Because of the striking similarity, the discussion in this contribution is almost completely restricted to the optimal control problem.

9.3 Criterion for Convergence

In this section a criterion on the terminal condition S for the (RDE) (or the final state penalty matrix for the optimal control problem) is given such that the solution of (RDE) converges to a solution of (ARE) for increasing time interval; moreover an algorithm is formulated for the identification of the limiting value. For simplicity of the discussion it is assumed that the undetectable eigenvalues of A are simple, and, except for pairs of complex conjugate eigenvalues, have different real parts. Otherwise the criteria become much more complex, without really providing more insight. It is assumed that (A, B) is stabilizable and moreover that the Hamiltonian matrix, defined in Appendix 9.5.2, has no eigenvalues on the imaginary axis. The reason for the latter assumption is commented upon in the sequel of the contribution.

Let the eigenvectors v_j corresponding to the m *undetectable* eigenvalues μ_j of A with respect to the output matrix C, be ordered according to decreasing real parts of the eigenvalue:

$$v_1, v_2, \ldots, v_k, v_{k+1}, \ldots, v_m \ .$$

If μ_j is real, then v_j is a real eigenvector. If μ_k and μ_{k+1} are complex conjugate eigenvalues, then v_k and v_{k+1} are complex conjugate eigenvectors; the ordering of the latter eigenvectors with respect to each other is arbitrary.

Consider the A-invariant subspaces in R^n:

$$\begin{aligned} W_0 &:= \{0\} \\ &\vdots \\ W_k &:= \operatorname{span}(v_1, \ldots, v_k) \\ &\vdots \\ W_m &:= \operatorname{span}(v_1, \ldots, v_m) \ . \end{aligned}$$

Note that W_m equals the undetectable subspace $ND(C, A)$. Consider also the associated subspaces $V_0, \ldots, V_m$ in $\mathbb{R}^n$:

$$V_k := SW_k \ .$$

With these concepts the criterion for the convergence of the solution of the Riccati differential equation can be formulated in a form which is somewhat more compact than the criterion formulation in our original paper [5, Theorem 4]. For the proof of the criterion the reader is referred to that paper.

Criterion 1. For a given terminal state penalty matrix S, the solution of (RDE) converges to a constant solution of (ARE) if and only if for any pair μ_k, μ_{k+1} of complex conjugate undetectable eigenvalues

$$\dim(V_{k+1}) = \dim(V_{k-1}) \text{ or } \dim(V_{k-1}) + 2$$

where dim denotes the dimension of a subspace. The limiting value is the solution of (ARE) supported by the subspace spanned by the eigenvectors corresponding to the real eigenvalues μ_j such that

$$\dim(V_j) = \dim(V_{j-1})$$

and to the complex conjugate eigenvalues μ_k and μ_{k+1}, such that

$$\dim(V_{k+1}) = \dim(V_{k-1})$$

(i.e., the eigenvectors in the so-called rejection set [5]).

Comment 1. If

$$\Omega = N(S) \cap ND(C, A)$$

is A-invariant, then the condition of Criterion 1 for convergence is certainly satisfied. The limit is the solution of (ARE) supported by the eigenvectors of A in Ω. Note also that in this case

$$\Omega = ND([\,C^T \quad F^T\,]^T, A)$$

where F satisfies $F^T F = S$, such that rank(F) = rank(S); hence F is unique up to a multiplication on the left by an orthogonal matrix.

Comment 2. The solution of (RDE) converges to the maximal solution P_+, i.e., the unique stabilizing solution of (ARE), if and only if

$$\dim(V_m) = m \quad \text{or} \quad ND(C, A) \cap N(S) = \{0\} \ .$$

In that case all modes are stabilized by the state feedback control, corresponding to P_+, i.e., all eigenvalues of the matrix $A - BB^T P_+$ have negative real parts. P_+ is the solution of (ARE) supported by the null vector. Note that this case occurs in particular if $S \geq P_+$.

Comment 3. The solution of (RDE) converges to the minimal positive semi-definite solution P_0 of (ARE) if and only if

$$\dim(V_m) = 0 \quad \text{or} \quad ND(C,A) \subset N(S)\,.$$

This means that the undetectable subspace $ND(C,A)$ lies completely in the null space of S. P_0 is such that no *undetectable* eigenvalues of A are stabilized in $A - BB^T P_0$. Particular cases are $S = 0$ (no terminal output penalization) and $S = C^T C$ (the terminal output penalization and the current output penalization agree). P_0 is the solution of (ARE) supported by the undetectable subspace itself. The smallest positive semi-definite solution of (ARE) corresponds to a maximally unstable closed loop system, having $ND(C,A)$ as unstable subspace. If (C,A) is detectable then the smallest and the largest positive semi-definite solutions of (ARE) coincide and the closed loop is asymptotically stable.

Comment 4. All positive semi-definite solutions of (ARE) can clearly be generated as limiting solutions of (RDE) for appropriate choices of the terminal cost matrix S. The most convenient choices of S are those where the subspace Ω is A-invariant, as discussed in Comment 1.

Comment 5. If all eigenvalues of A are real, then the solution of (RDE) converges for any positive semi-definite matrix S.

9.3.1 Examples

Example 1 below illustrates a case where the solution of (RDE) does not converge to a constant solution. Example 2 shows the dependence of the limiting value on the terminal condition S.

Example 1. Let

$$A = \begin{bmatrix} 1 & -1 \\ 1 & 1 \end{bmatrix}, B = \begin{bmatrix} 1 & 0 \\ 0 & 1 \end{bmatrix}$$

$$C = [0 \quad 0]\ , \ S = \begin{bmatrix} 1 & 0 \\ 0 & 0 \end{bmatrix}.$$

In this case A has two undetectable eigenvalues $1 + \mathrm{j}$ and $1 - \mathrm{j}$. The subspaces considered in Criterion 1 are in this example:

$$V_0 = \{0\}\ , \ V_2 = \{\begin{bmatrix} k \\ 0 \end{bmatrix} \mid k \in \mathbf{R}\}\,.$$

Criterion 1 shows that the solution of (RDE) does not converge for the given terminal condition S. It can easily be checked that the solution is

$$P(t) = f(t) \begin{bmatrix} 2\cos^2(t - t_1) & -\sin(2t - 2t_1) \\ -\sin(2t - 2t_1) & 2\sin^2(t - t_1) \end{bmatrix}$$

with

$$f(t) = 1/[1 + \exp(2t - 2t_1)] .$$

The solution tends to a periodic solution if the final time t_1 tends to infinity along multiples of π.

Example 2. Let

$$A = \begin{bmatrix} 1 & 0 \\ 0 & 2 \end{bmatrix} , \; B = \begin{bmatrix} 1 & 0 \\ 0 & 1 \end{bmatrix}$$

$$C = [0 \quad 0] .$$

The algebraic Riccati equation has four positive semi-definite solutions:

$$P_+ = \begin{bmatrix} 2 & 0 \\ 0 & 4 \end{bmatrix} , \; P_1 = \begin{bmatrix} 0 & 0 \\ 0 & 4 \end{bmatrix}$$

$$P_2 = \begin{bmatrix} 2 & 0 \\ 0 & 0 \end{bmatrix} , \; P_0 = \begin{bmatrix} 0 & 0 \\ 0 & 0 \end{bmatrix} .$$

Criterion 1 shows that the solution of (RDE) converges for any positive semi-definite matrix S. The limiting value is:

(a) P_0 if and only if $S = 0$;

(b) P_2 if and only if

$$S = \begin{bmatrix} a & 0 \\ 0 & 0 \end{bmatrix} ,$$

for positive a;

(c) P_1 if and only if

$$S = \begin{bmatrix} a \\ b \end{bmatrix} [a \quad b] ,$$

for non-zero b;

(d) P_+ for any positive definite matrix S.

9.4 Finite and Infinite Horizon Control Problems

Let us reconsider the statement of the LQ-optimal control problems introduced in Section 2. The statement of the infinite horizon optimal control problem is as follows:

Problem (P_1). With J defined in (9.7), find

$$J^* = \inf_{u(.)} J \tag{9.10}$$

and the corresponding optimal control $u^*(.)$ achieving this minimum cost.

This problem is to be compared to the limiting case of the finite horizon problem for increasing time interval:

Problem (P_2). With $J(t_1)$ defined in (9.2), let

$$J^*(t_1) = \inf_{u(.)} J(t_1) \tag{9.11}$$

and let $u_a(.)$ denote the corresponding optimal control. Determine

$$J^{**} = \lim_{t_1 \to \infty} J^*(t_1) \tag{9.12}$$

and, if the limit exists, find the limiting behavior $u^{**}(.)$ of the optimal control $u_a(.)$ for increasing t_1.

The question is whether J^* is equal to J^{**} and whether u^* coincides with u^{**}. If so, the optimal control for the infinite horizon problem can be approximated by the optimal control of the finite horizon problem for a sufficiently large time interval, which is also called the "receding horizon control problem".

One may consider an alternative way of expressing that one wants Sx or Fx to become small in the future by introducing the constraint:

$$Sx(t_1) = 0\,.$$

For the infinite horizon problem this corresponds to a stabilization constraint, expressing that we want to steer $Sx(t)$ to zero ultimately. This leads to two additional LQ-problems:

Problem (P_3). Find

$$I^* = \inf_{u(.)} I \tag{9.13}$$

subject to the stabilization constraint

$$Sx(t) \to 0 \quad \text{as} \quad t \to \infty$$

where

$$I = \lim_{t_1 \to \infty} I(t_1)$$

with $I(t_1)$ given by (9.3), and determine the optimal control $u^*(.)$ achieving this minimum cost.

Here again it is interesting to investigate whether this problem can be solved as the limiting case of the finite time interval control problem with constraint on the terminal state:

Problem (P_4). Let

$$I^*(t_1) = \inf_{u(.)} \{I(t_1) \mid Sx(t_1) = 0\} \tag{9.14}$$

and let $u_b(.)$ denote the control achieving the minimum cost. Determine

$$I^{**} = \lim_{t_1 \to \infty} I^*(t_1) \tag{9.15}$$

and, if the limit exists, find the limiting behavior $\underline{u}^{**}(.)$ of the optimal control $u_b(.)$ for increasing t_1.

Here also the question is whether I^* is equal to I^{**} and whether $\underline{u}^{**}$ coincides with $\underline{u}^*$. Of course for the finite horizon problem involved in (P_4) to be solvable, the controllability of the output Sx is required; this means that $R(S)$ should be a subspace of $C(A, B)$.

Interesting results on the relations between the four LQ-problems can be derived from the analysis of the convergence properties of the solutions of the Riccati differential equation and from the analysis of the solution of the infinite horizon problem. The following properties were derived by the authors [26].

Lemma 1. *The solution of the infinite horizon problem (P_1) is such that along the solutions of the system with optimal control, $SA^k x(t)$ tends to zero as t tends to infinity, for $k = 0, 1, 2, \ldots$.*

This lemma [26, Lemma 3] immediately yields [26, pp. 36–40]:

Equivalence Result 1. The infinite horizon problems (P_1) and (P_3) are equivalent, and the limits involved in the problem statements exist simultaneously.

The main result on the relation between the receding horizon problems (P_2) and (P_4) is [26, Theorem 2]:

Equivalence Result 2. Problems (P_2) and (P_4) are equivalent in the sense that the limits involved in both problem statements exist simultaneously, and if they exist, they are equal.

Comment 6. It is interesting to note that the proof of the above result shows that even if the limits involved in the problem statements (P_2) and (P_4) do not exist, the limiting behavior for large $t_1 - t_0$ is the same in both cases.

Finally, the relationship between the infinite horizon problems (P_1) and (P_3) on the one hand, and the receding horizon problems (P_2) and (P_4) on the other hand, follows from the criteria for the convergence of the solution of (RDE) discussed in the previous section.

It follows from Lemma 1 that the solution of (P_1) and (P_3) is related to the positive semi-definite solution of (ARE) supported by the largest A-invariant subspace in

$$\Omega = N(S) \cap ND(C, A) .$$

On the other hand the analysis of the convergence properties of the solution of (RDE) in Section 3 yields a criterion for the existence of a solution of (P_2) and (P_4). Moreover it reveals that the solution of (P_2) and (P_4), if it exists, is related to the positive semi-definite solution of (ARE) supported by a subspace of the same dimension as the subspace Ω, and which is exactly identified in Criterion 1. Hence [26, Theorem 4]:

Equivalence Result 3. The solutions of the infinite horizon problems (P_1) and (P_3) are exactly the limiting cases of the receding horizon problems (P_2) and (P_4) if and only if the subspace Ω is A-invariant.

Comment 7. If the above condition is not satisfied, the largest A-invariant subspace contained in Ω is of smaller dimension than Ω itself. In that case the solution of the infinite horizon problems (P_1) and (P_3) corresponds to the stabilization of more undetectable system eigenvalues than the limiting case of the receding horizon problems (P_2) and (P_4).

Comment 8. If the subspace Ω is A-invariant, then the solution of the infinite horizon problems (P_1) and (P_3) leads to the stabilization of the same modes if the output Fx is included in the final state penalty or the stabilization constraint, as if this output would be contained in the integral of the cost function. Note however that the allocation of the closed loop modes is different. In particular, the solution of problems (P_1) and (P_3) corresponds to an asymptotically stable closed loop system if and only if the system

$$\dot{x}(t) = Ax(t)$$

with output

$$y(t) = \begin{bmatrix} C \\ F \end{bmatrix} x(t)$$

is detectable. This case occurs in particular if S is strictly positive definite; this corresponds to a result discussed by Brunovsky and Komornik [3, Theorem 4]. In that case the stabilization constraint for problem (P_1) actually requires stabilization of the system. The receding horizon problems (P_2) and (P_4) and the infinite horizon problems (P_1) and (P_3) yield the same result, corresponding to the feedback control

$$u(t) = -B^T P_+ x(t)$$

when P_+ is the maximal (stabilizing) solution of (ARE). A weaker condition for convergence of the solution of (RDE) to the stabilizing solution is $S \geq P_+$.

Comment 9. The solution of problems (P_1) and (P_3) leads to the smallest positive semi-definite solution P_0 of the ARE if and only if the undetectable subspace $ND(C, A)$ is a subspace of the kernel $N(S)$. In this case the final state penalization or the stabilization constraint does not affect the solution. If there is no terminal state penalization or final state constraint ($S = 0$), the condition is automatically satisfied; this corresponds also to a result derived by Brunovsky and Komornik [3, Theorem 2]. In that case the solution agrees with the limiting cases of problems (P_2) and (P_4).

Comment 10. The main differences between the infinite time problems (P_1) and (P_3) and the receding horizon problems (P_2) and (P_4), if the limits exist, can be characterized as follows:

- For the infinite time problem an undetectable mode of A is stabilized if its corresponding eigenvector is not in the null space of the final state penalty matrix. In general the number of undetectable modes in the null space of S is smaller than the dimension of the subspace Ω, unless Ω is A-invariant (i.e., spanned by eigenvectors of A).
- For the receding horizon problem only q undetectable eigenvalues of A are stabilized, where q is the dimension of the subspace V_m defined in Section 3 (i.e., the image of $ND(C, A)$ under S).
- Let v_1 and v_2 be two eigenvectors, corresponding to the undetectable eigenvalues μ_1 and μ_2, which for simplicity are assumed to be real. Let $\mu_1 > \mu_2$. Then

$$S(v_1 + v_2) = 0 \quad \text{and} \quad Sv_1 \neq 0$$

imply that in the infinite horizon problem the eigenvalues μ_1 and μ_2 will *both* be stabilized. In the receding horizon problem the eigenvalue μ_2 will certainly not be stabilized, because Sv_1 and Sv_2 are linearly dependent. Whether μ_1 is stabilized or not, depends on whether Sv_1 is dependent on Sv_k for eigenvectors v_k whose corresponding undetectable eigenvalues have larger real parts than μ_1.

Example 3. We reconsider Example 1 of Section 3 for various choices of the terminal state penalty matrix. The algebraic Riccati equation has two positive definite solutions:

- The smallest solution P_0 is equal to the zero matrix. If the state is neither penalized in the integral of the cost function nor in the final state penalization, then the optimal control is "no control". In that case

$$A_0 = A - BB^T P_0 = A$$

has two unstable eigenvalues $(1 + \mathrm{j})$ and $(1 - \mathrm{j})$.
- The stabilizing solution P_+ is

$$P_+ = 2I_2$$

where I_2 denotes the identity matrix of order two. The corresponding closed loop plant matrix

$$A_+ = A - BB^T P_+ = \begin{bmatrix} -1 & 1 \\ -1 & -1 \end{bmatrix}$$

has two stable eigenvalues $(-1 + \mathrm{j})$ and $(-1 - \mathrm{j})$. The *cheapest* stabilizing control shifts the unstable open loop eigenvalues to their mirror images with respect to the imaginary axis, as was proved by Kwakernaak and Sivan [18, Theorem 3.12].

By means of this example the three modes of behavior for large time intervals can nicely be illustrated:

- If S is the zero matrix (no final state penalization), then the solution of (RDE) is

$$P(t) = 0\,.$$

The finite horizon optimal control corresponds to "no control"; the limiting case of the receding horizon problem clearly corresponds to the solution of the infinite horizon problem without stabilization constraint.

- If S is positive semi-definite and has rank one, e.g.

$$S = \begin{bmatrix} 1 & 0 \\ 0 & 0 \end{bmatrix}$$

then the receding horizon problem yields a non-convergent solution of (RDE), as was shown in Section 1. The infinite horizon problem (P_1) yields a solution which stabilizes *both* modes; the same is true for the infinite horizon problem (P_3).

- If S is positive definite, e.g. S is the identity matrix, then the solution of (RDE) is

$$P(t_0) = 2[1 + \exp(2t_0 - 2t_1)]^{-1} I_2$$

where I_2 denotes the identity matrix of order 2. Hence the limiting cases of the receding horizon problems (P_2) and (P_4) as well as the infinite horizon problems (P_1) and (P_3) correspond to a stabilization of both modes, with

$$u(t) = -2x(t)\,.$$

Comment 11. It is rather surprising that the limiting case of (P_4) may yield a state feedback control which does not stabilize the output Sx. Indeed in the controlled system some components of Sx may go off to infinity for increasing time! This is illustrated in the next example.

Example 4. Consider the system of Example 2, with the constraint

$$x_1(t_1) + x_2(t_1) = 0$$

where x_1 and x_2 denote the two state variables. The LQ-optimal control satisfying the constraint is

$$u(t) = -\begin{bmatrix} \exp(-2\tau)x_1(t) + \exp(-3\tau)x_2(t) \\ \exp(-3\tau)x_1(t) + \exp(-4\tau)x_2(t) \end{bmatrix} f(\tau)$$

with $\tau = t - t_1$ and

$$f(\tau) = 4/[\exp(-4\tau) + 2\exp(-2\tau) - 3].$$

For very large $t_1 - t$ the control tends to

$$u(t) = -\begin{bmatrix} 0 \\ 4x_2(t) \end{bmatrix}.$$

This control stabilizes the eigenvalue 2, but not the eigenvalue 1 of the matrix A. Hence the control is such that the output $x_1 + x_2$ is not stabilized, i.e.,

$$\lim_{t\to\infty} [x_1(t) + x_2(t)] = 0$$

does not hold. Even worse, the state x_1 blows up as t increases.

Comment 12. Example 4 shows that one should be very careful in introducing the infinite horizon problem as the limiting case of a finite horizon problem, although it is justified for the standard regulator problem (stabilizable system with detectable output in the integral term of the cost function). In particular it is clear from the example that a stabilization constraint in the infinite horizon problem can not always be simulated by the corresponding terminal state constraint or terminal state penalization for the receding horizon problem. It is therefore interesting to investigate under what conditions it can be done. There are two possibilities:

1. A first procedure [26, p. 38] consists of replacing the stabilization constraint on Sx in the infinite horizon problem by a terminal constraint or terminal penalization of Wx instead of Sx in the receding horizon problem, where

   $$W := S + A^T SA + \cdots + (A^T)^{n-1} SA^{n-1} .$$

   The validity of this approach readily follows from Equivalence Result 3.
2. A second procedure corresponds to replacing the state penalization

   $$x^T Qx = x^T C^T Cx$$

   in the integral term of the cost function by

   $$x^T Qx + \alpha x^T Sx = x^T C^T Cx + \alpha x^T F^T Fx$$

   for small positive α (or considering the limiting case for α decreasing to zero).

Comment 13. If (A, B) is stabilizable and H has eigenvalues on the imaginary axis, which means that A has undetectable eigenvalues on the imaginary axis, then (ARE) has no stabilizing solution; however, there is still a largest solution P_s (also called the strong solution) which is such that the eigenvalues of the matrix $A - BB^T P_s$ have a negative real part, except for the undetectable imaginary axis eigenvalues of A [6], [9]. The convergence analysis and the relation between the solutions of (ARE) and the limiting values of (RDE) remains practically unaltered, except that the stabilizing solution P_+ of (ARE) should be replaced by the strong solution Ps. The different positive semi-definite solutions of (ARE) correspond to different sets of undetectable eigenvalues with positive real part that are not stabilized, i. e. not affected by the feedback associated with the particular solution of (ARE). In addition all unobservable eigenvalues with negative real part are not affected by the feedback corresponding to *any* positive semi-definite solution of (ARE); the same is true for the undetectable eigenvalues with zero real part. If the eigenvectors corresponding to the latter eigenvalues are in $N(S)$, then the infinite horizon optimal control problem has a solution; the results discussed above remain practically unaltered. Otherwise the infinite horizon control problem has no solution. This is illustrated in the next example. The above discussion shows that the assumption that H has no imaginary axis eigenvalues is not only introduced for mathematical reasons, but also for the discussion of Section 4 to be practically relevant.

Example 5. Consider the scalar system

$$\dot{x}(t) = u(t)$$

with

$$I(t_1) = \int_0^{t_1} u(t)^2 \mathrm{d}t \, .$$

One readily checks that this system has an undetectable mode at the origin. The following results are readily derived:

1. The limiting case of the finite horizon problem with stabilization constraint $x(t_1) = 0$ or final state penalization $x(t_1)^2$ corresponds to a zero control. The closed loop is not asymptotically stable. The corresponding trajectory is: $x(t) = x(t_0)$. The undetectable system eigenvalue is not relocated. The corresponding limiting value of the cost is zero. For this example the solution of the (scalar) Riccati differential equation is given by

$$p(t) = 1/(t_1 - t)$$

for the stabilization constraint $x(t_1) = 0$ and

$$p(t) = \alpha/[1 + \alpha(t_1 - t)]$$

in the case of final state penalization, with the positive constant α denoting the weighting factor of $x(t_1)^2$ in the cost function. These expressions show that the convergence of $p(t)$ towards zero as t_1 tends to infinity, is not exponential. This is a consequence of the presence of an undetectable mode or an eigenvalue of the Hamilton matrix on the imaginary axis.
2. The infinite horizon problem with stabilization constraint has no solution. Indeed consider all possible stabilizing feedback controls:

$$u(t) = -kx(t)$$

with k positive. The infinite horizon cost equals $k/2$. Hence the cost decreases with k; since k must be positive, there is no optimal feedback.

Comment 14. The assumption of stabilizability could be relaxed to output stabilizability, that is the property that there exists a control input for (9.1) such that $Qx(t)$ and $Sx(t)$ tend to zero for increasing t. This is discussed by Geerts [11] for the case $S = 0$.

Comment 15. Very similar remarks can be made with respect to the filtering problem and with respect to discrete-time systems. The reader is referred to the literature [6], [9] for an explicit discussion.

9.5 Appendices

9.5.1 Notations and Definitions

The following notations are used throughout this chapter. For a square matrix A, $L^-(A)$, $L^+(A)$, $L^0(A)$ respectively denote the A-invariant subspaces spanned by the (generalized) eigenvectors corresponding to the eigenvalues with negative real part, positive real part, and zero real part, respectively. The superscript T denotes matrix transposition. The spectrum, the null space and the range of the matrix A are denoted by $\sigma(A)$, $N(A)$ and $R(A)$.

Let

$$L^{0+}(A) := L^+(A) \oplus L^0(A)$$

where $\oplus$ denotes the direct sum of subspaces. Let A be an $(n \times n)$-matrix, B an $(n \times r)$-matrix and C an $(m \times n)$-matrix. The unobservable subspace of the pair (C, A) is given by

$$NO(C,A) = N(C) \cap N(CA) \cap \ldots \cap N(CA^{n-1}).$$

The undetectable subspace of (C, A) is

$$ND(C,A) := NO(C,A) \cap L^{0+}(A)\,.$$

The pair (C, A) is said to be observable or detectable if respectively $NO(C, A)$ or $ND(C, A)$ equals $\{0\}$. It is clear that the subspaces $NO(C, A)$ and $ND(C, A)$ are A-invariant, and hence spanned by (generalized) eigenvectors of A. A (generalized) eigenvector of A in $NO(C, A)$ is called unobservable; its corresponding eigenvalue is said to be an unobservable eigenvalue. If in addition the eigenvalue has a nonnegative real part, the eigenvalue and the eigenvector are said to be undetectable.

The (A, B)-controllable subspace is given by

$$C(A,B) := R([B \quad AB \quad \ldots \quad A^{n-1}B])\,.$$

The pair (A, B) is said to be controllable if the subspace $C(A, B)$ is R^n. A weaker property is stabilizability of the pair (A, B). This property requires that $C(A, B)$ contains $L^{0+}(A)$, or equivalently the existence of a suitable matrix K such that $A+BK$ has only eigenvalues with negative real part. Another equivalent condition is that the stabilizable subspace

$$S(A,B) = L^-(A) + C(A,B)$$

is $\mathbf{R}^n$. For the dynamical and feedback control interpretation of these properties, the reader is referred to the relevant literature [15], [18].

9.5.2 Properties of the Solution of the Algebraic Riccati Equation

An extensive literature [7], [16], [17], [19], [23], [24], [25] exists on the existence and the properties of the solutions of the algebraic Riccati equation (9.9). In this appendix only those features are summarized which are relevant for the present chapter. A more general and detailed discussion of the solutions of (ARE) is given in other chapters of this book.

Closely related to (ARE) is the Hamiltonian matrix

$$H := \begin{bmatrix} A & -BB^T \\ -Q & -A^T \end{bmatrix} .$$

Let $Q = C^TC$. The following properties of the solutions of (ARE) are used in the present contribution.

1. If (A, B) is stabilizable and the eigenvalues of A on the imaginary axis are detectable (which is certainly true if (C, A) is detectable), then H has no imaginary axis eigenvalues.
2. If H has imaginary axis eigenvalues, then (ARE) has no stabilizing solution, i.e., no solution P such that all eigenvalues of $A - BB^TP$ have negative real part.
3. If (A, B) is stabilizable and (C, A) detectable, then (ARE) has a unique positive semi-definite solution. This solution is moreover the unique stabilizing solution. If (C, A) is observable, this solution is positive definite.
4. If (A, B) is controllable, then the symmetric solutions of (ARE) form a lattice with a common largest element and a common smallest element, which are positive and negative semi-definite, respectively.
5. If (A, B) is stabilizable, (C, A) is not detectable, but there are no undetectable eigenvalues on the imaginary axis, then (ARE) has multiple positive semi-definite symmetric solutions. These solutions also form a lattice. The common largest element P_+ is such that $A - BB^TP_+$ is stable (i.e., has only eigenvalues with negative real parts). All other positive semi-definite solutions P are such that $A - BB^TP$ has some eigenvalues with positive real part; these eigenvalues are undetectable eigenvalues of A. The corresponding (generalized) eigenvectors correspond with these of A, and are hence in $N(P)$. All positive semi-definite solutions can be characterized by the set of unstable eigenvalues of the matrix $A - BB^TP$, or equivalently by the undetectable eigenvalues of A which are not stabilized. Another characterization is the A-invariant subspace corresponding to these eigenvalues; this subspace is contained in $ND(C, A)$ and is given by

$$\Omega(P) = N(P) \cap ND(C, A) = L^+(A - BB^TP) .$$

The A-invariant subspace $\Omega(P)$ is said to be the supporting subspace of the solution P of (ARE). Conversely, if Ω is a real A-invariant subspace of $ND(C,A)$, there is a positive semi-definite solution P of (ARE) such that

$$\Omega = L^+(A - BB^T P) = N(P) \cap ND(C,A) .$$

If P and P' are two positive semi-definite solutions, with corresponding supporting spaces Ω and Ω', then $P \geq P'$ if and only if Ω is a subspace of Ω'. Explicitly for the largest solution P_+ the subspace is the null space

$$\Omega(P_+) = \{0\} .$$

The smallest element P_o of the lattice of positive semi-definite solutions is such that all undetectable eigenvalues are eigenvalues of $A - BB^T P_o$, or, otherwise said, no undetectable eigenvalues are stabilized. The A-invariant subspace supporting P_o is hence $ND(C,A)$. Note that the unobservable eigenvalues of A with negative real part are not affected by the feedback corresponding to the positive semi-definite solutions of (ARE) and are thus eigenvalues of $A - BB^T P$ for any positive semi-definite solution P of (ARE). The corresponding eigenvectors are in the null space of any positive semi-definite solution of (ARE).

6. If (A, B) is stabilizable, (C, A) not detectable, and there are undetectable eigenvalues on the imaginary axis (which means that the Hamiltonian matrix H has imaginary axis eigenvalues), the situation is somewhat different. The positive semi-definite solutions of (ARE) still form a lattice. The largest element P_s, called the strong solution, stabilizes all eigenvalues, except the undetectable eigenvalues on the imaginary axis; the latter are eigenvalues of $A - BB^T P$ for all solutions of (ARE). There is no solution of (ARE) such that $A - BB^T P$ has only eigenvalues with negative real part. Otherwise the ordering features mentioned in 5. remain unaltered, except that $ND(C,A)$ should be replaced by

$$ND(C,A) \cap L^+(A) .$$

Acknowledgment. The first author (JLW) gratefully acknowledges research support from the National Fund for Scientific Research in Belgium (FKFO-Grant). The authors thank Ir. Koenraad Audenaert for converting the text file from Wordperfect into TEX.

References

1. Anderson, B.D.O.; Moore, J.B.: Linear Optimal Control. Prentice Hall, Englewood Cliffs, N.J. 1971
2. Anderson, B.D.O.; Moore, J.B.: Optimal Filtering. Prentice-Hall, Englewood Cliffs, N.J. 1979
3. Brunovsky, P.; Komornik, J.: The Riccati equation solution of the linear-quadratic problem with constrained terminal state. IEEE Transactions on Automatic Control AC-26 (1981) 398–402

4. Bucy, R.S.; Joseph, P.D.: Filtering for Stochastic Processes with Applications to Guidance. Wiley, New-York 1968
5. Callier, F.M.; Willems, J.L.: Criterion for the convergence of the solution of the Riccati differential equation. IEEE Transactions on Automatic Control AC-26 (1981) 1232–1242
6. Chan, W.; Goodwin, G.C.; Sin, K.S.: Convergence properties of the Riccati difference equation in optimal filtering of nonstabilizable systems. IEEE Transactions on Automatic Control AC-29 (1984) 110–118
7. Coppel, W.A.: Matrix quadratic equations. Bulletin of the Australian Mathematical Society 10 (1974) 377–401
8. Davis, M.H.A.: Linear Estimation and Stochastic Control. Chapman and Hall, London 1977
9. De Souza, C.E.; Gevers, M.R.; Goodwin, G.C.: Riccati equations in optimal filtering of nonstabilizable systems having singular state transition matrices. IEEE Transactions on Automatic Control AC-31 (1986) 831–838
10. Faurre, P.; Clerget, M.; Germain, F.: Opérateurs Rationnels Positifs. Dunod, Paris 1979
11. Geerts, A.H.W.: A necessary and sufficient condition for solvability of the linear-quadratic control problem without stability. Systems and Control Letters 11 (1988) 47–52
12. Kalman, R.E.; Falb, P.L.; Arbib, M.L.: Topics in Mathematical System Theory. McGraw-Hill, New York 1969
13. Kalman, R.E.; Bucy, R.S.: New results in linear filtering and prediction theory. Transactions of the ASME, Ser.D. Journal of Basic Engineering 83D (1961) 95–108
14. Kalman, R.E.: Contributions to the theory of optimal control. Bol. Soc. Mat. Mexicana 5 (1960) 102–119
15. Knobloch, H.W.; Kwakernaak, H.; Lineare Kontrolltheorie. Springer-Verlag, Berlin 1985
16. Kucera, V.: On nonnegative definite solutions to matrix quadratic equations. Automatica 8 (1972) 413–423
17. Kucera, V.: A contribution to matrix quadratic equations. IEEE Transactions on Automatic Control AC-17 (1972) 344–347
18. Kwakernaak, H.; Sivan, R.: Linear Optimal Control Systems. Wiley-Interscience, New York 1972
19. Molinari, B.P.: The time-invariant linear-quadratic optimal control problem. Automatica 13 (1977) 347–357
20. Poubelle, M.A.; Petersen, I.R.; Gevers, M.R.; Bitmead, R.R.: A miscellany of results on an equation of count J.F. Riccati. IEEE Transactions on Automatic Control AC-31 (1986) 651–653
21. Sasagawa, T.: A necessary and sufficient condition for the solution of the Riccati equation to be periodic. IEEE Transactions on Automatic Control AC-25 (1980) 564–566
22. Schweppe, F.C.: Uncertain Dynamic Systems. Prentice-Hall, Englewood Cliffs 1973
23. Shayman, M.A.: Geometry of the algebraic Riccati equation, Part I and II. SIAM Journal on Control and Optimization 21 (1983) 375–409
24. Shayman, M.A.: On the periodic solutions of the matrix Riccati equation. Mathematical Systems Theory 16 (1983) 267–287
25. Willems, J.C.: Least-squares stationary optimal control and the algebraic Riccati equation. IEEE Transactions on Automatic Control AC-16 (1971) 621–634
26. Willems, J.L.; Callier, F.M.: Large finite horizon and infinite horizon LQ-optimal control problems. Optimal Control Applications and Methods 4 (1983) 31–45
27. Wimmer, H.K.: The algebraic Riccati equation: conditions for the existence and uniqueness of solutions. Linear Algebra and Its Applications 58 (1984) 441–460
28. Lehtomaki, N.A.; Sandell, N.R. Jr.; Athans, M.: Robustness results in linear-quadratic Gaussian based multivariable control design. IEEE Transactions on Automatic Control AC-26 (1981) 75–93

10 Riccati Difference and Differential Equations: Convergence, Monotonicity and Stability

Robert R. Bitmead and Michel Gevers

10.1 Introduction

The main theme of this Chapter will be the connections between various Riccati equations and the closed loop stability of control schemes based on Linear Quadratic (LQ) optimal methods for control and estimation. Our presentation will encompass methods applicable both for discrete time and continuous time, and so we discuss concurrently the difference equations (discrete time) and the differential equations (continuous time) – the intellectual machinery necessary for the one suffices for the other and so it makes sense to dispense with both cases in one fell swoop.

Our strategy for the exploration of this subject is as follows:

- The connection between particular Riccati equations and their associated LQ optimal control and estimation problems is established. This is done for
  - Finite horizon LQ optimal control and least squares state estimation,
  - Infinite horizon LQ control and estimation,
  - Receding horizon LQ control and estimation.
- Asymptotic stability problems are posed for infinite and receding horizon variants of these closed loop systems.
- General Lyapunov stability methods for linear difference and differential equations are then presented in a form amenable to LQ applications using the Riccati equation.
- Convergence and monotonicity properties of Riccati equation solutions are then treated.
- Stability inherent to the infinite horizon problems is investigated using the Lyapunov methods.
- We then expand proceedings using monotonicity methods to establish stability for receding horizon LQ solutions.
- Two brief examples of the application of these methods are presented from adaptive control, where receding horizon LQ controllers have gained considerable currency because of their computational simplicity, and harmonic analysis, where Kalman filtering methods may be applied to reject noise in signals.

Bittanti, Laub, Willems (Eds.)
The Riccati Equation

The examples illustrate both the extent of applicability of these optimal methods and the facility of the tools connected with so-called Fake Algebraic Riccati Techniques, which are our central theme here.

10.2 Linear Quadratic Optimal Control Problems and Riccati Equations

10.2.1 Discrete Time

Our focus here will be on the feedback control of the linear system

$$x_{t+1} = Fx_t + Gu_t, \tag{10.1}$$

where x_t is the n-vector state process and u_t the m-vector control process. The matrices F and G are constant (and therefore the system (10.1) is time-invariant) and have appropriate dimensions $n \times n$ and $n \times m$.

We pose three LQ optimization problems distinguished by their criterion horizons: finite horizon LQ optimal control, infinite horizon LQ optimal control and receding horizon LQ optimal control. Each of these problems involves the specification of a quadratic performance objective for the linear system (10.1) – hence 'Linear Quadratic'. This is a mature subfield of Control Systems and the reader is referred to the excellent texts [1-7] for further material and derivation of the earlier results.

In many circumstances it is desirable to consider time-varying systems and time-varying LQ optimization criteria. This yields (expectedly) time-varying control laws. Our study here is, in general, restricted to the behaviour of LQ time-invariant control laws and so we consider initially an assortment of optimal control problems with constant, nonnegative definite weighting (penalty) matrices Q and R. This time-invariance of weighting matrices will then be relaxed at certain instances later connected with receding horizon problems. Throughout we will assume that the control weighting matrix, R or R_t, is positive definite and so is invertible.

Finite Horizon Linear Quadratic Optimal Control. Find the input sequence $\{u_t | t = 0, \ldots, N-1\}$ which minimizes the criterion

$$J(N, x_0, u) = x_N^T P_0 x_N + \sum_{j=0}^{N-1} \left\{ x_j^T Q_{N-j-1} x_j + u_j^T R_{N-j-1} u_j \right\}. \tag{10.2}$$

Here x_0 is the initial state, Q_j is a sequence of nonnegative definite matrices penalizing the excursions of the state from zero, R_j is a sequence of positive definite matrices penalizing the control energy, and P_0 is the nonnegative definite penalty matrix on the terminal state x_N.

The solution to this problem is given by a feedback control law as follows:

$$\begin{aligned} u_{N-j} &= -(G^T P_{j-1} G + R_{j-1})^{-1} G^T P_{j-1} F x_{N-j}, \\ &\triangleq K_{j-1} x_{N-j}, \quad j = 1, \ldots, N, \end{aligned} \tag{10.3}$$

where the control gain, K_j, is given by

$$K_j = -(G^T P_j G + R_j)^{-1} G^T P_j F \tag{10.4}$$

and P_j is the solution sequence of the following *Riccati Difference Equation* (RDE):

$$P_{j+1} = F^T P_j F - F^T P_j G (G^T P_j G + R_j)^{-1} G^T P_j F + Q_j, \tag{10.5}$$

solved forwards in j from the initial condition P_0.

We make several remarks concerning this problem and its solution.

- The control signal sequence, $\{u_j\}$, is a causal feedback of the system state sequence, $\{x_j\}$.
- The feedback gain sequence, $\{K_j\}$, is computed from the Riccati difference equation (10.5) solution sequence, $\{P_j\}$, which in turn is computed from P_0, $\{Q_j\}$, $\{R_j\}$, F and G via the iteration of the RDE.
- Given the above matrices, one iterates the RDE for the sequence $\{P_j\}$ from the initial condition P_0 (which is the final state penalty) effectively backwards in time relative to the evolution of the plant state. Hence the peculiar indexing in (10.3). These P_j and their corresponding gains K_j may be precomputed since they depend only on the model information and on the given penalty weightings. This contrasts to the state, x_t, which needs to be measured on-line. Alternatively, the state too can be precomputed but this then yields an open loop control law.
- The control sequence $\{u_t\}$ associated with the finite horizon LQ problem is defined only over a fixed time interval. It is easily demonstrated that the RDE has no finite escape properties and so the P_j are always bounded. Thus there is no sensible notion of stability which can be attributed to this LQ formulation.
- The minimal value of the criterion $J(N, x_0, u)$ is given by

  $$J^\star(N, x_0, u) = x_0^T P_N x_0.$$

Infinite Horizon Linear Quadratic Optimal Control. We next consider a variant of the finite horizon problem as we allow the horizon, N, to increase without bound. We now consider the case where the weighting matrices Q_j and R_j are constants, Q and R respectively, and examine the convergence of the finite horizon solution to a time-invariant control law.

Specifically, we consider the problem: Find the input sequence $\{u_t | t = 0, 1, \ldots\}$ which minimizes the criterion

$$\begin{aligned} J_\infty(x_0, u) &= \lim_{N \to \infty} J(N, x_0, u) \\ &= \lim_{N \to \infty} \sum_{j=0}^{N-1} \left\{ x_j^T Q x_j + u_j^T R u_j \right\}. \end{aligned} \tag{10.6}$$

Let us study this problem briefly.

- If $[F, G]$ is a stabilizable pair, i.e., all modes of F which are uncontrollable are associated with eigenvalues of magnitude strictly less than one, then a feedback control law, $u_t^s = K^s x_t$, exists which causes $x_t \to 0$ as $t \to \infty$.
- Evaluating the criterion $J(N, x_0, u_t^s)$ for $s = 0, \ldots, N$, we see from the optimality property that this value of $J(N, x_0, x_t^s)$ exceeds the optimal value, $J^\star(N, x_0, u)$, whatever the particular choice of N. Further, since u_t^s causes the closed-loop plant to be exponentially stable, $J(N, x_0, u_t^s)$ converges to $J(\infty, x_0, u_t^s)$ as $N \to \infty$ and this latter quantity is finite. Thus the sequence of optimal costs, $\{J^\star(N, x_0, u) | N = 1, 2, \ldots\}$, belongs to a compact set and so possesses a convergent subsequence with a finite limit point, $J_\infty^\star(x_0)$. The problem stationarity plus optimality combine to ensure the uniqueness of this limit point.
- Since $J_\infty^\star(x_0)$ exists and is finite and $R > 0$, the optimal control law tends to zero as $t \to \infty$. If $[F, Q^{1/2}]$ is detectable then this implies that $x_t \to 0$, i.e., asymptotic stability is achieved. This aspect of infinite horizon LQ control will be more formally treated shortly.

Just as this optimal control problem is posed as the limit of a finite horizon LQ optimal control problem, so too is the solution the limit of the finite horizon solution:

$$\begin{aligned} u_j &= -(G^T P_\infty G + R)^{-1} G^T P_\infty F x_j, \\ &\triangleq K_\infty x_j, \quad j = 1, 2, \ldots, \end{aligned} \tag{10.7}$$

where the constant control gain, K_∞, is given by

$$K_\infty = -(G^T P_\infty G + R)^{-1} G^T P_\infty F \tag{10.8}$$

and P_∞ is the maximal nonnegative definite solution of the following *Algebraic Riccati Equation* (ARE):

$$P_\infty = F^T P_\infty F - F^T P_\infty G (G^T P_\infty G + R)^{-1} G^T P_\infty F + Q. \tag{10.9}$$

Comparing this with the earlier remarks on the RDE,

- The control law (10.7) is still a causal linear state variable feedback.
- The optimal cost is $J^\star(x_0, u) = x_0^T P_\infty x_0$.
- The control gain is now constant over all time. Therefore the feedback law plus the state process define a time-invariant closed loop for which asymptotic stability questions may readily be posed.
- The ARE (10.9) usually possesses many solutions, as may be seen by examining the scalar case. Our requirement here is that the maximal nonnegative definite solution be chosen.
- Under reasonable conditions on the LQ problem to be discussed later, one has that $P_N \to P_\infty$ as $N \to \infty$. That is, the maximal solution is also the limiting solution.

- The feedback gain, K_∞, is computed from the ARE solution, P_∞, but now this ARE solution is fixed in time. The ARE is an algebraic matrix equation, which is typically solved by symplectic eigenvector methods. Considerable effort was devoted to alternative procedures [8], [9] by which approximate ARE solutions or large N RDE solutions were computed. This computational difficulty of solving the ARE used to be rather burdensome in applications and so led to the proposal of a mixed-mode LQ optimal control problem between finite and infinite horizon. This we now discuss.

Receding Horizon Linear Quadratic Optimal Control. Historically speaking, the computational solution of the ARE posed a serious problem for the application of infinite horizon LQ control. Nowadays with better algorithms and hardware, this is less the case or, to draw on Yasser Arafat[1],

C'est caduque!

except in the important and recently resurgent field of LQ adaptive control where an instance of infinite horizon LQ control masquerading as finite horizon LQ control appears – *Receding Horizon LQ Control.* The numerical problem for infinite horizon LQ control hinged on the ability to solve the ARE (a symplectic eigenvector problem) reliably. In the adaptive control context this solution needs to be constructed at each adaptation step.

Receding horizon LQ control appears to be attributable to Thomas [10] and involves the following:

- At time t the plant is in state x_t and an N-step finite horizon LQ optimal control problem is posed. Find $\{u_{t+s} | s = 0, \dots, N-1\}$ which minimizes

$$J(N, x_t, u) = x_{t+N}^T P_0 x_{t+N} + \sum_{j=0}^{N-1} \left\{ x_{t+j}^T Q_{N-j-1} x_{t+j} + u_{t+j}^T R_{N-j-1} u_{t+j} \right\}. \tag{10.10}$$

- The feedback control signal u_t only is applied.
- The N-step finite horizon problem is re-solved for time $t+1$ from state x_{t+1}.

The appellation 'Receding Horizon' refers to the fact that at each time a finite horizon problem is solved N steps into the future but that this horizon remains N steps distant. In the adaptive control context this translates into designing a controller with a fixed look-ahead or 'prediction horizon'.

Being but an N-step finite horizon with a sliding initial condition, the control solution here is given from (10.3) as

$$\begin{aligned} u_t &= -(G^T P_{N-1} G + R_{N-1})^{-1} G^T P_{N-1} F x_t \\ &\triangleq K_{N-1} x_t \end{aligned} \tag{10.11}$$

where P_{N-1} is the Nth element in the solution sequence of the RDE (10.5), commencing with initial condition P_0.

1 Paris, May 1989.

Predictably, some remarks are in order.

- The control law (10.11) is a causal linear state variable feedback. That is, the optimal control may be written as a causal feedback of the state process. It may also be written as an explicit (open-loop) function of time.
- The control gain, K_{N-1}, is constant. This is so in spite of the time variation of the gain in the finite horizon solution, and is due to the recession of the horizon.
- Computationally the solution of the receding horizon problem entails the iteration of the RDE for N steps.
- A major feature of this receding horizon solution is that, with relatively little increase in complexity, control constraints may be directly handled.
- Asymptotic stability questions arise here because, even though a finite horizon problem is involved, the control law (10.11) is applied on the infinite horizon.
- The LQ performance of the closed loop system is not easily related to the RDE solution applied, even with constant Q and R, except if N is taken very large.
- When the control horizon N is allowed to become large and Q and R are fixed, we know (and shall see) that $P_{N-1} \to P_\infty$ and the resolution of the closed loop stability and LQ performance is easily inferred from the infinite horizon LQ problem.

It is one of our aims here to focus on the stability and performance aspects of receding horizon LQ schemes.

10.2.2. Continuous Time

We consider the linear, time-invariant, continuous-time system

$$\dot{x}(t) = Ax(t) + Bu(t), \tag{10.12}$$

where $x(t)$ is the n-vector state process and $u(t)$ an admissible m-vector control process. The matrices A and B have appropriate dimensions. Again we pose three LQ optimal control problems and provide solutions, although we do not dwell too long where matters correspond fully with discrete time.

Finite Horizon Linear Quadratic Optimal Control. Find the input function $\{u(t)|t \in (0,T)\}$ which minimizes the criterion

$$J(T, x(0), u) = x(T)^T P(0) x(T) \tag{10.13}$$

$$+ \int_0^T \{x(t)^T Q(T-t) x(t) + u(t)^T R(T-t) u(t)\}\, dt. \tag{10.14}$$

This represents an integral version of the sum appearing in (10.2). The solution is given by

$$\begin{aligned} u(t) &= -R(T-t)^{-1} B^T P(T-t) x(t) \\ &\triangleq K(T-t) x(t), \end{aligned} \tag{10.15}$$

where the control gain, $K(\tau)$, is given by

$$K(\tau) = -R(\tau)^{-1}B^T P(\tau), \tag{10.16}$$

and $P(\tau)$ is the solution of the *Riccati Differential Equation* (also denoted RDE)

$$\dot{P}(t) = A^T P(t) + P(t)A - P(t)BR(t)^{-1}B^T P(t) + Q(t) \tag{10.17}$$

with initial condition $P(0) \geq 0$.

The similarity to the discrete-time case is mostly self-evident and so many of the earlier comments carry over *mutatis mutandis* to continuous time. Thus

- The optimal control may be written as the causal feedback of the state process via control gain $K(T-\tau)$, computed from $P(T-\tau)$, the solution of the RDE (10.17) with initial condition $P(0)$.
- An asymptotic stability question is not sensibly posed with respect to this finite horizon LQ problem.
- The optimal value of the performance criterion, $J(T, x(0), u)$, is given by

  $$J^*(T, x(0), u) = x(0)^T P(T)x(0).$$

Infinite Horizon Linear Quadratic Optimal Control. Parallel to the discrete case, we introduce an infinite horizon, stationary LQ problem: Find the control function $\{u(t) | t \in (0, \infty)\}$ which minimizes the criterion

$$J_\infty(x(0), u) = \int_0^\infty \{x(t)^T Qx(t) + u(t)^T Ru(t)\}\, dt. \tag{10.18}$$

(One may alternatively consider the case where $T \to \infty$ in (10.13) as was done in (10.6).) The solution is given by the feedback law

$$\begin{aligned} u(t) &= -R^{-1}B^T Px(t) \\ &\triangleq Kx(t), \end{aligned} \tag{10.19}$$

where the constant control gain, K, is given by

$$K = -R^{-1}B^T P, \tag{10.20}$$

and P is the constant, maximally nonnegative definite solution of the (continuous) *Algebraic Riccati Equation* (also denoted ARE)

$$0 = A^T P + PA - PBR^{-1}B^T P + Q. \tag{10.21}$$

Again the earlier remarks still apply especially those directed towards the asymptotic stability problem. We shall address these shortly but firstly move on to consider the receding horizon problem.

Receding Horizon Linear Quadratic Optimal Control. Because of the increased reliance on digital computer solutions to control problems and the computational difficulties associated with the solution of the (continuous) ARE (a Hamiltonian eigenvector problem is involved), these continuous receding horizon problems are perhaps even more *caduque* than their discrete counterparts. Nevertheless and given some recent interest [11] in specifying a continuous version of the Adaptive Generalized Predictive Control we complete our catalogue by providing an appropriate specification.

- At time t the plant is in state $x(t)$ and a T-ahead finite horizon LQ optimal control problem is posed. Find $\{u(t+s)|s \in (0,T)\}$ which minimizes $J(T, x(t), u)$ defined by (10.13).
- Apply $u(t)$ and then re-solve for the next instant t^+.

Clearly, in parallel to earlier development, the solution to this problem is given equivalently by

$$u(t) = -R(T)^{-1}B^T P(T)x(t), \tag{10.22}$$

with $P(T)$ the solution of the continuous RDE (10.17) at time T with initial condition $P(0) \geq 0$.

The similarity with the discrete case is apparent, especially inasmuch as the asymptotic stability question is raised once again since (10.22) represents a time-invariant control law to be applied on the infinite timescale. We shall address means of resolving this very issue shortly but make the observation at this juncture that the discrete finite horizon problem, and hence the receding horizon problem, differ from the continuous versions in one key aspect. That is, the discrete finite horizon LQ problem is a finite dimensional optimization over the space of all possible N-step control m-vectors, i.e., $R^{N\times m}$, while the continuous version is an infinite dimensional problem. The conclusion that we draw from this is that, in continuous time, receding horizon LQ strategies are likely to be less appealing than in discrete time because of the inherent difficulty of incorporating constraints into the former – the ability easily to impose conditions on, for example, control signal magnitudes or slewing rates is a major support for the use of receding horizon LQ controllers in discrete-time adaptive control.

10.3 Linear Optimal State Estimation and Riccati Equations

The material presented so far has concentrated on connections between LQ optimal control problems in either continuous time or discrete time and their respective Riccati equations (10.5), (10.9), (10.17), (10.21). As is widely known, these state control problems are dual to certain state estimation problems. Now we shall briefly treat these state estimation, or Kalman Filtering, ideas.

The correspondence between LQ control and least squares linear state estimation is strict duality but, for our needs here of addressing the asymptotic stability of LQ-based but suboptimal feedback control strategies (such as the receding horizon LQ control), we shall find that certain issues arise differently in each circumstance. These distinctions will be noted here.

10.3.1 Discrete Time

We consider the following system

$$x_{t+1} = Fx_t + Gw_t \tag{10.23}$$

$$y_t = Hx_t + v_t \tag{10.24}$$

where x_t is the state (no longer assumed measurable), y_t the measured output process, and w_t and v_t are zero mean, white, mutually independent gaussian noise processes with covariance

$$\mathrm{E}\begin{pmatrix} w_t \\ v_t \end{pmatrix}\begin{pmatrix} w_s^T & v_s^T \end{pmatrix} = \begin{pmatrix} Q_t & 0 \\ 0 & R_t \end{pmatrix} \delta_{ts}.$$

Our goal is to predict optimally the state, x_t, from output measurements $\{y_s | s = 1, \ldots, t-1\}$ from the initial data that our knowledge of x_0 is distributed as a gaussian random variable with mean x_0 and covariance Σ_0. The least squares solution, computed recursively, is the Kalman Predictor:

$$\hat{x}_{t+1|t} = F\hat{x}_{t|t-1} + K_t(y_t - H\hat{x}_{t|t-1}) \tag{10.25}$$

where the Kalman gain, K_t, is given by

$$K_t = F\Sigma_t H^T (H\Sigma_t H^T + R_t)^{-1}, \tag{10.26}$$

and Σ_t is the solution of the discrete filtering RDE

$$\Sigma_{t+1} = F\Sigma_t F^T - F\Sigma_t H^T (H\Sigma_t H^T + R_t)^{-1} H\Sigma_t F^T + Q_t, \tag{10.27}$$

solved forwards in time from initial conditions x_0, Σ_0. The state estimate is gaussian and has the following moments,

$$E(\hat{x}_{t|t-1}) = x_t,$$

$$E\left((x_t - \hat{x}_{t|t-1})(x_t - \hat{x}_{t|t-1})^T\right) = \Sigma_t.$$

The formulation of infinite horizon Kalman filtering problems follows directly from here by allowing $t \to \infty$ above. For stationary plants and noise covariance matrices Q and R, this infinite time or stationary solution is given by the following

$$\hat{x}_{t+1|t} = F\hat{x}_{t|t-1} + K(y_t - H\hat{x}_{t|t-1}) \tag{10.28}$$

where the fixed Kalman gain, K, is given by

$$K = F\Sigma_\infty H^T (H\Sigma_\infty H^T + R)^{-1}, \tag{10.29}$$

and Σ_∞ is the solution of the discrete filtering ARE

$$\Sigma_\infty = F\Sigma_\infty F^T - F\Sigma_\infty H^T (H\Sigma_\infty H^T + R)^{-1} H\Sigma_\infty F^T + Q. \tag{10.30}$$

Comparison of the control RDE (10.5) with the filtering RDE (10.27) establishes the duality. The major distinction between the control and filtering problems in practical terms is that the RDE of the latter is iterated forwards in time as opposed to the reverse iteration of (10.5). Issues of asymptotic stability of the closed loop control system translate into questions on the asymptotic stability of the undriven filters

$$\hat{x}_{t+1|t} = (F - K_t H)\hat{x}_{t|t-1}$$

or, in the stationary case,

$$\hat{x}_{t+1|t} = (F - KH)\hat{x}_{t|t-1}.$$

We shall center our attention on these stability questions shortly.

10.3.2 Continuous Time

Since the passage from LQ control to least squares state estimation involves the introduction of stochastic signals, in continuous time this is accompanied by the need also to convert to the Itô calculus. The state is presumed to evolve according to

$$dx(t) = Fx(t)dt + Gdw(t) \tag{10.31}$$
$$dy(t) = Hx(t)dt + dv(t), \tag{10.32}$$

where conditions above apply except that now $w(t)$ and $v(t)$ are mutually independent Brownian Motions with intensities $Q(t)$ and $R(t)$ respectively.

The least squares optimal state estimate now is given by the Kalman filter

$$d\hat{x}(t) = (F - K(t)H)\,\hat{x}(t)dt + K(t)dy(t) \tag{33}$$
$$K(t) = \Sigma(t)H^T R(t)^{-1} \tag{10.34}$$
$$\dot{\Sigma}(t) = F\Sigma(t) + \Sigma(t)F^T - \Sigma(t)H^T R(t)^{-1} H\Sigma(t) + Q(t), \tag{10.35}$$

with the obvious choice of initial conditions.

We shall not return explicitly to this Itô formulation of this problem since our task will be to analyse the stability properties of the unforced filter

$$d\hat{x}(t) = (F - K(t)H)\,\hat{x}(t)dt,$$

which is equally well treated by normal differential calculus.

10.4 Time Out

At this stage we have presented a collection of optimal control and optimal estimation problems in which Riccati difference equations, Riccati differential equations and Algebraic Riccati Equations arise. We shall be addressing the issues of asymptotic stability of the associated closed loop and unforced systems in various circumstances. These we summarize here:

1. Asymptotic stability of the stationary infinite horizon LQ closed loop,

$$x_{t+1} = \left(F - G(G^T P_\infty G + R)^{-1} G^T P_\infty F\right) x_t, \tag{10.36}$$
$$\dot{x}(t) = \left(F - G R^{-1} G^T P(\infty)\right) x(t) \tag{10.37}$$

2. Asymptotic stability of the stationary infinite horizon Kalman filter,

$$\hat{x}_{t+1} = \left(F - F\Sigma_\infty H^T (H\Sigma_\infty H^T + R)^{-1} H\right) \hat{x}_t, \tag{10.38}$$
$$\dot{\hat{x}}(t) = \left(F - H^T R^{-1} \Sigma(\infty) H\right) \hat{x}(t) \tag{10.39}$$

3. Asymptotic stability of the infinite horizon Kalman filter with arbitrary initial condition,

$$\hat{x}_{t+1} = \left(F - F\Sigma_t H^T (H\Sigma_t H^T + R)^{-1} H\right) \hat{x}_t, \tag{10.40}$$
$$\dot{\hat{x}}(t) = \left(F - H^T R^{-1} \Sigma(t) H\right) \hat{x}(t) \tag{10.41}$$

4. Asymptotic stability of the receding horizon LQ closed loop, with fixed N

$$x_{t+1} = \left(F - G(G^T P_N G + R)^{-1} G^T P_N F\right) x_t, \tag{10.42}$$
$$\dot{x}(t) = \left(F - G R^{-1} G^T P(N)\right) x(t) \tag{10.43}$$

5. Asymptotic stability of the Kalman filter frozen at a particular iteration,

$$\hat{x}_{t+1} = \left(F - F\Sigma_N H^T (H\Sigma_N H^T + R)^{-1} H\right) \hat{x}_t, \tag{10.44}$$
$$\dot{\hat{x}}(t) = \left(F - H^T R^{-1} \Sigma(N) H\right) \hat{x}(t) \tag{10.45}$$

We remark here that this set of stability questions displays some of the appealing variety of possible optimal estimation and control problems. The time-varying stability problem for Kalman filtering, problem 3, is noteworthy for the difficulty of posing a sensible dual in LQ control. This has been advanced in [12] (Section 5), however, by including a final state weighting in infinite horizon LQ. For the other pairs of problems, 1 & 2 and 4 & 5, duality means that the resolution of the one element deals immediately with the other. This holds even though the latter two problems are deliberately derived as suboptimal strategies. It is these last two problems which possess the more novelty because of their recent applications, upon which we shall comment later.

10.5 Asymptotic Stability Methods for Linear Equations

The range of asymptotic stability problems raised above refers to linear time invariant systems 1(2) and 4(5) and to the time varying system 3. We now present some tools suitable for the study of these issues. This presentation will follow the development in [13].

We begin by stating the discrete time Lyapunov stability theorem.

Theorem 10.1. *Consider the vector difference equation*

$$x_{t+1} = F_t x_t, \tag{10.46}$$

with transition function

$$\Phi(t+N,t) = F_{t+N-1}F_{t+N-2}\dots F_{t+1}F_t.$$

Suppose there exists a positive definite matrix sequence, $\infty > \beta I \geq P_t \geq \alpha I > 0$ *such that*

$$F_{t+1}^T P_{t+1} F_{t+1} - P_t = -N_t N_t^T, \tag{10.47}$$

for some matrix sequence, N_t*, and all* t*. Then (10.46) is stable in the sense of Lyapunov.*

If further the pair $[F_t, N_t^T]$ *is uniformly completely observable, i.e., there exists constants* $T > 0$*,* $\gamma > 0$*,* $\delta > 0$ *such that for all* t*,*

$$\infty > \gamma I \geq \sum_{i=0}^{T-1} \Phi^T(t+i,t) N_{t+i} N_{t+i}^T \Phi(t+i,t) \geq \delta I > 0, \tag{10.48}$$

then (10.46) is exponentially asymptotically stable.

If F_t *is a constant,* F*, then* P_t *may also be chosen to be constant,* P*, as may be* N_t*,* N*, and the uniform observability condition (10.48) may be replaced by detectability of the pair* $[F, N^T]$*.*

Finally, the condition of uniform complete observability of $[F_t, N_t^T]$ *may be replaced by the same condition on* $[F_t - K_t N_t^T, N_t^T]$ *for any bounded* K_t *and the same conclusion holds.*

We do not provide a proof of this theorem here since it is available in [13], except for the final paragraph which is an easy extension. The key result to be drawn from this theorem is that a detectability condition arises in the assessment of stability and that this is needed to conclude the rate. The astute reader will have noticed the similarity to the earlier heuristic statements about infinite horizon LQ stability, where such a condition was foreshadowed.

In continuous time the result is the logical counterpart modulo the need to include regularization of the differential equation.

Theorem 10.2. *Consider the vector differential equation*

$$\dot{x}(t) = F(t)\,x(t), \tag{10.49}$$

with $F(\cdot)$ *bounded and locally integrable and with transition function* $\Phi(\tau, t)$*. Suppose there exists a positive definite matrix function* $\infty > \beta I \geq P(t) \geq \alpha I > 0$ *such that*

$$P(t)F(t) + F^T(t)P(t) = -N(t)N^T(t) \tag{10.50}$$

for some matrix function, $N(t)$*, and all* t*. Then (10.49) is stable in the sense of Lyapunov.*

If further the pair $[F(t), N^T(t)]$ *is uniformly completely observable, i.e., there exists constants* $T > 0$, $\gamma > 0$, $\delta > 0$ *such that for all* t,

$$\infty > \gamma I \geq \int_t^{t+T} \Phi^T(\tau,t)N(\tau)N^T(\tau)\Phi(\tau,t)\,d\tau \geq \delta I > 0, \tag{10.51}$$

then (10.49) is exponentially asymptotically stable.

If $F(t)$ *is a constant,* F*, then* $P(t)$ *may also be chosen to be constant,* P*, as may be* $N(t)$*,* N*, and the uniform observability condition (10.51) may be replaced by detectability of the pair* $[F, N^T]$.

Finally, the condition of uniform complete observability of $[F(t), N^T(t)]$ *may be replaced by the same condition on* $[F(t) - K(t)N^T(t), N^T(t)]$ *for bounded, locally integrable* $K(t)$ *and the same conclusion holds.*

Recall that detectability of $[F, N^T]$ corresponds to the condition that any unobservable modes of this pair be strictly stable. An algebraic test for detectability is that, for any eigenvector v of F with eigenvalue λ, i.e., $Fv = \lambda v$, if $N^T v = 0$ then $|\lambda| < 1$ (in the discrete case) or $\mathrm{Re}(\lambda) < 0$ (in the continuous case). Predictably, the development will next turn towards the application of these stability methods to the closed loop systems given by (10.36)–(10.45). This, in turn, will allow the connection to be made between problem specification and closed loop stability.

The next stage in our treatment of LQ stability problems will be to address the question of asymptotic stability for the stationary infinite horizon LQ optimal control problems (10.36), (10.37) or the dual stationary filtering problems (10.38), (10.39). As evidenced earlier, we need only consider explicitly either the control problems or the filtering problems to infer stability properties for the other. We begin with the discrete case and then present the continuous version. Before launching fully into the analysis of stability, we recognise the pivotal role of the ARE in these problems and consider firstly some properties of the Riccati equations and their solutions.

10.6 Riccati Equation Solution Properties: Convergence and Monotonicity

Now is the juncture in which those results concerned with the convergence and monotonicity properties of the Riccati equations will be presented. These results will, admittedly, tend to appear somewhat peripheral to our thrust toward stability but they provide the machinery underpinning the later work on stability. The issues in this section are to describe some pertinent properties of and connections between solutions of the Riccati equations and, further, to examine some dependencies of these solutions on parameters.

10.6.1 Discrete Time

Infinite horizon LQ and Kalman filtering problems are associated with algebraic Riccati equations (10.9), (10.30) and we have already commented briefly on the potential multiplicity of solutions. Indeed in specifying the desired solution we have referred to the maximal nonnegative definite solution. We now examine some properties of this solution.

Existence of Maximal Nonnegative ARE Solution.

Theorem 10.3. *[12] Consider the ARE associated with an infinite horizon LQ control problem,*

$$P = F^T PF - F^T PG(G^T PG + R)^{-1} G^T PF + Q \tag{10.52}$$

where

- $[F, G]$ *is stabilizable,*
- $[F, Q^{1/2}]$ *has no unobservable modes on the unit circle,*
- $Q \geq 0$ *and* $R > 0$.

Then there exists a unique, maximal, positive definite symmetric solution $\bar{P}$.

This theorem specifies conditions necessary for the existence of a positive definite maximal solution. If unobservable modes of $[F, Q^{1/2}]$ are permitted on the unit circle then the strict positivity of $\bar{P}$ gives way only to nonnegativity. The stabilizability condition is critical to the sense of an infinite horizon LQ problem. This result makes formal that which is heuristically reasonable. We next consider the extent to which the (infinite horizon) ARE solution might properly be regarded as the limiting value of the solution of the (finite horizon) RDE as the finite horizon grows without bound.

Convergence of RDE Solution to ARE Solution.

Theorem 10.4 *Consider the ARE (10.52) above and its maximal solution* $\bar{P}$, *and consider the RDE*

$$P_{t+1} = F^T P_t F - F^T P_t G(G^T P_t G + R)^{-1} G^T P_t F + Q. \tag{10.53}$$

Then, provided $[F, G]$ *is stabilizable,* $R > 0$, $[F, Q^{1/2}]$ *is detectable and* $P_0 \geq 0$, $P_t \to \bar{P}$ *as* $t \to \infty$.

Notice here, once again, that the theorem statement reinforces those earlier heuristics by which the validity of the infinite horizon solution was justified. The key condition in this theorem is that detectability on $[F, Q^{1/2}]$ is introduced. This is stronger than the 'no unobservable modes on the unit circle' condition of Theorem 10.3. Alternative complementary theorems may also be developed which trade such detectability strictions for more severe constraints on the initial condition matrix P_0, typically that $P_0 \geq \bar{P}$. For the moment, however, this form is best suited to our purposes.

It is worth remarking here that very revealing examples of the sufficiency of these theorem conditions may be simply developed by considering the scalar case with unstable $F = 2$, say. In this case, the ARE is a simple scalar quadratic equation for which existence of real solutions and their positivity are easily examined.

Comparison of RDE Solutions. deSouza [14] has recently provided a lovely extension of the results of Nishimura [15] and Poubelle [16], [17] on the comparative properties between solutions of like RDE's. An earlier version is attributed to Claude Samson. The proof is by substitution into the RDE.

Lemma 10.1. *[14] Consider two RDEs (10.53) with the same F, G and R matrices but possibly different Q's, Q^1 and Q^2 respectively. Denote their solution matrices P_t^1 and P_t^2 respectively. Then, the difference between the two solutions $\tilde{P}_t = P_t^2 - P_t^1$ satisfies the following equation*

$$\tilde{P}_{t+1} = \bar{F}_t^{1T} \tilde{P}_t \bar{F}_t^1 - \bar{F}_t^{1T} \tilde{P}_t G (G^T P_t^2 G + R)^{-1} G^T \tilde{P}_t \bar{F}_t^1 + \tilde{Q} \tag{10.54}$$

or,

$$\tilde{P}_{t+1} = \bar{F}_t^{1T} \tilde{P}_t \bar{F}_t^1 - \bar{F}_t^{1T} \tilde{P}_t G (G^T \tilde{P}_t G + \tilde{R}_t)^{-1} G^T \tilde{P}_t \bar{F}_t^1 + \tilde{Q} \tag{10.55}$$

where

$$\bar{F}_t^1 = F - G(G^T P_t^1 G + R)^{-1} G^T P_t^1 F$$

$$\tilde{Q} = Q_2 - Q_1$$

$$\tilde{R}_t = G^T P_t^1 G + R.$$

A wealth of useful results stems easily from this astute algebraic observation by deSouza. For example,

Theorem 10.5. *Under the conditions of Lemma 10.1, suppose that*

$$Q^1 \geq Q^2,$$

and, for some t we have

$$P_t^1 \geq P_t^2,$$

then for all $k > 0$

$$P_{t+k}^1 \geq P_{t+k}^2.$$

Proof. In (10.54) we have, under the theorem conditions, that $\tilde{P}_t \leq 0$, $\tilde{Q}_t \leq 0$ and $P_t^2 \geq 0$. Thus $\tilde{P}_{t+1} \leq 0$ and the result is established for $k = 1$. By induction the theorem follows for all positive k.

Note that the arbitrariness of the assignment of superscripts to solutions P_t^1 and P_t^2 means that complementary results are directly established with, say, $P_{t+k}^2 \geq P_{t+k}^1$. This feature will be seen again in the immediately following results.

Monotonicity Properties of RDE Solutions. We are now in a position to apply the deSouza Lemma 10.1 to derive far reaching monotonicity properties of the RDE solution which play a central part in stability analyses to follow. We have the following cascade of results flowing from clever application of the above lemma to a single RDE solution sequence but with differing time indices.

Theorem 10.6. *[16] If the non-negative definite solution P_t of the RDE (10.53) is nonincreasing at one time, i.e.,*

$$P_{t+1} \leq P_t \quad , \textit{for some } t,$$

then P_t is monotonically nonincreasing for all subsequent times,

$$P_{t+k+1} \leq P_{t+k} \quad , \textit{for all } k \geq 0.$$

Proof. Identify P_t^1 with P_t, P_t^2 with P_{t+1}, and take $Q^1 = Q^2$ in Lemma 10.1 or Theorem 10.5 where now $\tilde{Q}_t = 0$.

Theorem 10.7. *[16] If the non-negative definite solution P_t of the RDE (10.53) is nondecreasing at one time, i.e.,*

$$P_{t+1} \geq P_t \quad , \textit{for some } t,$$

then P_t is monotonically nondecreasing for all subsequent times,

$$P_{t+k+1} \geq P_{t+k} \quad , \textit{for all } k \geq 0.$$

Proof. Identify P_t^1 with P_{t+1}, P_t^2 with P_t, and take $Q^1 = Q^2$ in Lemma 10.1 or Theorem 10.5 where now $\tilde{Q}_t = 0$.

These two monotonicity theorems describe the effective sign definiteness of the change in successive solution values. The deSouza Lemma 10.1 may be equally well applied to derive a similar property of second differences of the RDE solutions.

Theorem 10.8. *[14] If the solution P_t of the RDE (10.53) has a nonpositive definite second difference at time t, i.e.,*

$$P_{t+2} - 2P_{t+1} + P_t \leq 0,$$

then for all $k \geq 0$,

$$P_{t+k+2} - 2P_{t+k+1} + P_{t+k} \leq 0.$$

Proof. Equation (10.55) states that $\Delta P_t \triangleq P_{t+1} - P_t$ satisfies an RDE with state weighting $\tilde{Q}_t = 0$ and control weighting $\tilde{R}_t$, which is greater than the original R.

Therefore ΔP_t itself obeys the monotonicity properties of any RDE solution. One then recognises that

$$\Delta P_{t+1} - \Delta P_t = P_{t+2} - 2P_{t+1} + P_t.$$

10.6.2 Continuous Time

The above results on discrete time Riccati equation properties carry over with but little alteration to the realm of continuous time. Indeed, the discrete case has historically been the more difficult from which to extract hard results because of issues such as transition function noninvertibility etc. We shall see, for example, that the continuous version of the deSouza Lemma is very much more easily established and that the history of monotonicity results is longer, going back at least to Kailath [18]. Because of this simplicity and similarity we shall attempt to be briefer.

Existence of Maximal Nonnegative ARE Solution.

Theorem 10.9. *Consider the ARE associated with an infinite horizon LQ control problem,*

$$0 = A^T P + PA - PBR^{-1}B^T P + Q. \tag{10.56}$$

where

- $[A, B]$ *is stabilizable,*
- $[A, Q^{1/2}]$ *has no unobservable modes on the imaginary axis,*
- $Q \geq 0$ *and* $R > 0$.

Then there exists a unique, maximal, positive definite symmetric solution $\bar{P}$.

Convergence of RDE Solution to ARE Solution.

Theorem 10.10. *Consider the ARE (10.56) above and its maximal solution* $\bar{P}$*, and consider the RDE*

$$\dot{P}(t) = A^T P(t) + P(t)A - P(t)BR^{-1}B^T P(t) + Q \tag{10.57}$$

Then, provided $[A, B]$ *is stabilizable,* $R > 0$, $[A, Q^{1/2}]$ *is detectable and* $P_0 \geq 0$, $P_t \to \bar{P}$ *as* $t \to \infty$.

Comparison of RDE Solutions, Monotonicity. Poubelle [17] now replaces deSouza as the fount of all wisdom as we move to continuous time. Here the central results pivot about the following lemma.

Lemma 10.2 *[17] Consider* $P(t)$*, the solution of the RDE*

$$\dot{P}(t) = A^T P(t) + P(t)A - P(t)BR^{-1}B^T P(t) + Q$$

with initial condition $P(0) \geq 0$ and denote the closed loop matrix

$$\bar{A}(t) = A - BR^{-1}B^T P(t).$$

Then $P(t)$ satisfies

$$\ddot{P}(t) = \dot{P}(t)[A - BR^{-1}B^T P(t)] + [A - BR^{-1}B^T P(t)]^T \dot{P}(t) \tag{10.58}$$
$$= \dot{P}(t)\bar{A}(t) + \bar{A}^T(t)\dot{P}(t). \tag{10.59}$$

Further, $\ddot{P}(t)$ satisfies

$$\dddot{P}(t) = \ddot{P}(t)[A - BR^{-1}B^T P(t)] + [A - BR^{-1}B^T P(t)]^T \ddot{P}(t) - 2\dot{P}(t)B^T P(t)B\dot{P}(t) \tag{10.60}$$
$$= \ddot{P}(t)\bar{A}(t) + \bar{A}^T(t)\ddot{P}(t) - 2\dot{P}(t)B^T P(t)B\dot{P}(t) \tag{10.61}$$

Proof. By differentiating successively (10.17) and then differentiating (10.58).

The two higher order versions of the RDE (10.58) and (10.60) are themselves RDEs of sorts. In particular, their alternative descriptions (10.59) and (10.61) are deliberately displayed as Lyapunov equations. We have the following simple result concerning the solution of such equations.

Lemma 10.3. *Consider the time-varying Lyapunov equation*

$$\dot{S}(t) = S(t)M(t) + M^T(t)S(t) + W(t), \qquad S(0) = S_0. \tag{10.62}$$

Denote by $\Phi(t,\tau)$ the transition matrix associated with $M(t)$. Then the solution of (10.62) is given by

$$S(t) = \Phi^T(t,0)S_0\Phi(t,0) + \int_0^t \Phi^T(t,\tau)W(\tau)\Phi(t,\tau)\,d\tau. \tag{10.63}$$

One immediately derives the following monotonicity theorems.

Theorem 10.11. *If the non-negative definite solution $P(t)$ of the RDE (10.57) is nonincreasing at one time, i.e.,*

$$\dot{P}(t) \leq 0 \quad \text{, for some } t,$$

then $P(t)$ is monotonically nonincreasing for all subsequent times,

$$\dot{P}(t+s) \leq 0 \quad \text{, for all } s \geq 0.$$

Proof. Since $P(t)$ is the solution of the RDE (10.57), by Lemma 10.2, $P(t)$ satisfies (10.59), which is a Lyapunov equation with zero driving term. Appealing to Lemma 10.3, we have

$$\dot{P}(t+s) = \Phi^T(t+s,t)\dot{P}(t)\Phi(t+s,t).$$

Theorem 10.12. *If the non-negative definite solution $P(t)$ of the RDE (10.57) is nondecreasing at one time, i.e.,*

$$\dot{P}(t) \geq 0 \quad , \textit{for some } t,$$

then $P(t)$ is monotonically nondecreasing for all subsequent times,

$$\dot{P}(t+s) \geq 0 \quad , \textit{for all } s \geq 0.$$

Theorem 10.13. *If the solution $P(t)$ of the RDE (10.57) has a nonpositive definite second derivative at time t, i.e., $\ddot{P}(t) \leq 0$, then for all $s \geq 0$, $\ddot{P}(t+s) \leq 0$.*

Proof. Write $W(t) = -2\dot{P}(t)B^T P(t)B\dot{P}(t)$, which is clearly nonpositive definite, then (10.61) is a Lyapunov equation and has a solution $\ddot{P}(t)$ which is given by

$$\ddot{P}(t+s) = \Phi^T(t+s,t)\ddot{P}(t)\Phi(t+s,t) + \int_t^{t+s} \Phi^T(t,\tau)W(\tau)\Phi(t,\tau)\, d\tau.$$

The result follows.

10.6.3 Summary

With this section we have derived and presented a combination of convergence, comparison and monotonicity results for AREs and RDEs in discrete and continuous time, which will provide the technical machinery with which to assault the asymptotic stability problems when coupled to the Lyapunov stability techniques of the preceding section. These monotonicity results are delightfully general and, naturally enough, reflect the structural properties of optimal control and estimation problems. The truly remarkable feature of these results and their associates is that, despite the proclaimed similarity between continuous time and discrete time and the duality between filtering and control, distinct methodologies of proof are often required from each of these specific areas to establish readily many results.

10.7 Stability in Infinite Horizon LQ Problems

Our treatment here will concentrate on the infinite horizon stationary discrete LQ and Kalman filtering stability Problems 1 and 2, (10.36) and (10.38), and their continuous variants, (10.37) and (10.39). Additionally, we consider also the stability of the time-varying infinite horizon Kalman filter Problem 3 (10.40) and (10.41). Thereafter, in the next section, we shall move on to deal with the receding horizon Problems 4 and 5 (10.42)–(10.45). Some tools, such as the recasting of the RDE/ARE as a Lyapunov equation will be common and so we commence with this aspect.

We will appeal directly to our earlier Lyapunov analysis to derive the following results using a standard device of rewriting the RDE or ARE as a Lyapunov equation. Denote the LQ gain by

$$K_j = -(G^T P_j G + R)^{-1} G^T P_j F. \tag{10.64}$$

Then the RDE (10.5) may be written, following some simple arithmetic, as

$$\begin{aligned} P_{j+1} &= F^T P_j F - F^T P_j G(G^T P_j G + R_j)^{-1} G^T P_j F + Q_j \\ &= (F + GK_j)^T P_j (F + GK_j) + K_j^T R_j K_j + Q_j. \end{aligned} \tag{10.65}$$

The astute reader will have picked (10.65) as a Lyapunov equation, with P_j serving the role of the element with the same symbol in the stability theory for linear equations. This we shall exploit.

In continuous time we have the equivalent version of the RDE or ARE. Denote the LQ control gain

$$K(t) = -R(t)^{-1} B^T P(t). \tag{10.66}$$

Now rewrite the RDE as

$$\begin{aligned} \dot{P}(t) &= A^T P(t) + P(t)A - P(t)BR(t)^{-1} B^T P(t) + Q(t) \\ &= (A + BK(t))^T P(t) + P(t)(A + BK(t)) + K^T(t) R K(t). \end{aligned} \tag{10.67}$$

10.7.1 Discrete Problems

Our root problem here is to examine the asymptotic stability of

$$x_{t+1} = \left(F - G(G^T P_\infty G + R)^{-1} G^T P_\infty F\right) x_t, \tag{10.68}$$

where P_∞ is the maximal nonnegative definite solution of the ARE,

$$P_\infty = F^T P\infty F - F^T P\infty G(G^T P_\infty G + R)^{-1} G^T P_\infty F + Q. \tag{10.69}$$

Theorem 10.14. *Consider the time-invariant linear vector difference equation (10.68) representing the closed loop of an infinite horizon LQ controlled system, where P_∞ is the maximal nonnegative definite solution, $\bar{P}$, of the ARE (10.69). Subject to the conditions:*

- *$[F, G]$ is stabilizable,*
- *$[F, Q^{1/2}]$ is detectable,*
- *$Q \geq 0$ and $R > 0$,*

then (10.68) is exponentially asymptotically stable.

Proof. From the theorem conditions and Theorem 10.3, we see that the ARE possesses a positive definite maximal solution, $\bar{P}$. Recognising the correspondence between (10.68) and (10.46) in Theorem 10.1, comparing the writing (10.65) of the ARE as a Lyapunov equation and then invoking Theorem 10.1, we see that, provided $[F + GK_\infty, (K_\infty^T R K_\infty + Q)^{1/2}]$ is a detectable pair, (10.68) will be stable. Since

$$K_\infty^T R K_\infty + Q = (\, K_\infty^T R^{1/2} \quad Q^{1/2}\,) \begin{pmatrix} R^{1/2} K_\infty \\ Q^{1/2} \end{pmatrix},$$

and

$$F + GK_\infty = F + (\, GR^{-1/2} \quad 0\,) \begin{pmatrix} R^{1/2} K_\infty \\ Q^{1/2} \end{pmatrix},$$

detectability of $[F, Q^{1/2}]$ suffices to prove stability.

This is the fundamental discrete infinite horizon stability result which shall form the basis of our successive analysis. The key feature is that through the writing of the ARE as a Lyapunov equation (10.65) involving the closed loop matrix of (10.68), the positive definite ARE solution $\bar{P}$ now serves to define a quadratic Lyapunov function. The critical theorem condition is the detectability requirement – the other conditions are better associated with the well posedness of the LQ problem. It is this detectability which shall reappear as pivotal to the development of the receding horizon results.

10.7.2 Continuous Problems

We now consider the asymptotic stability of

$$\dot{x}(t) = \left(F - GR^{-1}G^T P(\infty)\right) x(t) \tag{10.70}$$

where $P(\infty)$ is the maximal nonnegative definite solution of the ARE,

$$0 = A^T P(\infty) + P(\infty)A - P(\infty)BR^{-1}B^T P(\infty) + Q. \tag{10.71}$$

Theorem 10.15. *Consider the time-invariant linear vector differential equation (10.70) representing the closed loop of an infinite horizon LQ controlled system, where $P(\infty)$ is the maximal nonnegative definite solution, $\bar{P}$, of the ARE (10.71). Subject to the conditions:*

- $[F, G]$ *is stabilizable,*
- $[F, Q^{1/2}]$ *is detectable,*
- $Q \geq 0$ *and* $R > 0$,

then (10.70) is exponentially asymptotically stable.

Proof. Parallels the discrete case with the appeal to Theorem 10.2 using the (10.67) reformulation of the ARE.

10.7.3 Asymptotic Stability of the Time-varying Kalman Filter

We treat only the discrete case here since the continuous version is identical in form. Consider the time-varying Kalman filter

$$\hat{x}_{t+1} = \left(F - F\Sigma_t H^T (H\Sigma_t H^T + R)^{-1} H\right) \hat{x}_t, \tag{10.72}$$

with Σ_t being the solution of the filtering RDE (10.27)

$$\Sigma_{t+1} = F\Sigma_t F^T - F\Sigma_t H^T (H\Sigma_t H^T + R_t)^{-1} H\Sigma_t F^T + Q_t. \tag{10.73}$$

We have the following.

Theorem 10.16. *Consider the time-varying difference equation (10.72) representing the time-varying Kalman filter operating from initial condition,* $\Sigma_0 \geq 0$. *Subject to the conditions:*

- $[F, HR_t^{-1/2}]$ *is uniformly observable,*
- $[F, Q_t^{1/2}]$ *is uniformly controllable,*
- $Q_t \geq 0$ *and* $R_t > 0$,

then (10.72) is exponentially asymptotically stable.

Proof. Under the theorem conditions, the RDE solution Σ_t is a positive definite matrix sequence bounded above and below. The filtering version of (10.65)

$$\Sigma_{t+1} = (F - K_t H)\Sigma_t (F - K_t H)^T + K_t R_t K_t^T + Q_t \tag{10.74}$$

then admits direct appeal to Theorem 10.1 for asymptotic stability of (10.72) to follow from the uniform observability of the pair $[F^T, Q_t^{1/2}]$. This corresponds to the stated controllability condition.

The upshot of this section has been to enunciate the conditions for the asymptotic stability of infinite horizon closed loop optimal solutions. The key features, stabilizability and detectability essentially, concur with those presaged earlier in our heuristic development. We next turn to deal with sufficient conditions for receding horizon stability.

10.8 Stability in Receding Horizon LQ Problems

Recall from our discussion in earlier sections that the genealogy of receding horizon LQ problems is the application of finite horizon LQ methods in the infinite horizon context, with an aim to achieving computational savings. In stationary circumstances, i.e., Q and R constant, with sufficiently large value of horizon N, the RDE Convergence Theorem 10.4 and the ARE Stability Theorem 10.14 combine to produce an obvious asymptotic stability result.

Theorem 10.17. *Consider the receding horizon LQ closed loop system*

$$x_{t+1} = \left(F - G(G^T P_N G + R)^{-1} G^T P_N F\right) x_t, \tag{10.75}$$

where P_N is the Nth term in the solution sequence of the RDE (10.5) with constant weighting matrices $Q \geq 0$ and $R > 0$ with initial condition $P_0 \geq 0$. Then, provided $[F, G]$ is stabilizable and $[F, Q^{1/2}]$ is detectable, there exists an N_0 such that (10.75) is exponentially asymptotically stable for all $N \geq N_0$.

While this result gives hope for the eventual stability of receding horizon based control systems, it delivers no guide to selection of a suitable N. To study the stability of receding horizon closed loops with arbitrary N we introduce a new tool.

10.8.1 The Fake Algebraic Riccati Equation

Guided by the ease of establishing stability for infinite horizon LQ controllers directly from the ARE we harken back to the RDE and attempt to have it masquerade as a fictitious or frozen or fake ARE. The RDE (10.5)

$$P_{j+1} = F^T P_j F - F^T P_j G(G^T P_j G + R_j)^{-1} G^T P_j F + Q_j,$$

is a recursion for P_{j+1} given P_j. We rewrite this as

$$P_j = F^T P_j F - F^T P_j G(G^T P_j G + R_j)^{-1} G^T P_j F + \bar{Q}_j, \tag{10.76}$$

$$\bar{Q}_j = Q + P_j - P_{j+1}. \tag{10.77}$$

Here (10.76) appears no longer to be a recursion for P_{j+1} but rather to be an algebraic equation for P_j. Specifically, we have recast the RDE so that P_j satisifes an algebraic Riccati equation with the original value of R_j but with a different value of Q given by (10.77).

The continuous version of this construction yields

$$0 = A^T P(t) + P(t)A - P(t)BR(t)^{-1}B^T P(t) + \bar{Q}(t) \tag{10.78}$$

$$\bar{Q}(t) = Q - \dot{P}(t). \tag{10.79}$$

We shall not focus too greatly on continuous time where the results do not deviate significantly from their discrete time counterparts. These above elementary modifications to the RDEs will play a critical role in the stability arguments to follow.

10.8.2 Receding Horizon Stability via FARE and Monotonicity

The RDE reformed as an ARE (10.76) or (10.78) is known as the *Fake Algebraic Riccati Equation* (FARE). Reference to the infinite horizon stability Theorem 10.14, immediately yields the following receding horizon stability result.

Theorem 10.18. *Consider the receding horizon LQ closed loop system (10.75) with P_N being the solution of the FARE (10.76) (the RDE (10.5)). Provided:*

- $[F,G]$ *is stabilizable,*
- $\bar{Q}_N \geq 0$ *and* $R_N > 0$,
- $[F,\bar{Q}_N^{1/2}]$ *is detectable,*

then (10.75) is exponentially asymptotically stable.

Proof. The FARE with $\bar{Q}_N \geq 0$ and $R_N > 0$ is an ARE with these Q and R which is associated with an infinite horizon LQ problem with such weighting matrices. According to Theorem 10.14, the closed loop of this artificial infinite horizon problem is asymptotically stable. But this closed loop is, in fact, also that of the original receding horizon problem.

The means of achieving the stability of these receding horizon control loops is to recast them so that they appear as infinite horizon problem solutions. This, in itself, is not too surprising a feature and does not indicate too great a leap forward except that the construction required for ARE solution is replaced by a simpler test on the FARE. A closer analysis of $\bar{Q}_t$ and $\bar{Q}(t)$ indicates further beneficial properties. Recall,

$$\bar{Q}_t = Q + P_t - P_{t+1}$$
$$\bar{Q}(t) = Q - \dot{P}(t).$$

We may now pose and answer some questions concerning the potential detectability properties of these $\bar{Q}$.

Lemma 10.4. *Suppose that* $[F, Q^{1/2}]$ *is detectable. Then*

$$P_{N+1} \leq P_N \quad \Rightarrow \quad \bar{Q}_N \geq Q \quad \Rightarrow \quad [F,\bar{Q}_N^{1/2}] \text{ is detectable},$$

or, in continuous time,

$$\dot{P}(t) \leq 0 \quad \Rightarrow \quad \bar{Q}(t) \geq Q \quad \Rightarrow \quad [F,\bar{Q}^{1/2}(t)] \text{ is detectable}.$$

This result establishes the connection between nonincreasing solutions of the RDE and $[F,\bar{Q}^{1/2}]$ detectability. All that remains is to invoke the monotonicity properties of the RDE already derived in Theorems 10.6 and 10.1 pertaining to the constant weighting RDEs (10.53) and (10.57). This we now do.

Theorem 10.19. *Consider the discrete time closed loop (10.75) derived from a receding horizon LQ problem with constant weighting matrices,* $Q \geq 0$ *and* $R > 0$, *and horizon N. Suppose that* $[F,G]$ *is stabilizable and* $[F,Q^{1/2}]$ *is detectable. Then if, for some N_0,*

$$P_{N_0+1} \leq P_{N_0},$$

(10.75) is exponentially asymptotically stable for any $N \geq N_0$.

Theorem 10.20. *Consider the continuous time closed loop derived from a receding horizon LQ problem with constant weighting matrices, $Q \geq 0$ and $R > 0$, and horizon T,*

$$\dot{x}(t) = \left(F - GR^{-1}G^T P(T)\right) x(t). \tag{10.80}$$

Suppose that $[F, G]$ is stabilizable and $[F, Q^{1/2}]$ is detectable. Then if, for some T_0,

$$\dot{P}(T_0) \leq 0,$$

(10.80) is exponentially asymptotically stable for any $T \geq T_0$.

These two theorems rely upon the montonicity properties of the RDE to establish that $\bar{Q}$ of the FARE always produces $[F, \bar{Q}^{1/2}]$ detectable and, thereby, a stable closed loop. The theorem statements rely upon the monotonic nonincreasing aspects of the RDE solutions. As a consequence of this, if P_j is always nonincreasing and converges to a constant nonnegative definite value, P_∞, then clearly one must satisfy $P_j \geq P_\infty$ for all j and $\bar{Q}_j$ exceeds Q for all j.

To admit the application of FARE results in a similar fashion in the circumstance of $P_j \leq P_\infty$, one may appeal to the second difference results on the RDE solution, Theorems 10.8 and 10.13, to ensure that $\bar{Q}$ never is permitted to become nonpositive definite even though $\bar{Q}_j \leq Q$. Specifically,

Theorem 10.21. *[14] Consider the discrete time closed loop (10.75) derived from a receding horizon LQ problem with constant weighting matrices, $Q \geq 0$ and $R > 0$, and horizon N. Suppose that, for some N_0:*

- *$[F, G]$ is stabilizable,*
- *$\bar{Q}_{N_0} \geq 0$,*
- *$[F, \bar{Q}_{N_0}^{1/2}]$ is detectable,*
- *$P_{N_0+2} - 2P_{N_0+1} + P_{N_0} \leq 0$,*

then (10.75) is asymptotically stable for any $N \geq N_0$.

Proof. We appeal to Theorem 10.8 to show that the final theorem condition above implies that the second difference of P_N is nonpositive for $N \geq N_0$. This, in turn, implies that the first difference, $P_{N+1} - P_N$, is nonincreasing for such N. Thus $\bar{Q}_N$ is a nondecreasing function of N.

The continuous version is older but similar.

Theorem 10.22. *[17] Consider the continuous time closed loop (10.80) derived from a receding horizon LQ problem with constant weighting matrices, $Q \geq 0$ and $R > 0$, and horizon T. Suppose that, for some T_0:*

- *$[F, G]$ is stabilizable,*
- *$\bar{Q}(T_0) \geq 0$,*
- *$[F, \bar{Q}(T_0)^{1/2}]$ is detectable,*
- *$\ddot{P}(T_0) \leq 0$,*

then (10.75) is asymptotically stable for any $T \geq T_0$.

We have now wended our way through an expanse of results on the stability of LQ optimal control and filtering systems combining: Lyapunov stability, monotonicity properties of RDEs, the FARE etc. The key observation has been to use the FARE coupled with monotonicity arguments to establish sensible stability requirements for the receding horizon LQ strategy. Further, more technically demanding variations on this theme are contained in the recent book [20]. We shall now present two examples of the application of these methods.

10.9 Examples

10.9.1 Stability of Generalized Predictive Control

Generalized Predictive Control (GPC) is a very popular process control design procedure due (*inter alia*) to Clarke, Mohtadi and Tuffs [19]. This control design especially has been successful in the field of Adaptive Control where many practical applications have been reported. At the heart of GPC is a receding horizon LQ problem. At time t the LQ criterion

$$J = \sum_{0}^{N} y_{t+j}^T y_{t+j} + \lambda \sum_{0}^{N_u} u_{t+j}^T u_{t+j}, \tag{10.81}$$

subject to $\{u_{t+j} = 0 | j = N_u + 1, \ldots, N\}$, is minimized and u_t is applied. Here N is the prediction horizon, N_u is the control horizon and $\lambda > 0$ is a control weighting.

For the case $N_u = N$, the GPC formulation may be viewed as receding horizon LQ with

$$Q = H^T H, \quad P_0 = H^T H, \quad R = \lambda I.$$

After identifying the GPC specification as receding horizon LQ, one appeals to the foregoing theory to determine stability of the controlled system. It is here that one strikes a snag – The above choiced for Q, R and P_0 do not admit simple affirmation of stability. Indeed, by inspection, one may replace the initial condition $P_0 = H^T H$ by the equivalent initial condition $P_{-1} = 0$. Thus $P_0 \geq P_{-1}$ and, by Theorem 10.7, P_j is then destined always to increase. This makes it more problematic to ensure that $\bar{Q}_j \geq 0$. It is then not surprising that GPC exhibits difficulties in assuring designed closed loop stability. Further modifications to GPC are evaluated in [20] using some more technical monotonicity devices.

Given that the GPC receding horizon LQ control meets stability difficulties because the P_j sequence is always monotonic nondecreasing, one might ask whether other general strategies have more success in forcing P_j to decrease. In [20], techniques which effectively select P_0 infinite are examined using methods of this chapter.

10.9.2 Harmonic Analysis in Noise

It is frequently the case in many signal processing problems that one wishes to evaluate or estimate the harmonic components of a (slowly varying) periodic signal in noise. Here we shall consider the case where the period of the signal is known. In such circumstances, one may write a signal model for the measured signal, y_t, as follows.

$$x_{t+1} = Fx_t + Gw_t \tag{10.82}$$
$$y_t = Hx_t + v_t, \tag{10.83}$$

where x_t is a vector of harmonic cosine and sine terms,

$$F = 1 \oplus \left\{ \begin{matrix} \text{block diag} \\ k = 1, 2, \ldots, n/2 - 1 \end{matrix} \begin{pmatrix} \cos k\theta & \sin k\theta \\ -\sin k\theta & \cos k\theta \end{pmatrix} \right\} \oplus (-1),$$
$$H = (\, 1/\sqrt{2} \quad 1 \quad 0 \quad 1 \quad 0 \quad \ldots \quad 1 \quad 0 \quad -1/\sqrt{2})^T,$$

w_t and v_t are independent noise processes with covariances Q and R. Note that F here has all of its eigenvalues on the unit cirlce. This formulation is examined in [21]. We consider two cases; the matrix Q is zero, and the matrix Q is nonzero. Harmonic analysis is identically equivalent to state estimation for this model.

When there is no state noise in the model (10.82) ($Q = 0$), the above set of equations describes a truly periodic signal corrupted by noise. If one attempts to set up the Kalman filter to estimate the state x_t optimally, then the limiting covariance Σ_∞ for the problem is zero, and thus the limiting Kalman gain is zero. This concurs with the intuitive solution to optimal estimation of a purely periodic signal in noise, i.e., average the answers over many periods. When F has all its eigenvalues on the unit circle and $Q = 0$, the stability results of earlier fail because $[F, Q^{1/2}]$ possesses many uncontrollable modes on the unit circle.

One technique to force stability into the solution of such problems is to set the input noise variance Q to a nonzero value arbitrarily. There are many other methods as well, see [2]. The summary choice of nonzero Q still obliges the designer to solve an ARE for a positive definite Σ_∞ to design the Kalman filter. An alternative approach explored in [21] is to select a positive value ϵ and determine a Kalman-like gain

$$K = F\epsilon IH^T(H\epsilon IH^T + R)^{-1},$$

associated with a $\Sigma_\infty = \epsilon I$. The question then arises: Does such a choice yield filter stability?

The answer is affirmative because, by construction of the FARE

$$\epsilon I - F\epsilon IF^T + F\epsilon IH^T(H\epsilon IH^T + R_t)^{-1}H\epsilon IF^T = \bar{Q},$$

one easily establishes that $\bar{Q} > 0$ for any $\epsilon > 0$. As a side remark, we mention that these filters degenerate to the Discrete Fourier Transform when $\epsilon \to \infty$.

10.10 Conclusion

We have led the reader through a tutorial development of the stability properties of Linear Quadratic feedback control systems and least squares state estimation in several guises: continuous and discrete time; finite, infinite and receding horizon. Our aim has been to reveal the connections between Lyapunov stability theory and the Riccati equations. The additional disclosures have then stemmed from these connections when coupled to properties purely of the Riccati equations themselves, namely convergence and monotonicity. These new tools provide simple stability tests for large classes of receding horizon LQ problems which are finding increasing practical application. The novelty here is the construction and application of Fake Algebraic Riccati Techniques.

References

1. B.D.O. Anderson and J.B. Moore, *Linear Optimal Control*, Prentice-Hall, Englewood Cliffs NJ, USA, 1971.
2. B.D.O. Anderson and J.B. Moore, *Optimal Filtering*, Prentice-Hall, Englewood Cliffs NJ, USA, 1979.
3. B.D.O. Anderson and J.B. Moore, *Optimal Control: Linear Quadratic Methods*, Prentice-Hall, Englewood Cliffs NJ, USA, 1990.
4. H. Kwakernaak and R. Sivan, *Linear Optimal Control Systems*, Wiley Interscience, New York, USA, 1972.
5. K.J. Åström and B. Wittenmark, *Computer Controlled Systems: Theory and Design*, Prentice-Hall, Englewood Cliffs NJ, USA, 1984.
6. F.L. Lewis, *Optimal Control*, Wiley Interscience, New York, USA, 1986.
7. F.L. Lewis, *Optimal Estimation*, Wiley Interscience, New York, USA, 1986.
8. W.H. Kwon and A.E. Pearson, 'On Feedback Stabilization of Time-varying Discrete Linear Systems,' *I.E.E.E. Transactions on Automatic Control*, vol. AC–23, pp. 479–481, 1978.
9. D.L. Kleinman, 'Stabilizing a Discrete Constant Linear System, with Application to Iterative Methods for Solving the Riccati Equation,' *I.E.E.E. Transactions on Automatic Control*, vol. AC–19, pp. 252–254, 1974.
10. Y.A. Thomas, 'Linear Quadratic Optimal Estimation and Control with Receding Horizon,' *Electronics Letters*, vol. 11, pp. 19–21, 1975.
11. P.J. Gawthrop and H. Demircioğlu, 'Continuous-time Generalized Predictive Control (CGPC)', *Proc. IFAC Symposium on Adaptive Systems in Control and Signal Processing*, Glasgow UK, pp. 123–128, 1989.
12. C.E. de Souza, M. Gevers and G.C. Goodwin, 'Riccati Equations in Optimal Filtering of Nonstabilizable Systems having Singular State Transition Matrices,' *I.E.E.E. Transactions on Automatic Control*, vol. AC–31, pp. 831–838, 1986.
13. B.D.O. Anderson, R.R. Bitmead, C.R. Johnson Jr, P.V. Kokotovic, R.L. Kosut, I.M.Y. Mareels, L. Praly and B.D.Riedle, *Stability of Adaptive Systems: Passivity and Averaging Analysis*, MIT Press, Cambridge MA, USA, 1986.
14. C.E. de Souza, 'Monotonicity and Stabilizability results for the Solution of the Riccati Difference Equation,' *Proc. Workshop on the Riccati Equation in Control, Systems and Signals*, S. Bittanti (ed.), Como, Italy, pp. 38–41, 1989.
15. T. Nishimura, 'On the *a priori* Information in Sequential Estimation Problems,' *I.E.E.E. Transactions on Automatic Control*, vol. AC–12, pp. 123–125, 1967.

16. R.R. Bitmead, M. Gevers, I.R. Petersen and R.J. Kaye, 'Monotonicity and Stabilizability Properties of Solutions of the Riccati Difference Equation: Propositions, Lemmas, Theorems, Fallacious Conjectures and Counterexamples,' *Systems and Control Letters*, vol. 5, pp. 309–315, 1985.
17. M.A. Poubelle, R.R. Bitmead and M. Gevers, 'Fake Algebraic Riccati Techniques and Stability,' *I.E.E.E. Transactions on Automatic Control*, vol. AC–33, pp. 379–381, 1988.
18. T. Kailath, 'Some New Algorithms for Recursive Estimation in Constant Linear Systems,' *I.E.E.E. Transactions on Information Theory*, vol. IT–21, pp. 750–760, 1975.
19. D.W. Clarke, C. Mohtadi and P.S. Tuffs, 'Generalized Predictive Control: Parts i and ii,' *Automatica*, vol. 23, pp. 137–160, 1987.
20. R.R. Bitmead, M. Gevers and V. Wertz, *Adaptive Optimal Control: the Thinking Man's GPC*, Prentice-Hall Australia, Sydney, 1990.
21. R.R. Bitmead, A.C. Tsoi and P.J. Parker, 'A Kalman Filtering Approach to Short Time Fourier Analysis,' *I.E.E.E. Transactions on Acoustics Speech and Signal Processing*, vol. ASSP-34, pp. 1493–1501, 1986.

11 Generalized Riccati Equations in Dynamic Games

Tamer Başar

11.1 Introduction

The discrete- and continuous-time Riccati equations, which play a prominent role in linear-quadratic control and filtering theory (as discussed extensively in other chapters of this book), appear also in discrete- and continuous-time dynamic games, albeit in more general forms. Both the existence and the characterization of noncooperative equilibria in zero-sum and nonzero-sum linear-quadratic dynamic games, under saddle-point, Nash and Stackelberg equilibrium concepts, involve the solutions of these generalized matrix Riccati (differential or algebraic) equations.

This chapter provides a unified survey of some of the available results in this area, for both finite and infinite horizon formulations, and by emphasizing the variations brought about by different information patterns, such as open-loop, closed-loop, delayed, sampled-data and their combinations. Section 2 concerns zero-sum linear-quadratic (LQ) dynamic/differential games, providing a detailed development of some key results on the topic. Section 3 shows the equivalence between H^{∞}-optimal control problems (in discrete and continuous time) and a subclass of zero-sum LQ games with zero initial states, which makes the theory of Section 2 directly applicable to such worst-case controller design problems. Section 4 presents some coupled Riccati equations that arise in connection with the Nash equilibria of LQ nonzero-sum dynamic games, and the chapter ends with the concluding remarks of Section 5.

It should be pointed out that there is voluminous literature on the role of Riccati equations in control and filtering, and also on dynamic/differential games that lead to generalized Riccati equations. It is clearly impossible to list all relevant references and discuss their individual contributions in this limited space. Therefore with the exception of some specific or very recent results, we have chosen to provide just a few book references, where needed, such as [16], [11] for regular LQ control problems and Riccati equations arising in this context, and [8] for dynamic/differential games. Further detailed references can be found in these texts.

Bittanti, Laub, Willems (Eds.)
The Riccati Equation

11.2 Zero-Sum Dynamic Games

In this section we derive and discuss properties of generalized Riccati equations (GRE'S) that arise in the saddle-point solutions of linear-quadratic zero-sum dynamic games, in both discrete and continuous time – the latter class also known as zero-sum differential games.

In the discrete time, the system equation (describing the evolution of the game dynamics) is given by

$$x_{k+1} = A_k x_k + B_k^1 u_k^1 + B_k^2 u_k^2, \quad k \in \mathcal{K} := \{1, \ldots, K\} \tag{11.1}$$

and the finite-horizon cost function is

$$L(u^1, u^2) = \sum_{k=1}^{K} \left\{ |x_{k+1}|^2_{Q_{k+1}} + \left|u_k^1\right|^2 - r_k \left|u_k^2\right|^2 \right\} \tag{11.2}$$

which is to be minimized by Player 1 and maximized by Player 2. Here, r_k is a positive scalar, Q_{k+1} is a nonnegative definite matrix, for each $k \in \mathcal{K}$, and $|x|_Q$ denotes the Euclidean (semi-) norm of x weighted by $Q \geq 0$, i.e., $|x|_Q^2 = x'Qx$. Note that we have taken the weighting matrices associated with u_k^1 and u_k^2 as identity matrices, without any loss of generality, since any nonidentity matrix can easily be absorbed into B_k^1 or B_k^2; of course, $r_k > 0$ could also be absorbed into B_k^2, but we prefer to keep it as a parameter since we will later have an occasion to study the saddle-point solution as a function of r. The state x_k has dimension n, and the control vector of Player i, u_k^i, has dimension m_i, $i = 1, 2$. It is assumed that the initial state x_1 is known to both players.

The formulation above will not be complete, unless we specify the information structure of the problem, that is the nature of the dependence of the control variables on the state. Here we will be interested primarily in three different types of information:

(i) *Closed-loop* (CL): The control is allowed to depend on the current as well as the entire past values of the state, i.e.,

$$\begin{aligned} u_k &= \mu(x_{[1,k]}), \quad k \in \mathcal{K} \\ x_{[1,k]} &:= (x_1, \ldots, x_k) \end{aligned} \tag{11.3}$$

where $\mu_{[1,k]} := (\mu_1, \ldots, \mu_k)$ is known as a (control) policy. We let $\mathcal{M}_{\text{CL}}$ denote the set of all such (Borel measurable) policies.

(ii) *Closed-loop 1-step delay* (CLD): The control is not allowed to depend on the current value of the state, but it can depend on the entire past values of the state, i.e.,

$$\begin{aligned} u_k &= \mu_k(x_{[1,k-1]}), \quad k \in \mathcal{K}, k \neq 1 \\ &= \mu_1(x_1), \qquad\qquad k = 1. \end{aligned} \tag{11.4}$$

We denote the corresponding policy space by $\mathcal{M}_{\text{CLD}}$.

(iii) *Open-loop* (OL): The control depends only on the initial state, i.e.,

$$u_k = \mu_k(x_1), \quad k \in \mathcal{K}. \tag{11.5}$$

The corresponding policy space is denoted by $\mathcal{M}_{\text{OL}}$.

Let $\mathcal{M}^i$ denote the policy space of Player i under any one of the information patterns introduced above. Further, introduce the function $J : \mathcal{M}^1 \times \mathcal{M}^2 \to \mathbf{R}$ by

$$\begin{aligned} &J(\mu^1, \mu^2) = L(u^1, u^2), \\ &\quad u_k^i = \mu_k^i(\cdot), k \in \mathcal{K}; \quad \mu_{[1,K]}^i \in \mathcal{M}^i, \quad i = 1, 2, \end{aligned} \tag{11.6}$$

where we have suppressed the dependence on the initial state x_1. The triple $\{J; \mathcal{M}^1, \mathcal{M}^2\}$ constitutes the **normal form** of the zero-sum dynamic game, in the context of which we can introduce the notion of a *saddle-point equilibrium.*

Definition 2.1. Given a zero-sum dynamic game $\{J; \mathcal{M}^1, \mathcal{M}^2\}$ in normal form, a pair of policies $(\mu^{1^*}, \mu^{2^*}) \in \mathcal{M}^1 \times \mathcal{M}^2$ constitutes a saddle-point solution if, for all $(\mu^1, \mu^2) \in \mathcal{M}^1 \times \mathcal{M}^2$,

$$J(\mu^{1^*}, \mu^2) \leq J^* := J(\mu^{1^*}, \mu^{2^*}) \leq J(\mu^1, \mu^{2^*}) \quad \square \tag{11.7a}$$

The quantity J^* above is called the *value* of the game, which is defined even if a saddle-point solution does not exist, as

$$\bar{J} := \inf_{\mu^1 \in \mathcal{M}^1} \sup_{\mu^2 \in \mathcal{M}^2} J(\mu^1, \mu^2) = J^* = \sup_{\mu^2 \in \mathcal{M}^2} \inf_{\mu^1 \in \mathcal{M}^1} J(\mu^1, \mu^2) =: \underline{J}. \tag{11.7b}$$

Here $\bar{J}$ and $\underline{J}$ are the *upper value* and the *lower value*, respectively, and generally we have the inequality $\bar{J} \geq \underline{J}$. Only when they are equal, as in (11.7b), that the value J^* of the game is defined.

In the continuous time, (11.1) and (11.2) are replaced, respectively, by

$$\frac{d}{dt}x(t) =: \dot{x} = A(t)x + B^1(t)u^1(t) + B^2(t)u^2(t), \quad t \geq 0 \tag{11.8}$$

$$L(u^1, u^2) = |x(t_f)|_{Q_f}^2 + \int_0^{t_f} \left\{ |x(t)|_{Q(t)}^2 + \left|u^1(t)\right|^2 - r(t)\left|u^2(t)\right|^2 \right\} dt \tag{11.9}$$

where $Q_f \geq 0$, $Q(t) \geq 0$, $t \in [0, t_f]$, $r(t) > 0$, t_f is the terminal time, all matrices have piecewise continuous entries, and so do u^1 and u^2. The initial state x_0 is known to both players.

Here we will consider again three types of information structure, two of which are direct counterparts of the two (CL and OL) introduced for discrete time:

(i) *Closed-loop* (CL):

$$u(t) = \mu(t; x(s), s \leq t), \quad t \geq 0. \tag{11.10}$$

Here we have to impose some additional (other than measurability) conditions on μ so that the state differential equation (11.8) admits a unique solution. One such condition is to assume that $\mu(\cdot)$ is Lipschitz-continuous in x. We will not discuss these conditions in detail here, because they are not explicitly used in the derivation of the saddle-point solution which turns out to be linear. Nevertheless, let us introduce the policy space $\mathcal{M}_{\text{CL}}$ to denote the general class of smooth closed-loop policies.

(ii) *Sampled-data* (SD): Here we have a partitioning of the time interval $[0, t_f]$, as

$$0 < t_1 < t_2 < \ldots < t_{K-1} < t_f$$

where t_k, $k \in \mathcal{K}$, is a sampling point. The controller has access to the values of the state only at the past (and present, if any) sampling points, i.e.,

$$u(t) = \mu(t; x(t_k), \ldots, x(t_1), x_0), \quad t_k \leq t < t_{k+1}, k \in \mathcal{K}. \tag{11.11}$$

where $\mu(\cdot)$ is piecewise continuous in t, and measurable in $x(t_k), x(t_{k-1}), \ldots$ $\ldots, x_0$. We denote the class of such policies by $\mathcal{M}_{\text{SD}}$.

(iii) *Open-loop* (OL):

$$u(t) = \mu(t; x_0), \quad t \geq 0 \tag{11.12}$$

where $\mu(\cdot)$ is piecewise continuous in t and measurable in x_0. The corresponding policy space is denoted by $\mathcal{M}_{\text{OL}}$.

The normal form for a continuous-time game can be defined as in (11.6), with Definition 2.1 being equally valid here for a saddle point. In this general setting we now introduce a simple, but very useful property of multiple saddle points, valid for both discrete and continuous-time formulations:

Property 2.1: Ordered interchangeability. *For a dynamic game in normal form, $\{J; \mathcal{M}^1, \mathcal{M}^2\}$, if $(\bar{\mu}^1, \bar{\mu}^2) \in \mathcal{M}^1 \times \mathcal{M}^2$ and $(\tilde{\mu}^1, \tilde{\mu}^2) \in \mathcal{M}^1 \times \mathcal{M}^2$ are two saddle-point pairs, then the pairs $(\bar{\mu}^1, \tilde{\mu}^2)$ and $(\tilde{\mu}^1, \bar{\mu}^2)$ also constitute saddle points.* □

We are now in a position to present the saddle-point solutions to the discrete and continuous time games introduced above, along with the associated generalized Riccati equations. For the finite-horizon case, we discuss only the main points of the derivation: for details the reader is referred to [8]. For the infinite horizon case we provide more details.

11.2.1 The Discrete-time LQ Dynamic Game

11.2.1.1 Open-loop information structure for both players The procedure is first to fix the open-loop policy of Player 2, say $\bar{u}^2_{[1,K]}$, and minimize $L(u^1, \bar{u}^2_{[1,K]})$ given by (11.2) with respect to $u^1 = u^1_{[1,K]}$. From LQ optimal control theory (see, e.g., [16]) it is well-known that the solution exists and is unique (since L is strictly convex in u^1), and is characterized in terms of the standard discrete-time Riccati equation. Furthermore, the optimal u^1 is linear in $u^2_{[1,K]}$ and x_1, i.e., for some linear function $\ell^1(\cdot)$, $u^{1*}_{[1,K]} = \ell^1(\bar{u}^2_{[1,K]}, x_1)$. Now, conversely, if we fix the open-loop policy of Player 1, say $\bar{u}^1_{[1,K]}$, and maximize $L(u^1_{[1,K]}, u^2)$ with respect to $u^2 = u^2_{[1,K]}$, the solution will again be linear, i.e., $u^{2^*}_{[1,K]} = \ell^2(\bar{u}^1_{[1,K]}, x_1)$, but existence and uniqueness will depend on whether L is strictly concave in u^2 or not. This requirement translates into the following condition:

Lemma 2.1. *The quadratic objective functional $L(u^1, u^2)$ given by (11.2), and under the state equation (11.1), is strictly concave in u^2 for every open-loop policy u^1 of Player 1, if, and only if,*

$$r_k I - B^{2'}_k S_{k+1} B^2_k > 0, \quad k \in \mathcal{K} \tag{11.13a}$$

where the sequence S_{k+1}, $k \in \mathcal{K}$, is generated by the Riccati equation

$$S_k = Q_k + A'_k S_{k+1} A_k + A'_k S_{k+1} B^2_k \left[r_k I - B^{2'}_k S_{k+1} B^2_k \right]^{-1} B^{2'}_k S_{k+1} A_k;$$

$$S_{K+1} = Q_{K+1}. \quad \square \tag{213b}$$

Under condition (11.13a), the quadratic open-loop game becomes convex-concave, and it follows from the general theory of static quadratic games ([8], p. 168) that it admits a unique saddle point – which has to be the unique fixed point of the linear mappings:

$$u^1_{[1,K]} = \ell^1\left(u^2_{[1,K]}, x_1\right); u^2_{[1,K]} = \ell^2\left(u^1_{[1,K]}, x_1\right).$$

Certain manipulations, details of which can be found in ([8], p. 248), lead to the conclusion that this unique fixed point is characterized in terms of a matrix sequence generated by another (other than (11.13b)) discrete-time Riccati equation. The result is given in the following theorem:

Theorem 2.1. *For the discrete-time linear-quadratic zero-sum dynamic game with open-loop information structure, let (11.13a) be satisfied and M_k, $k \in \mathcal{K}$, be a sequence of matrices generated by*

$$M_k = Q_k + A'_k M_{k+1} \Lambda^{-1}_k A_k; \quad M_{K+1} = Q_{K+1} \tag{11.14a}$$

where

$$\Lambda_k := I + \left(B^1_k B^{1'}_k - \frac{1}{r_k} B^2_k B^{2'}_k \right) M_{k+1}. \tag{11.14b}$$

Then,

(i) Λ_k, $k \in \mathcal{K}$, *are invertible.*
(ii) The game admits a unique saddle-point solution, given by

$$u^{1^*}_k = \mu^{1^*}_k(x_1) = -B^{1'}_k M_{k+1} \Lambda^{-1}_k A_k x^*_k \tag{11.15a}$$

$$u^{2^*}_k = \mu^{2^*}_k(x_1) = \frac{1}{r_k} B^{2'}_k M_{k+1} \Lambda^{-1}_k A_k x^*_k, \quad k \in \mathcal{K} \tag{11.15b}$$

where $\{x^*_{k+1}, k \in \mathcal{K}\}$ is the corresponding state trajectory, generated by

$$x^*_{k+1} = \Lambda^{-1}_k A_k x^*_k, \quad x^*_1 = x_1. \tag{11.15c}$$

(iii) The saddle-point value of the game is

$$L^* = L(u^{1^*}, u^{2^*}) = x_1' A_1' M_2 \Lambda_1^{-1} A_1 x_1. \tag{11.16}$$

(iv) If the matrix in (11.13a) has at least one negative eigenvalue, then the upper value becomes unbounded. □

11.2.1.2 Closed-loop Information structure for both players. When players have access to closed-loop state information (with memory), then the inherent redundancy in information gives rise to a multiplicity of saddle-point equilibria – each one leading to the same value (in view of Property 2.1) but not necessarily requiring the same existence conditions. Hence, in this case one has to bring in a further refinement on the saddle-point solution, such as *strong time consistency*, or *noise insensitivity*. We call a (saddle-point) solution "strongly time consistent" if it provides a solution to any truncated version of the original game, regardless of the values of the new initial states. More precisely,

Definition 2.2: Strong time consistency [3]. From the original game defined on the time interval $[1, K]$, construct a new game on a shorter time interval $[\ell, K]$, by setting $\mu^1_{[0,\ell-1]} = \beta^1_{[0,\ell-1]}$, $\mu^2_{[0,\ell-1]} = \beta^2_{[0,\ell-1]}$, where $\beta^i_{[0,\ell-1]}$, $i = 1, 2$, are fixed but arbitrarily chosen. Let $(\mu^{1^*} \in \mathcal{M}^1_{\mathrm{CL}}, \mu^{2^*} \in \mathcal{M}^2_{\mathrm{CL}})$ be a saddle-point solution for the original game. Then, it is "*strongly time consistent*", if the pair $(\mu^{1^*}_{[\ell,K]}, \mu^{2^*}_{[\ell,K]})$ is a saddle-point solution of the new game (on the interval $[\ell, K]$), regardless of the choices for $\beta^i_{[0,\ell-1]}$, $i = 1, 2$, and for every ℓ, $2 \le \ell \le K$. □

By "noise insensitivity" we mean that the saddle-point solution retains its equilibrium property even if the state equation (11.1) has an additional additive term which is a zero-mean white noise sequence. It turns out that both refinement schemes lead to the same solution, which is the one obtained by solving in retrograde time a sequence of static games – in the spirit of dynamic programming. Such a special closed-loop saddle-point solution is known as a *feedback* saddle-point solution, which is given in the following theorem [8].

Theorem 2.2. *For the discrete-time linear-quadratic zero-sum dynamic game with closed-loop information structure,*

(i) There exists a unique feedback saddle-point solution if, and only if,

$$G_k := r_k I - B_k^{2'} M_{k+1} B_k^2 > 0, \quad k \in \mathcal{K} \tag{11.17}$$

where the sequence M_{k+1}, $k \in \mathcal{K}$, is generated by (11.14a).

(ii) Under condition (11.17), Λ_k, $k \in \mathcal{K}$, are invertible, and the unique feedback saddle-point policies are

$$u_k^{1^*} = \mu_k^{1^*}(x_k) = -B_k^{1'} M_{k+1} \Lambda_k^{-1} A_k x_k \tag{11.18a}$$

$$u_k^{2^*} = \mu_k^{2^*}(x_k) = \frac{1}{r_k} B_k^{2'} M_{k+1} \Lambda_k^{-1} A_k x_k, \quad k \in \mathcal{K}, \tag{11.18b}$$

with the corresponding trajectory still generated by (11.15c). Furthermore the saddle-point value is given by (11.16).

(iii) If the matrix G_k in (11.17) has a negative eigenvalue for some $k \in \mathcal{K}$, then the game does not admit a saddle point under any information structure, and its upper value becomes unbounded. □

Proof. We prove here only part (iii), since it is so important for the development to follow. Proofs for other parts of the theorem can be found in [8] where the basic tool is dynamic programming, which involves the solution of a sequence of quadratic games in retrograde time. For each k, existence of a unique saddle point to the corresponding static game is guaranteed by the positive definiteness of G_k, and if there is some $\bar{k} \in \mathcal{K}$ such that $G_{\bar{k}}$ has a negative eigenvalue, then the corresponding static game in the sequence does not admit a saddle point, and being a quadratic game this implies that the upper value of the game is unbounded. If G_k has a zero (but no negative) eigenvalue, then whether the corresponding game admits a saddle point or not depends on the precise value of the initial state x_1, and in particular if $x_1 = 0$ one can allow G_k to have a zero eigenvalue and still preserve the existence of a saddle point. Since the "zero-eigenvalue case" can be recovered as the limit of the "positive-eigenvalue case", we will henceforth not address the "zero-eigenvalue case".

Now, contrary to the statement of (iii), assume that the dynamic game admits a saddle point, say $(\hat{\mu}^1, \hat{\mu}^2)$, even though G_k has a negative eigenvalue for some $k \in \mathcal{K}$; clearly, this saddle-point solution cannot be the feedback solution (2.18). Let $\bar{k}$ be the largest integer in $\mathcal{K}$ for which G_k has a negative eigenvalue, and $G_{\bar{k}+1} > 0$. In the original game, fix the policies of the players during the first $\bar{k}$ stages as

$$\mu^1_{[1,\bar{k}]} = \hat{\mu}^1_{[1,\bar{k}]}, \quad \mu^2_{[1,\bar{k}]} = \hat{\mu}^2_{[1,\bar{k}]}$$

and let the corresponding state trajectory be $\hat{x}_{[1,\bar{k}]}$. Then, we essentially have the same game defined on the shorter time interval $[\bar{k}+1, K]$, with initial state $\hat{x}_{\bar{k}+1}$. By one of the two hypotheses, the pair $(\mu^{1*}_{[\bar{k}+1,K]}, \mu^{2*}_{[\bar{k}+1,K]})$, defined by (2.18), constitutes a saddle point for this game (actually for all $x_{\bar{k}+1}$, and not only for $\hat{x}_{\bar{k}+1}$), and by the other hypothesis the truncated version of $(\hat{\mu}^1, \hat{\mu}^2)$ on the trajectory $\hat{x}_{[1,\bar{k}]}$, i.e.,

$$\{\hat{\mu}^1_k(x_{[\bar{k}+2,k]}, \hat{x}_{[1,\bar{k}+1]}), \hat{\mu}^2_k(x_{[\bar{k}+2,k]}, \hat{x}_{[1,\bar{k}+1]}); k \geq \bar{k}+1\}$$

constitutes a saddle point for the same game. By the ordered interchangeability of multiple saddle points (Property 2.1), the pair $\{\hat{\mu}^1_k(x_{[\bar{k}+2,k]}, \hat{x}_{[1,\bar{k}+1]}), \mu^{2*}_k(x_k); k \geq \bar{k}+1\}$ also constitutes a saddle point.

Let us now consider the same game, but with one additional stage. That is, now we fix the policies during the first $\bar{k}-1$ stages, as $\mu^1_{[1,\bar{k}-1]} = \hat{\mu}^1_{[1,\bar{k}-1]}$, $\mu^2_{[1,\bar{k}-1]} = \hat{\mu}^2_{[1,\bar{k}-1]}$. The resulting game is another linear-quadratic game, defined on the time interval $[\bar{k}, K]$, and with initial state $\hat{x}_{\bar{k}}$. Let us denote the kernel of this game by

$L^{(\bar{k})}(u^1_{[\bar{k},K]}, \mu^2_{[\bar{k},K]}; \hat{x}_{\bar{k}})$, which we know admits the saddle-point solution

$$\{\hat{\mu}^1_{[\bar{k},K]}(x_{[\bar{k}+1,k]}, \hat{x}_{[1,\bar{k}]}), \hat{\mu}^2_k(x_{[\bar{k}+1,k]}, \hat{x}_{[1,\bar{k}]}); \quad k \geq \bar{k}\}.$$

Now, by this saddle-point property,

$$\begin{aligned}
\infty > & \min_{\mu^1_{[\bar{k},K]}} \max_{u^2_{[\bar{k},K]}} L^{(\bar{k})}(\mu^1_{[\bar{k},K]}, u^2_{[\bar{k},K]}; \hat{x}_{\bar{k}}) \\
&= \min_{\mu^1_{\bar{k}}} \max_{u^2_{\bar{k}}} \min_{\mu^1_{[\bar{k}+1,K]}} \max_{u^2_{[\bar{k}+1,K]}} L^{(\bar{k})}(\mu^1_{[\bar{k},K]}, u^2_{[\bar{k},K]}; \hat{x}_{\bar{k}}) \\
&\geq \min_{\mu^1_{\bar{k}}} \max_{u^2_{\bar{k}}} \min_{\mu^1_{[\bar{k}+1,K]}} L^{(\bar{k})}(\mu^1_{[\bar{k},K]}, u^2_{\bar{k}}, \hat{\mu}^2_{[\bar{k}+1,K]}; \hat{x}_{\bar{k}}) \qquad (\circ) \\
&= \min_{\mu^1_{\bar{k}}} \max_{u^2_{\bar{k}}} [x'_{\bar{k}+1} M_{\bar{k}+1} x_{\bar{k}+1} + u^{1'}_{\bar{k}} u^1_{\bar{k}} - r_k u^{2'}_k u^2_k + c_{\bar{k}}]
\end{aligned}$$

where $c_{\bar{k}}$ is some constant, which depends on the previously chosen policies and the initial state x_1 (and not on $u^1_{\bar{k}}$ or $u^2_{\bar{k}}$). In the above, the inequality in the third line has followed because $\mu^{2*}_{[\bar{k}+1,K]}$ is not necessarily the maximizing $u^2_{[\bar{k}+1,K]}$. The last line has followed because, as mentioned earlier, the pair $(\mu^{1*}_{[\bar{k}+1,K]}, \hat{\mu}^2_{[\bar{k}+1,K]})$ provides a saddle point for the game defined on $[\bar{k}+1, K]$, with the value given by (11.16), with subscript "1" replaced by "$\bar{k}+1$". Since, by hypothesis, $G_{\bar{k}+1}$ has a negative eigenvalue, the function

$$x'_{\bar{k}+1} M_{\bar{k}+1} x_{\bar{k}+1} - r_k u^{2'}_{\bar{k}} u^2_{\bar{k}}$$

is not bounded above (in $u^2_{\bar{k}}$), and hence ($\circ$) is unbounded, which is a contradiction to the other hypothesis that $L^{(\bar{k})}$ has a saddle point. □

Note that the implication of the last statement of the theorem, just proved, is the property that under the closed-loop information pattern the feedback saddle-point solution requires the least stringent existence conditions; in other words, if the matrix in (11.17) has a negative eigenvalue for some $k \in \mathcal{K}$, then a saddle point will not exist even if the policies are allowed to depend on past values of the state. This implies that saddle-point solutions under other information structures (provided that they exist) can be constructed from the feedback saddle-point solution (2.18) on the trajectory (11.15c). The open-loop solution of Theorem 2.1 is one such example, where for existence we require an additional (more restrictive) condition (11.13a). Two other such cases are discussed in the following subsections.

11.2.1.3 Closed-loop/open-loop information structure. If Player 2 has access to only open-loop information (and Player 1 still has closed-loop information), the existence condition of Theorem 2.2 will remain intact, with a corresponding saddle-point solution given by (11.18a) and (11.15b), leading to the same saddle-point value (11.16). The game in fact admits more than one equilibrium, but this particular one requires the least stringent existence conditions. However, as opposed to the case of Theorem 2.2, this solution will neither be strongly time consistent (in fact "strong time consistency" is not a meaningful notion in this case) nor be noise insensitive. There is, however, a (unique) noise-insensitive pure saddle-point solution for the game, given for Player 1 by

$$\tilde{u}_k^1 = \tilde{\mu}_k^1(x_k, x_1) = \mu_k^{1^*}(x_k^*) + P_k^1[x_k - x_k^*(x_1)] \tag{11.19a}$$

where $x_k^*(x_1)$ is generated by (11.15c), and P_k^1, $k \in \mathcal{K}$ is a matrix sequence defined by

$$P_k^1 = -B_k^{1'} Z_{k+1} \left[I + B_k^1 B_k^{1'} Z_{k+1}\right]^{-1} A_k \tag{11.19b}$$

$$Z_k = Q_k + A_k' Z_{k+1} \left[I + B_k^1 B_k^{1'} Z_{k+1}\right]^{-1} A_k; \quad Z_{K+1} = Q_{K+1}. \tag{11.19c}$$

For Player 2, on the other hand, the open-loop saddle-point policy is still (11.15b), which will be the unique L-maximizing solution under (11.19a) if, and only if,

$$r_k I - B_k^{2'} \tilde{S}_{k+1} B_k^2 > 0, \quad k \in \mathcal{K}, \tag{11.20a}$$

where $\tilde{S}_{k+1}$, $k \in \mathcal{K}$, is generated by

$$\tilde{S}_{k+1} = Q_k + P_k^{1'} P_k^1 + \tilde{A}_k' \tilde{S}_{k+1} \tilde{A}_k + \tilde{S}_{k+1} B_k^2 \left[r_k I - B_k^{2'} \tilde{S}_{k+1} B_k^2\right]^{-1} B_k^{2'} \tilde{S}_{k+1} \tilde{A}_k;$$
$$\tilde{S}_{K+1} = Q_{K+1}, \tag{11.20b}$$

and

$$\tilde{A}_k := A_k + B_k^1 P_k^1. \tag{11.20c}$$

Condition (11.20a) is in fact the associated strict concavity condition, which directly follows from Lemma 2.1. If this condition is not satisfied, but condition (11.17) is, the upper value of the stochastic game with the additive noise will still be bounded, but the saddle-point solution will involve a mixed policy for Player 2. The corresponding policy for Player 1 will still be linear, in the current and initial values of the state, with the gain coefficients, in general, depending on the second (and higher) order moments of the additive noise. We do not discuss the details of this case here since Riccati equations do not play a role in the characterization of these controllers. Interested reader is referred to [2] for an illustrative example on the characterization of such mixed policies; a complete theory can be found in [7].

Now let us reverse the roles of the players, and endow Player 2 with closed-loop information, and Player 1 with open-loop information. Since Player 1 cannot influence the strict concavity of L with respect to u^2, again Lemma 2.1 is useful here, which means that we have to require condition (11.13a) to hold. Under this condition, it follows from Theorem 2.1 and using Property 2.1 that there exists a saddle point to the game, with Player 1's policy given by (11.15a). For Player 2, on the other hand, it will be any representation of (11.18b), using the current and initial values of the state, on the state trajectory (11.15c). These all will require the same existence condition, which is (11.13a). Only one of these representations will have the noise insensitivity property, which is the one obtained by maximizing over $\mu^2 \in \mathcal{M}^2_{\mathrm{CL}}$ the function

$$E\left\{L(\mu^{1^*}(x_1), \mu^2(\cdot))\right\}$$

where μ^{1^*} is given by (11.15a), the state equation (11.1) has an additional additive zero-mean white noise term, and $E\{\cdot\}$ denotes the expectation over the statistics of this noise sequence. This stochastic control problem admits the unique maximizing solution:

$$\tilde{u}^2_k = \tilde{\mu}^2_k(x_k, x_1) = \mu^{2^*}_k(x_1) + P^2_k\left[x_k - x^*_k(x_1)\right] \tag{11.21a}$$

where $\mu^{2^*}_k$ is given by (11.15b), and $x^*_k(x_1)$ is generated by (11.15c). P^2_k is a gain matrix given by

$$P^2_k = \frac{1}{r_k} B^{2'}_k S_{k+1}\left[r_k I - B^{2^*}_k B^{2'}_k S_{k+1}\right]^{-1} A_k \tag{11.21b}$$

where S_{k+1} is generated by (11.13b), $k \in \mathcal{K}$. Hence the solution of the OL/CL game again involves two generalized Riccati equations, which are precisely the ones introduced in the context of Theorem 2.1 – whereas only one of them, i.e., (11.14a), was used in the construction of the saddle-point policies in the OL/OL case, both are needed in the OL/CL case. The *noise-insensitive* pure saddle-point solution in the CL/OL case, however, involves one regular and two generalized Riccati equations, as we have seen earlier, with only two of them actually used in the construction of the solution (namely, (11.14a) and (11.19c)), with the third one, (11.20b), used only in the existence conditions. These results are now summarized in the following theorem.

Theorem 2.3. *For the discrete-time linear-quadratic zero-sum dynamic game,*

(i) There exists a unique noise-insensitive saddle-point solution under the OL/CL information pattern, given by (11.15a) and (11.21a), provided that condition (11.13a) is satisfied. If the matrix in (11.13a) has at least one negative eigenvalue for some $k \in \mathcal{K}$, then the upper value of the game under this information structure becomes unbounded.

(ii) There exists a saddle-point solution under the CL/OL information pattern, given by (11.18a) and (11.15b), provided that condition (11.17) holds. If the matrix in (11.17) has at least one negative eigenvalue for some $k \in \mathcal{K}$, then the upper value of the game becomes unbounded.

(iii) The saddle-point value in both cases is the same (whenever they both exist), and is given by (11.16).

(iv) The saddle-point solution in (ii) is not noise-insensitive. The unique noise-insensitive pure-strategy saddle-point solution is given by (11.19a) and (11.15b), which however is valid under the more stringent condition (11.20a). □

11.2.1.4 Closed-loop 1-step delay information for both players If both players' permissible policies are as given by (11.4), the saddle-point solution is again not unique, in view of Property 2.1. However, requiring the solution to be additive noise-insensitive leads to a unique solution by using Theorem 2.2, Property 2.1 and a dynamic programming type argument. The solution is essentially a particular representation of the feedback solution (2.18) on the state trajectory (11.15c). Here, instead of providing a rigorous proof of the result, we simply motivate the construction, starting with the feedback policies (2.18). A proof can be found in [5]. First, we remark that if the control policies were restricted to depend only on the most recent value of the state, the unique saddle-point solution would be given by (using the notation of Theorem 2.2):

$$\begin{aligned} u_k^1 = \hat{\mu}_k^1(x_{k-1}) &= -B_k^{1'} M_{k+1} \Lambda_k^{-1} A_k \Lambda_{k-1}^{-1} A_{k-1} x_{k-1}, \quad k > 1 \\ &= -B_1^{1'} M_2 \Lambda_1^{-1} A_1 x_1, \quad k = 1 \end{aligned} \tag{11.22a}$$

$$\begin{aligned} u_k^2 = \hat{\mu}_k^2(x_{k-1}) &= (1/r_k) B_k^{2'} M_{k+1} \Lambda_k^{-1} A_k \Lambda_{k-1}^{-1} A_{k-1} x_{k-1}, \quad k > 1 \\ &= (1/r_1) B_1^{2'} M_2 \Lambda_1^{-1} A_1 x_1, \quad k = 1 \end{aligned} \tag{11.22b}$$

which is a valid solution provided that the cost function $L(\hat{\mu}^1, u^2)$ is strictly concave in u^2. This result readily follows from Theorem 2.2 in view of Property 2.1. It is not difficult to see that this is not a noise-insensitive solution. Now, to construct one (in fact the unique one) that is noise-insensitive and requires the least stringent existence condition, we rewrite (11.18a) as

$$u_k^1 = -B_k^{1'} M_{k+1} \Lambda_k^{-1} A_k \underbrace{(A_{k-1} x_{k-1} + B_{k-1}^1 u_{k-1}^1 + B_{k-1}^2 u_{k-1}^2)}_{\xi_k}$$

where ξ_k will have to be expressed in terms of not only x_{k-1}, but also $x_{[1,k-2]}$, through u_{k-1}^1 and u_{k-1}^2. Likewise, (11.18b) is rewritten as

$$u_k^2 = \frac{1}{r_k} B_k^{2'} M_{k+1} \Lambda_k^{-1} A_k \xi_k$$

in view of which, for all $k \in \mathcal{K}$,

$$\xi_{k+1} = A_k x_k - \left(B_k^1 B_k^{1'} - \frac{1}{r_k} B_k^2 B_k^{2'} \right) M_{k+1} \Lambda_k^{-1} A_k \xi_k, \quad \xi_1 = x_1$$

$\Leftrightarrow$

$$\xi_{k+1} = A_k x_k + (\Lambda_k^{-1} - I) A_k \xi_k, \quad \xi_1 = x_1. \tag{11.23}$$

This is an n-dimensional compensator, the output of which at stage k replaces x_k in (2.18). The condition for the resulting set of policies to be in saddle-point equilibrium is strict concavity of $L(u^1, u^2)$ in u^2 when u^1 is given by

$$u_k^{1^*} = \mu_k^{1^*}(\xi_k) = -B_k^{1'} M_{k+1} \Lambda_k^{-1} A_k \xi_k, \quad k \in \mathcal{K}. \tag{11.24}$$

Let $\zeta_k := (x_k', \xi_k')'$, which is generated by

$$\zeta_{k+1} = \hat{A}_k \zeta_k + \hat{B}_k^2 u_k^2, \quad \zeta_1 = (x_1', x_1')' \tag{11.25a}$$

where

$$\hat{A}_k := \begin{pmatrix} I & -B_k^1 B_k^{1'} M_{k+1} \Lambda_k^{-1} \\ I & \Lambda_k^{-1} - I \end{pmatrix} A_k; \quad \hat{B}_k^2 := \begin{pmatrix} B_k^2 \\ 0 \end{pmatrix}. \tag{11.25b}$$

In terms of ζ_k, $k \in \mathcal{K}$, $L(u^{1^*}, u^2)$ can be written as

$$L(u^{1^*} u^2) = \sum_{k=1}^{K} \left\{ |\zeta_{k+1}|^2_{\hat{Q}_{k+1}} - r_k \left| u_k^2 \right|^2 \right\} \tag{11.25c}$$

where

$$\begin{aligned} \hat{Q}_{K+1} &= \text{diag } (Q_{K+1}, 0); \\ \hat{Q}_k &= \text{diag } \left(Q_k, A_k'(\Lambda_k^{-1})' M_{k+1} B_k^1 B_k^{1'} M_{k+1} \Lambda_k^{-1} A_k \right), \quad k \in \mathcal{K}. \end{aligned} \tag{11.25d}$$

To determine the condition of strict concavity of (11.25c) in u^2, we can now use Lemma 2.1 to arrive at

$$r_k I - \hat{B}_k^{2'} \hat{S}_{k+1} \hat{B}_k^2 > 0, \quad k \in \mathcal{K} \tag{11.26a}$$

where $\hat{S}_{k+1}$, $k \in \mathcal{K}$, is generated by

$$\begin{aligned} \hat{S}_k &= \hat{Q}_k + \hat{A}_k' \hat{S}_{k+1} \hat{A}_k + \hat{A}_k' \hat{S}_{k+1} \hat{B}_k^2 \left[r_k I - \hat{B}_k^{2'} \hat{S}_{k+1} \hat{B}_k^2 \right]^{-1} \hat{B}_k^{2'} \hat{S}_{k+1} \hat{A}_k; \\ \hat{S}_{K+1} &= \hat{Q}_{K+1}. \end{aligned} \tag{11.26b}$$

We are now in a position to state the following theorem.

Theorem 2.5. *For the discrete-time linear-quadratic zero-sum dynamic game with closed-loop 1-step delay information for both players, let condition (11.26a) hold. Then,*

(i) The game admits a unique noise-insensitive saddle-point solution given by (2.18) with x_k replaced by ξ_k generated by (11.23). The saddle-point value is again given by (11.16).

(ii) If the matrix in (11.26a) has a negative eigenvalue for at least one $k \in \mathcal{K}$, then the game does not admit a saddle point, and its upper value is unbounded. □

Note again that in the above we need the solution of a single n-dimensional generalized Riccati equation (11.14a) in the construction of the saddle-point policies; however, the solution of another ($2n$-dimensional) Riccati equation (11.26b) is needed to check the condition of existence.

This condition would not change even if Player 2 has access to full state information, so that the saddle-point solution of the theorem would remain a saddle point (though no longer noise-insensitive) under the CLD/CL information pattern. Finally it is worth noting at this point that, by Theorem 2.2 (iii) and Property 2.1, condition (11.26a) implies condition (11.17), so that under the former the matrix Λ_k is invertible for all $k \in \mathcal{K}$.

11.2.1.5 The infinite-horizon game with closed-loop information. We now consider the time-invariant version of the discrete-time game, as $K \to \infty$ and under closed-loop information. It is natural to expect the saddle-point solution to be in the form (from (2.18))

$$u_k^{1\infty} = \mu^{1\infty}(x_k) = -B^{1'}\bar{M}\bar{\Lambda}^{-1}Ax_k, \quad k \geq 1 \tag{11.27a}$$

$$u_k^{2\infty} = \mu^{2\infty}(x_k) = \frac{1}{r}B^{2'}\bar{M}\bar{\Lambda}^{-1}Ax_k, \quad k \geq 1 \tag{11.27b}$$

where the subscripts on all matrices have been dropped because they are taken to be constants. In the above,

$$\bar{\Lambda} := I + \left(B^1B^{1'} - \frac{1}{r}B^2B^{2'}\right)\bar{M} \tag{11.28a}$$

and $\bar{M}$ is the limiting solution of (11.14a) which (if exists) should satisfy the generalized algebraic Riccati equation (GARE)

$$\bar{M} = Q + A'\bar{M}\bar{\Lambda}^{-1}A. \tag{11.28b}$$

Of course in the context of Theorem 2.2 we also need a condition to replace (11.17) for the infinite horizon case. Toward this end we first establish the monotonicity of the sequence generated by (11.14a), a result presented earlier in [2].

Lemma 2.2. *Given an integer K, let the condition (11.17) of Theorem 2.2 be satisfied for all $k \leq K$. Let M_k, $k = K, K-1, \ldots, 1$ be the sequence generated by iteration (11.14a). Then,*

$$M_k \geq M_{k+1} \quad \textit{for all} \quad k \leq K. \tag{11.29}$$

(i.e., $M_k - M_{k+1}$ is a nonnegative definite matrix.)

Proof. The proof is similar to the proof of monotonicity of the solution of the discrete-time Riccati equation in linear-quadratic control, but now we use the saddle-point value function rather than the dynamic programming value function. Toward this end, we first note the following recursion associated with the "feedback" game covered by Theorem 2.2 (see [8], p. 254).

$$V_k(x) = \min_{u^1}\max_{u^2}\left\{V_{k+1}\left(Ax + B^1u^1 + B^2u^2\right) + |x|_Q^2 + \left|u^1\right|^2 - r|u^2|^2\right\},$$

$$V_{K+1}(x) = |x|_Q^2,$$

where $V_k(x)$ is the saddle-point value of the dynamic game with only $K - k + 1$ stages (i.e., starting at stage k at state x, and running through K). We know from Theorem 2.2 that $V_k(x) = |x|_{M_k}^2$, and

$$|x|_{M_k}^2 = \min_{u^1}\max_{u^2}\left\{|Ax + B^1u^1 + B^2u^2|_{M_{k+1}}^2 + |x|_Q^2 + |u^1|^2 - r|u^2|^2\right\} \tag{$*$}$$

for every $x \in \mathbb{R}^n$. Now, since for two functions $g(\cdot,\cdot)$ and $f(\cdot,\cdot)$,

$$g(u^1,u^2) \geq f(u^1,u^2) \Rightarrow \inf_{u^1}\sup_{u^2} g(u^1,u^2) \geq \inf_{u^1}\sup_{u^2} f(u^1,u^2),$$

it follows from $(*)$ that $M_{k+1} \geq M_{k+2}$ implies $M_k \geq M_{k+1}$. Hence, the proof of the Lemma will be completed (by induction) if we show that $M_K \geq M_{K+1} = Q$. Since we know that under condition (11.17), with $k = K$,

$$Q\Lambda_K^{-1} = Q\left[I + \left(B^1B^{1'} - \frac{1}{r}B_2B^{2'}\right)Q\right]^{-1} \geq 0,$$

it follows that

$$M_K = Q + A'Q\Lambda_K^{-1}A \geq Q = M_{K+1},$$

which completes the proof. □

The next lemma provides a set of conditions under which the sequence $\{M_{k,K+1}\}_{k=K+1}^{1}$ is bounded, for every $K > 0$. Here we use a double index on M to indicate explicitly the dependence of the sequence on the terminal (starting) time point K. Of course, for the time-invariant problem, the elements of the sequence will depend only on the difference, $K - k$, and not on k and K separately. Let us also introduce the counterpart of (11.17) in the time-invariant infinite-horizon case:

$$rI - B^{2'}\bar{M}B^2 > 0 \tag{11.30}$$

Lemma 2.3. *Let $\bar{M}$ be a positive definite solution of the GARE (11.28b), which also satisfies the condition (11.30). Then, for all $K > 0$,*

$$\bar{M} - M_{k,K+1} \geq 0, \quad k \leq K+1 \tag{11.31}$$

where $M_{k,K}$ is generated by (2.14).

Proof. First note that (11.30) implies that $\bar{\Lambda}$ is nonsingular, and since by hypothesis $\bar{M} > 0$, it follows that $\bar{M}\bar{\Lambda}^{-1} > 0$. If this property is used in (11.28b), we immediately have $\bar{M} \geq Q$, which shows (11.31) for $k = K+1$. We now show that the validity of (11.31) for $k+1$ implies its validity for k. Toward this end let us first assume that $M_{k+1,K+1} > 0$, and note from (11.14a) and (11.28b), that

$$\begin{aligned}\bar{M} - M_{k,K+1} &= A'\left(\bar{M}\bar{\Lambda}^{-1} - M_{k+1,K+1}\Lambda_{k,K+1}^{-1}\right)A \\ &= A'\left(\left[\bar{M}^{-1} + B^1B^{1'} - \frac{1}{r}B^2B^{2'}\right]^{-1} - \left[M_{k+1,K+1}^{-1} + B^1B^{1'} - \frac{1}{r}B^2B^{2'}\right]^{-1}\right)A\end{aligned}$$

which is nonnegative definite since $\bar{M} \geq M_{k+1,K+1}$ by the hypothesis of the inductive argument. By the continuous dependence of the eigenvalues of a matrix on its elements the result holds also for $M_{k+1,K+1} \geq 0$, since we can choose a matrix $N(\epsilon)$, and a sufficiently small positive parameter ϵ_0 such that $0 < N(\epsilon) \leq \bar{M}$, $0 < \epsilon < \epsilon_0$, and $N(0) = M_{k+1,K+1}$. This completes the proof of the lemma. □

The next lemma says that any nonnegative definite solution of (11.28b) has to be positive definite, if we take the pair (A, C) to be observable, where $C'C = Q$. The proof of this result is similar to that of the standard ARE which arises in linear regulatory theory [16]; it is given below for the sake of completeness.

Lemma 2.4. *Let (A, C) be observable, where $C'C = Q$. Then, if there exists a nonnegative definite solution to the GARE (11.28b), satisfying (11.30), it is positive definite.*

Proof. Assume that, to the contrary, GARE has a nonnegative definite solution with at least one zero eigenvalue. Let x be an eigenvector corresponding to this eigenvalue. Then,

$$x'\bar{M}x = x'C'Cx + x'A'\bar{M}\bar{\Lambda}^{-1}Ax = 0$$

$$\Rightarrow$$

$$Cx = 0 \text{ and } \bar{M}^{\frac{1}{2}}Ax = 0$$

where $\bar{M}^{\frac{1}{2}}$ is the unique nonnegative definite square-root of $\bar{M}$, and the last result follows because both terms on the right are nonnegative definite quadratic terms, and $\bar{M}\bar{\Lambda}^{-1}$ can equivalently be written as

$$\bar{M}^{\frac{1}{2}}(I + \bar{M}^{\frac{1}{2}}(B^1B^{1'} - \frac{1}{r}B^2B^{2'})\bar{M}^{\frac{1}{2}})^{-1}\bar{M}^{\frac{1}{2}}.$$

Next, multiply the GARE from left and right by $x'A'$ and Ax, respectively, to arrive at

$$x'A'\bar{M}Ax = x'A'C'CAx + x'(A')^2\bar{M}\bar{\Lambda}^{-1}A^2x.$$

The left-hand side is *zero* by the earlier result, and hence

$$CAx = 0 \quad \text{and} \quad \bar{M}^{\frac{1}{2}}A^2x = 0.$$

Now multiply the GARE from left and right by $x'(A^2)'$ and A^2x, respectively, and continue in the manner above to finally arrive at the relation

$$x'\left(C', A'C', (A^2)'C', \ldots, (A^{n-1})'C'\right) = 0$$

which holds, under the observability assumption, only if $x = 0$. Hence $\bar{M} > 0$. □

An important consequence of the above result, also in view of Lemmas 2.2 and 2.3, is that if there exist multiple positive-definite solutions to the GARE (11.28b), there is a minimal such solution (minimal in the sense of matrix partial ordering), say $\bar{M}^+$, and that

$$\lim_{K\to\infty} M_{k,K+1} = \bar{M}^+ > 0,$$

whenever (A, C) is an observable pair. Clearly, this minimal solution will also determine the value of the infinite-horizon game, in the sense that

$$\inf_{\mu^1\in\mathcal{M}^1_{CL}} \sup_{\mu^2\in\mathcal{M}^2_{CL}} J^\infty(\mu^1,\mu^2) = x_1'(\bar{M}^+ - Q)x_1 = \sup_{\mu^2\in\mathcal{M}^2_{CL}} \inf_{\mu^1\in\mathcal{M}^1_{CL}} J^\infty(\mu^1,\mu^2) \tag{11.32}$$

where J^∞ is (11.6) as $K \to \infty$.

The following lemma now says that the existence of a positive definite solution to the GARE (11.28b) is not only sufficient, but also necessary for the value of the game to be bounded.

Lemma 2.5. *Let (A, C) be observable. Then, if the GARE (11.28b) does not admit a positive definite solution, the upper value of the game $\{J^\infty; \mathcal{M}^1_{\rm CL}, \mathcal{M}^2_{\rm CL}\}$ is unbounded.*

Proof. Since the limit point of the monotonic nonnegative definite sequence $\{M_{k,K+1}\}$ has to constitute a solution to (11.28b), nonexistence of a positive definite solution to (11.28b) (which also implies nonexistence of a nonnegative definite solution, in view of Lemma 2.4) implies that for each fixed k, the sequence $\{M_{k,K+1}\}_{K>0}$ is unbounded. This means that given any (sufficiently large) $\alpha > 0$, there exists a $K > 0$, and an initial state $x_1 \in \mathbb{R}^n$, such that the value of the K stage game exceeds $\alpha|x_1|^2$. Now choose $u^2_k = 0$ for $k > K$. Then,

$$\begin{aligned}\inf_{\mu^1}\sup_{\mu^2} J^\infty(\mu^1, \mu^2) \geq \inf_{\mu^1} \Bigg\{ &\Bigg\{ \sum_{k=K+1}^{\infty} |x_{k+1}|^2_Q + |u^1_k|^2 \Bigg\} + \\ &\Bigg\{ \sum_{k=1}^{K} |x_{k+1}|^2_Q + |u^1_k|^2 - |\mu^{2*}_k(x_k)|^2 \Bigg\}\Bigg\} \\ &\geq x_1'(M_1 - Q)x_1 > \alpha|x_1|^2 - x_1'Qx_1\end{aligned}$$

which shows that the upper value can be made arbitrarily large. In the above, μ^{2*}_k in the first inequality is the feedback saddle-point controller for Player 2 in the K-stage game, and the second inequality follows because the summation from $K+1$ to ∞ is nonnegative and hence the quantity is bounded from below by the value of the K-stage game.

□

Theorem 2.6. *Consider the infinite-horizon discrete-time linear-quadratic dynamic game, with (A, C) constituting an observable pair, where $C'C = Q$. Then,*

(i) The game has equal upper and lower values if, and only if, the GARE (11.28b) admits a positive definite solution satisfying the condition (11.30).

(ii) If the GARE admits a positive definite solution, satisfying (11.30), then it admits a minimal such solution, to be denoted $\bar{M}^+$. Then, the finite value of the game is (11.32).

(iii) The upper (minimax) value of the game is finite if, and only if, the upper and lower values are equal.

(iv) If $\bar{M}^+ > 0$ exists, as given above, the controller $\mu^{1\infty}$ given by (11.27a), with $\bar{M}$ replaced by $\bar{M}^+$, attains the finite upper value, in the sense that

$$\sup_{\mu^2 \in \mathcal{M}_{CL}} J^\infty(\mu^{1\infty}, \mu^2) = x_1'(\bar{M}^+ - Q)x_1, \tag{11.33}$$

and the maximizing feedback solution above is given by (11.27b), again with $\bar{M}$ *replaced by* $\bar{M}^+$.

(v) *Whenever the upper value is bounded, the feedback matrix*

$$F^1 := (I - B^1 B^{1'} \bar{M}^+ \bar{\Lambda}^{+^{-1}})A \tag{11.34a}$$

is stable, that is it has all its eigenvalues inside the unit circle. Equivalently, the linear system

$$x_{k+1} = F^1 x_k + B^2 u_k^2 \tag{11.34b}$$

is input-output stable.

Proof. Parts (i)–(iii) follow from the sequence of Lemmas 2.2-2.5, as also discussed prior to the statement of the theorem. To prove part (iv), we first note that the optimization problem in (11.33) is the maximization of

$$\sum_{k=1}^{\infty} |x_{k+1}|_Q^2 + |B^{1'} \bar{M}^+ \bar{\Lambda}^{+^{-1}} A x_k|^2 - r|u_k^2|^2$$

over $u^2_{[1,\infty]}$, subject to the state equation constraint (11.34b). First consider the truncated (K-stage) version:

$$\begin{aligned}
&\max_{u^2_{[1,K]}} \sum_{k=1}^{K} \left\{ |x_{k+1}|_Q^2 + |B^{1'} \bar{M}^+ \bar{\Lambda}^{+^{-1}} A x_k|^2 - r|u_k^2|^2 \right\} \\
&\leq \max_{u^2_{[1,K]}} \left\{ |x_{k+1}|_{\bar{M}^+}^2 + \sum_{k=1}^{K} |x_k|_{\tilde{Q}_k}^2 - r|u_k^2|^2 \right\}
\end{aligned}$$

where

$$\tilde{Q}_k := \begin{cases} Q + A'(\bar{M}^+ \bar{\Lambda}^{+^{-1}})' B^1 B^{1'} \bar{M}^+ \bar{\Lambda}^{+-1} A, & k > 1 \\ A'(\bar{M}^+ \Lambda^{+^{-1}})' B^1 B^{1'} \bar{M}^+ \bar{\Lambda}^{+^{-1}} A, & k = 1 \end{cases}$$

with the inequality following because $\bar{M}^+ \geq Q$. Adopting a dynamic programming approach to solve this problem, we have, for the last stage, after some manipulations:

$$\max_u \left\{ |F^1 x_K + B^2 u|_{\bar{M}^+}^2 + |x_K|_{\tilde{Q}_K}^2 - r|u|^2 \right\} = |x_K|_{\bar{M}^+}^2,$$

which is uniquely attained by

$$\begin{aligned}
u^* &= (rI - B^{2'} \bar{M}^+ B^2)^{-1} B^{2'} \bar{M} F^1 x_k \\
&\equiv (1/r) B^{2'} \bar{M}^+ \bar{\Lambda}^{+^{-1}} A x_K
\end{aligned} \tag{$*$}$$

under the strict concavity condition

$$rI - B^{2'}\bar{M}^{+}B^{2} > 0,$$

which is (11.30) and is satisfied by hypothesis. Hence recursively we solve identical optimization problems at each stage, leading to

$$\max_{u^2[1,K]} \sum_{k=1}^{K} \left\{ |x_{k+1}|_Q^2 + |B^{1'}\bar{M}^{+}\bar{\Lambda}^{+^{-1}}Ax_k|^2 - r|u_k^2|^2 \right\} \leq x_1'(\bar{M}^{+} - Q)x_1$$

where the bound is independent of K. Since we already know from part (ii) that this bound is the value of the game, it readily follows that the control μ^{1^∞} attains it, and the steady state controller (∗) maximizes $J^\infty(\mu^{1\infty}, \mu^2)$.

Now, finally to prove part (v), we use an argument similar to that used in linear regulator theory. Toward this end, first note that boundedness of the upper value implies, with $u_k^2 \equiv 0$, $k \geq 1$, that

$$|x_k|_Q^2 + |\mu_k^{1^\infty}(x_k)|^2 \to 0 \quad \text{as} \quad k \to \infty$$

$$\Leftrightarrow \qquad Cx_k \to 0 \quad \text{and} \quad Sx_k \to 0, \quad S := B^{1'}\bar{M}^{+}\bar{\Lambda}^{+^{-1}}A$$

$$Cx_{k+1} \to 0 \Rightarrow C(A + B^1 S)x_k \to 0 \Rightarrow CAx_k \to 0$$

$$\cdots \quad \cdots \quad \cdots$$

$$Cx_{k+n-1} \to 0 \Rightarrow \quad \ldots \quad \Rightarrow CA^{(n-1)}x_k \to 0.$$

But $CA^i x_k \to 0$, $i = 0, \ldots, n-1$, implies by observability that $x_k \to 0$, and hence F^1 is stable since x_k is generated here by (11.34b), with $u_k^2 \equiv 0$. □

Note that the theorem above does *not* claim that the policies (11.27a) and (11.27b) with $\bar{M} = \bar{M}^{+}$ are in saddle-point equilibrium. Part (iv) says that with μ^1 fixed at $\mu^{1\infty}$, $\mu^{2\infty}$ maximizes $J^\infty(\mu^{1\infty}, \mu^2)$, which is only one side (left-hand side) of the saddle point inequality (11.7a). It is not necessarily true, however, that the minimum of $J^\infty(\mu^1, \mu^{2\infty})$ over $\mathcal{M}_{\mathrm{CL}}^1$ is attained by $\mu^{1\infty}$. An example to this effect will be provided later for the continuous-time case.

11.2.2 The Continuous-time LQ Dynamic (Differential) Game

We now present the counterpart of the results of the previous subsection in the continuous time, by paralleling the development there.

11.2.2.1 Open-loop information structure for both players. Here we can view the controls of the two players as elements of appropriate Hilbert spaces of square-integrable functions, and the differential game becomes a static quadratic game defined on infinite dimensional spaces [9]. For the game to admit a unique saddle point, we again require strict concavity of $L(u^1, u^2)$ in $u^2 \in \mathcal{M}_{\mathrm{OL}}^2$, for every fixed $u^1 \in \mathcal{M}_{\mathrm{OL}}^1$. This is precisely the condition of existence of a unique solution to the

LQ optimization problem

$$\max_{u^2 \in \mathcal{M}^2_{\mathrm{OL}}} L(u^1, u^2)$$

which is known to have an associated Riccati differential equation [11]. Since the objective function is indefinite, the Riccati equation may have "finite escape", in which case we say that "the Riccati equation has a *conjugate* point on the the given interval". This now lays the ground for the following counterpart of Lemma 2.1.

Lemma 2.6. *The quadratic objective functional $L(u^1, u^2)$ given by (11.9), and under the state equation (11.8), is strictly concave in u^2 for every open-loop policy u^1 of Player 1, if, and only if, the following Riccati differential equation does not have a conjugate point on the interval $[0, t_f]$:*

$$\dot{S} + A'S + SA + Q + \frac{1}{r} SB^2B^{2'} S = 0; \quad S(t_f) = Q_f. \quad \square \tag{11.35}$$

The discussion preceding Theorem 2.1 equally applies here, leading to the following result on the open-loop saddle-point solution [8]:

Theorem 2.7. *For the linear-quadratic differential game with open-loop information structure, let the condition of Lemma 2.6 be satisfied, and introduce the following Riccati differential equation:*

$$\dot{Z} + A'Z + ZA + Q - ZB^1B^{1'} Z + \frac{1}{r} ZB^2B^{2'} Z = 0; \quad Z(t_f) = Q_f. \tag{11.36}$$

Then,

(i) The Riccati differential equation (11.36) does not have a conjugate point on the interval $[0, t_f]$.

(ii) The game admits a unique saddle-point solution, given by

$$u^{1^*}(t) = \mu^{1^*}(t; x_0) = -B^1(t)'Z(t)x^*(t) \tag{11.37a}$$

$$u^{2^*}(t) = \mu^{2^*}(t; x_0) = \frac{1}{r(t)} B^2(t)'Z(t)x^*(t), \quad t \geq 0 \tag{11.37b}$$

where $x^*_{[0,t_f]}$ is the corresponding state trajectory, generated by

$$\dot{x}^* = (A - (B^1B^{1'} - \frac{1}{r}B^2B^{2'})Z(t))x^*; \quad x^*(0) = x_0. \tag{11.38}$$

(iii) The saddle-point value of the game is

$$L^* = L(u^{1^*}, u^{2^*}) = x_0'Z(0)x_0. \tag{11.39}$$

(iv) If the Riccati equation (11.35) has a conjugate point of the "odd" type in the open interval $(0, t_f)$, then the upper value of the game is unbounded. $\square$

The fact that nonexistence of a conjugate point for (11.35) is a more stringent condition than the nonexistence of a conjugate point to (11.36) can be demonstrated through a simple scalar example ([8], p. 292). Let the system dynamics be

$$\dot{x} = 2u^1 - u^2$$

and the performance index be given by

$$L(u^1, u^2) = |x(t_f)|^2 + \int_0^{t_f} [u^1(t)^2 - u^2(t)^2] dt.$$

The Riccati differential equation (11.36) admits the unique solution

$$Z(t) = \frac{1}{1 + t_f - t},$$

which is valid for all $t_f > 0$. Hence, there is no conjugate point for (11.36) on the interval $(0, \infty)$. For (11.35), however, a finite solution exists only if $t_f < 1$, in which case

$$S(t) = \frac{1}{1 + t - t_f}, \quad 0 \le t \le t_f < 1.$$

Hence, (11.35) has a conjugate point if the interval is of unit length or longer.

11.2.2.2 Closed-loop information structure for both players Here, as in the discrete-time case, there exist generally multiple equilibria, with again a refinement being possible by invoking *strong time consistency* or additive (Wiener) *noise-insensitivity*, where these notions can be introduced analogously (see [3], [8]). They lead to a unique saddle-point solution, also called *feedback* saddle-point solution since it satisfies a Hamilton-Jacobi-Bellman (Isaacs) equation which involves the solution of quadratic games pointwise in time. A simpler derivation of the feedback saddle-point solution, though, involves the "completing-the-squares" argument of quadratic optimization, where one can show, through routine manipulations, that (11.9), under (11.8), can equivalently be written as

$$\begin{aligned} L(u^1, u^2) = x_0' Z(0) x_0 + \int_0^{t_f} |u^1 + B^1(t)' Z(t) x(t)|^2 dt \\ - \int_0^{t_f} |u^2 - \frac{1}{r} B^2(t)' Z(t) x(t)|^2 r(t) dt, \end{aligned} \tag{11.40}$$

provided that there exists a continuously differentiable $Z(\cdot)$ solving (11.36) for all $t \in (0, t_f]$. Since this representation for L completely decouples the controls of the players, the saddle-point solution can readily be read off (11.40). This observation, together with some additional reasoning, leads to the following theorem:

Theorem 2.8. *For the linear-quadratic zero-sum differential game with closed-loop information structure, defined on the time interval* $[0, t_f]$,

(i) *There exists a unique feedback saddle-point solution, if and only if, the Riccati differential equation (11.36) does not have a conjugate point in the interval* $[0, t_f]$.

(ii) *The unique strongly time consistent (feedback) saddle-point policies are*

$$\mu^{1^*}(t, x(t)) = -B^1(t)'Z(t)x(t) \tag{11.41a}$$

$$\mu^{2^*}(t, x(t)) = \frac{1}{r(t)}B^2(t)'Z(t)x(t), \quad t \geq 0 \tag{11.41b}$$

where $Z(\cdot)$ *is the unique solution of (11.36). Furthermore, the saddle-point value is given by (11.39), and the corresponding state trajectory solves (11.38).*

(iii) *If the Riccati differential equation (11.36) has a conjugate point of the "odd" type in the open interval* $(0, t_f)$*, then the game does not admit a saddle point, and its upper value becomes unbounded.*

Proof. Part (ii) readily follows from the representation (11.40). Note that since the matrix Riccati equation admits a continuously differentiable solution in some neighborhood of the terminal time t_f, one can find a $\bar{t} < t_f$ such that the differential game defined on the interval $[\bar{t}, t_f]$ admits a saddle-point solution, as given in the theorem.

A proof for parts (i) and (iii) of the theorem can be found in [10], where it has also been shown that a saddle point can survive a conjugate point if it is of the "even" type, in which case the product $Z(t)x(t)$ remains bounded even though Z itself (and most likely the controller gain) becomes unbounded. Here we provide an alternate more intuitive proof for the *necessity* of nonexistence of a conjugate point, though under the assumption that every component of the state vector is uniformly (over time) bounded away from zero. Toward this end suppose that even though there exists a conjugate point in the interval $[0, t_f]$, say t_s, the game admits a saddle-point solution on $[0, t_f]$, say $(\hat{\mu}^1, \hat{\mu}^2)$, with a corresponding state trajectory $\hat{x}_{[0,t_f]}$. Let $\epsilon > 0$ be sufficiently small, so that $t_s + \epsilon < t_f$. Consider a new game on the shorter time interval $[t_s + \epsilon, t_f]$, obtained from the original one by setting

$$\mu^1_{[0,t_s+\epsilon]} = \hat{\mu}^1_{[0,t_s+\epsilon]}, \quad \mu^2_{[0,t_s+\epsilon]} = \hat{\mu}^2_{[0,t_s+\epsilon]}$$

and with a new initial state $\hat{x}(t_s + \epsilon)$. Since the original games does not have a conjugate point on the interval $[t_s + \epsilon, t_f]$, this new (truncated) game admits a saddle point, with value $\hat{x}(t_s+\epsilon)'Z(t_s+\epsilon)\hat{x}(t_s+\epsilon)$ + constant, where the quadratic term becomes arbitrarily large as $\epsilon \downarrow 0$, since t_s is a conjugate point, and we have assumed that every component of $\hat{x}(t_s + \epsilon)$ is uniformly bounded away from zero for some initial state of the original system. For the same truncated game, we have $(\hat{\mu}^1_{[t_s+\epsilon,t_f]}, \hat{\mu}^2_{[t_s+\epsilon,t_f]})$ also a saddle-point pair (by hypothesis) for which the value is bounded above (as $\epsilon \downarrow 0$) since it is part of a saddle point solution for the entire game (beyond the conjugate point). This leads to a contradiction, by Property 2.1, since the saddle-point value has to be unique. □

As in the discrete time case, the implication of the last statement of the theorem is that the feedback saddle-point solution requires the least stringent conditions on the parameters of the game, and hence the saddle-point solutions under other information structures (provided that they exist) can be constructed from the feedback saddle-point solution (2.41), on the state trajectory which is the solution of (11.38). The open-loop solution of Theorem 2.7 is one such case, where for existence we require the additional condition given in Lemma 2.6. Two other applications of this idea are given below, in the contexts of closed-loop/open loop and sampled-data information structures.

11.2.2.3 Closed-loop/open-loop information structure. If Player 2 has open-loop and Player 1 closed-loop information, the existence condition of Theorem 2.8 remains intact, with the saddle-point solution given this time by (11.41a) and (11.37b), leading to the same saddle-point value (11.39). Again we have multiplicity of saddle-point equilibria, but this particular one requires the least stringent existence conditions. It is, however, neither strongly time consistent nor noise-insensitive. Replacing the system equation (11.8) by the Itô stochastic differential equation

$$dx = \left[A(t)x + B^1(t)u^1(t) + B^2(t)u^2(t)\right] dt + dw_t,$$

where w_t, $t \geq 0$, is a vector Wiener process, and using the expected value of (2.7) as the performance index, there turns out to be a unique pure saddle-point solution (which is also independent of the covariance of the noise process), given for Player 1 (as the counterpart of (11.19a) in the continuous time) by

$$\tilde{u}^1(t) = \tilde{\mu}^1(t, x(t), x_0) = \mu^{1^*}(t, x^*(t; x_0)) - B^1(t)'Z^1(t)[x(t) - x^*(t; x_0)] \quad (11.42a)$$

where $x^*(t; x_0)$ is the solution to (11.38), and $Z^1(\cdot)$ is the unique nonnegative definite solution of the Riccati differential equation

$$\dot{Z}^1 + A'Z^1 + Z^1A + Q - Z^1B^1B^{1'}Z^1 = 0; \quad Z^1(t_f) = Q_f, \quad (11.42b)$$

which admits a continuously differentiable solution for all t, $0 \leq t \leq t_f$, without any restrictions on t_f. Note that this is a particular representation of the feedback policy (11.41a) on the saddle-point trajectory. For Player 2, the open-loop saddle-point policy is still (11.37b), which however is now a valid maximizing solution under a more restrictive condition, which is (from Lemma 2.6) nonexistence of a conjugate point to the following Riccati differential equation on the time interval $[0, t_f]$:

$$\dot{\tilde{S}} + \tilde{A}'\tilde{S} + \tilde{S}\tilde{A} + Q + Z^1B^1B^{1'}Z^1 + \frac{1}{r}\tilde{S}B^2B^{2'}\tilde{S} = 0; \quad \tilde{S}(t_f) = Q_f. \quad (11.43a)$$

where

$$\tilde{A} := A - B^1B^{1'}Z^1. \quad (11.43b)$$

This conjugate-point condition is more restrictive than that of Theorem 2.8, but less stringent than that of Theorem 2.7.

Now, if we instead endow Player 2 with closed-loop and Player 1 with open-loop information, the least stringent existence condition will be that of Lemma 2.6, that is it would be identical with the OL/OL case. Player 1's unique saddle-point policy will be (11.37b), whereas Player 2's policy will not be unique, as any representation of the feedback policy (11.41b) on the state trajectory (11.38) will yield the same saddle-point value, as long as strict convexity of L in u^1 is not lost. As in the discrete-time case, only one of these will be noise-insensitive, which (following the same reasoning as in the discrete-time case) can be obtained as

$$\tilde{u}^2(t) = \tilde{u}^2(t; x(t), x_0) = \mu^{2^*}(t, x^*(t; x_0)) + \frac{1}{r(t)} B^2(t)' S(t)[x(t) - x^*(t; x_0)], \quad t \geq 0 \tag{11.44}$$

where S is the solution of the Riccati equation (11.35). The following theorem now summarizes these conclusions.

Theorem 2.9. *For the linear-quadratic zero-sum differential game defined on the interval* $[0, t_f]$,

(i) *There exists a unique noise-insensitive saddle-point solution under the OL/CL information structure, given by (11.37a) and (11.44), provided that the condition of Lemma 2.6 is satisfied. If the Riccati differential equation (11.35) has a conjugate point of the odd type in the open interval* $(0, t_f)$, *then the upper value of the game is unbounded under the OL/CL information pattern.*
(ii) *There exists a saddle-point solution under the CL/OL information pattern, given by (11.41a) and (11.37b), provided that the condition of Theorem 2.8 holds. If the generalized Riccati equation (11.36) has a conjugate point of the odd type in the interval* $(0, t_f)$, *then the upper value is unbounded under the CL/OL information pattern.*
(iii) *In both cases, the saddle-point value is (11.39).*
(vi) *In (ii) the given saddle-point solution is not noise-insensitive. The unique noise-insensitive pure-strategy saddle-point solution is given by (11.42a) and (11.37b), which however is valid under the more stringent condition of nonexistence of a conjugate point to (11.43a) in the interval* $[0, t_f]$. □

11.2.2.4 Sampled-data information for both players Here we seek a saddle point in the class of controllers that have the structure (11.11), and in view of part (iii) of Theorem 2.8, together with Property 2.1, every sampled-data solution to the game has to be a representation of the feedback saddle-point solution (2.41) on the trajectory (11.38). One such representation, which is in fact the only one that depends only on the most recent sampled value of the state, is

$$\mu^{1s}(t, x(t_k)) = -B^1(t)' Z(t) \Phi(t, t_k) x(t_k), \quad t_k \leq t < t_{k+1}, \tag{11.45a}$$

$$\mu^{2s}(t, x(t_k)) = \frac{1}{r} B^2(t)' Z(t) \Phi(t, t_k) x(t_k), \quad t_k \leq t < t_{k+1} \tag{11.45b}$$

where Φ is the state transition function associated with the matrix

$$F(t) := A(t) - [B^1(t) B^1(t)' - \frac{1}{r(t)} B^2(t) B^2(t)'] Z(t) \tag{11.45c}$$

and $Z(\cdot)$ is the solution to (11.36). Clearly, for (11.45a)-(11.45b) to constitute a saddle-point solution, first we have to require that (11.36) have no conjugate points in the given interval, but in addition we have to make sure that the function $J(\mu^{1s}, \mu^2)$ is concave in μ^2. The corresponding condition is given in the theorem below, which also says that the solution (11.45a)-(11.45b) in fact requires the least stringent conditions, besides being strongly time consistent and noise-insensitive.

Theorem 2.10. *For the linear-quadratic zero-sum differential game (11.8)-(11.9), with sampled state information,*

(i) There exists a unique strongly time consistent and noise-insensitive saddle point solution if, and only if, the generalized Riccati equation (11.36) does not have a conjugate point in the interval $[0, t_f]$, and the following K matrix Riccati equations also do not have conjugate points in the given subintervals:

$$\dot{S}_k + A'S_k + S_k A + Q + \frac{1}{r} S_k B^2 B^{2'} S_k = 0;$$
$$S_k(t_{k+1}) = Z(t_{k+1}); \quad t_k \leq t < t_{k+1}, k = K-1, K-2, \ldots 0. \quad (11.46)$$

(ii) Under the conditions above, the corresponding saddle-point policies are (11.45a) and (11.45b), with the value of the game again given by (11.39).

(iii) If any one of the Riccati equations (11.46) has a conjugate point of the odd type in the corresponding interval, the differential game with the given sampling times does not admit a saddle point and has unbounded upper value.

Proof. We should first remark that the set of conjugate point conditions on the Riccati equations (11.46) is precisely the condition for strict concavity of the function $(J(\mu^{1s}, \mu^2)$ in μ^2. But simply showing this would not prove the theorem, since the policy (11.45a) was just one particular representation of the feedback saddle-point policy (11.41a). A more complete proof turns out to be one that follows basically the steps of the proof of Theorem 2.2, with the only difference being that in the present case the static games in between measurement points are continuous-time open-loop games (of the type covered by Theorem 2.7), while in the discrete-time game of Theorem 2.2 they were games defined on Euclidean spaces. Accordingly, under the sampled-data measurement scheme, we have the dynamic programming recursion

$$\min_{\mu^1 \in \mathcal{M}^1_{\rm SD}} \max_{\mu^2 \in \mathcal{M}^2_{\rm SD}} J(\mu^1, \mu^2) = \min_{\mu^1_{[0,t_1)}} \max_{\mu^2_{[0,t_1)}} \left\{ \int_0^{t_1} (|x|^2_{Q(t)} + |u^1|^2 - r|u^2|^2) dt + \ldots \right.$$

$$\left. \min_{\mu^1_{[t_{K-1},t_f]}} \max_{\mu^2_{[t_{K-1},t_f]}} \left\{ \int_{t_{K-1}}^{t_f} (|x|^2_{Q(t)} + |u^1|^2 - r|u^2|^2) dt + |x(t_f)|^2_{Q_f} \right\} \right\}$$

where on the time interval $[t_k, t_{k+1})$, $u^i(t) = \mu^i_{[t_k,t_{k+1})}(t, x(t_k), \ldots, x(t_1), x_0)$. This dynamic programming recursion involves the solution of a sequence of (nested) open-loop linear-quadratic zero-sum games, eachone having a bounded upper value if, and only if, (from Theorem 2.7) the Riccati equation (11.46) does not have a conjugate point on the corresponding subinterval. The boundary values are determined using Property 2.1, that is the fact that open-loop and closed-loop feedback saddle-point policies generate the same trajectory whenever both saddle points exist. It is also because of this reason that the saddle-point policies have the structure (2.45), which are also unique strongly time consistent saddle-point policies since they are derived using dynamic programming. This completes the proofs of parts (i) and (ii). Part (iii) follows from a reasoning similar to the one used in the proof of Theorem 2.2 (iii), under the observation that if the upper value of any one of the open-loop games encountered in the derivation above is unbounded, then the upper value of the original game will also be unbounded with sampled-data measurements. □

Note that in construction of the sample-data saddle-point strategies (2.45) we need the solution of a single n-dimensional generalized Riccati equation (11.36). However, to determine the existence of a saddle point we need, in addition, the solutions (or the conjugate point conditions) of K n-dimensional Riccati differential equations (11.46), one for each sampling subinterval. These conditions would still be the prevailing ones if Player 2 had access to full (continuous) state information, and the pair of policies (2.45) would still be in saddle-point equilibrium under this extended (closed-loop) information for Player 2. However, this solution will not retain its noise-insensitivity property; a noise-insensitive saddle-point policy for Player 2 will depend on the current as well as the most recent sampled state measurements, when he has CL information. Such a solution can easily be constructed by repeated application of Theorem 2.9(i) on each sampling subinterval. The existence condition will still be the same as the one in Theorem 2.10.

11.2.2.5 The infinite-horizon differential game with closed-loop and sampled-data information. We study here the time-invariant version of the differential game, as $t_f \to \infty$, first under closed-loop and then under sampled-data information. As in subsection 2.1.5, we would expect the feedback saddle-point solution to be in the form (from (2.41))

$$u^{1\infty}(t) = \mu^{1\infty}(x(t)) = -B^{1'}\bar{Z}x(t) \tag{11.47a}$$

$$u^{2\infty}(t) = \mu^{2\infty}(x(t)) = -\frac{1}{r}B^{2'}\bar{Z}x(t), \quad t \geq 0 \tag{11.47b}$$

where all the matrices and the scalar r are constants, and $\bar{Z}$ is the limiting solution of (11.36) which (if exists) should satisfy the continuous-time generalized algebraic Riccati equation (GARE)

$$A'\bar{Z} + \bar{Z}A + Q - \bar{Z}(B^1B^{1'} - \frac{1}{r}B^2B^{2'})\bar{Z} = 0. \tag{11.48}$$

Denote the solution of the generalized Riccati differential equation (11.36) by $Z(t;t_f)$, where dependence on the terminal time is explicitly recognized. Now, the first question that would be of interest is: If $Z(t;t_f)$ is bounded for all $t_f >$

0 (i.e., (11.36) does not have any conjugate points on $[0, t_f]$ for any t_f) and $\lim_{t_f \to \infty} Z(t, t_f) = \bar{Z}$ exists, does this necessarily imply that the pair (2.47) is in saddle-point equilibrium. The answer is "no", as the following (counter) example, taken from [18] clearly illustrates.

Example. Consider the scalar state dynamics

$$\dot{x} = x + u^1 + u^2, \quad x(0) = x_0$$

and the objective function

$$J^\infty = \int_0^\infty (x^2 + (u^1)^2 - 2(u^2)^2)dt.$$

The generalized Riccati differential equation (11.36) associated with this game is

$$\dot{Z} + 2Z + 1 - \frac{1}{2}Z^2 = 0; \quad Z(t_f; t_f) = 0$$

which admits the unique solution

$$Z(t; t_f) = \sqrt{6}\tanh\left[\sqrt{6}(t_f - t) + \tanh^{-1}(-2\sqrt{6})\right] + 2.$$

This function has a well-defined limit, as $t_f \to \infty$, which is independent of t:

$$\bar{Z} = 2 + \sqrt{6}.$$

Hence the policies (11.47a)-(11.47b) are:

$$\mu^{1\infty}(x) = -(2 + \sqrt{6})x; \quad \mu^{2\infty}(x) = \frac{2 + \sqrt{6}}{2}x.$$

Note that the closed-loop system under these policies is

$$\dot{x} = -\frac{\sqrt{6}}{2}x$$

which is asymptotically stable. Also, the system only under $\mu^{1\infty}$ is

$$\dot{x} = -(1 + \sqrt{6})x + u^2$$

which is input-output stable.

Now it can be shown that if $u^1 = \mu^{1\infty}(\cdot)$, the J^∞-maximizing solution for Player 2 is $\mu^{2\infty}$. However, when $u^2 = \mu^{2\infty}(\cdot)$, the J^∞-minimizing solution for Player 1 is not $\mu^{1\infty}$. The underlying minimization problem is

$$\int_0^\infty \left[-\left(4 + 2\sqrt{6}\right)x^2 + (u^1)^2\right] dt \to \text{minimum}$$

subject to

$$\dot{x} = \frac{4 + \sqrt{6}}{2}x + u^1.$$

Since the open-loop system is unstable, and there is a negative cost on state in the cost function, clearly the minimizing solution is $u^1 \equiv 0$, under which the cost is driven to $-\infty$.

Even though there is no continuity in the minimizer's policy at $t_f = \infty$, nevertheless the value (defined by (11.7b), even in the absence of a saddle point) is continuous. It has been shown in [18] that given an $\epsilon > 0$, there exists a $T(\epsilon) > 0$ such that by choosing the time-varying feedback policy

$$\mu_\epsilon^2(t, x(t)) = \begin{cases} (1/r)B^{2'} Z(t;T)x(t), & 0 \le t < T \\ 0, & t \ge T \end{cases}$$

Player 2 can assure that

$$\min_{u^1} J^\infty(u^1, \mu_\epsilon^2) \ge (2 + \sqrt{6})(x_0)^2 - \epsilon(x_0)^2$$

where

$$(2 + \sqrt{6})(x_0)^2 = \lim_{t_f \to \infty} Z(t; t_f)(x_0)^2$$

which is the limit of the value of the finite horizon game (as $t_f \to \infty$). Since $\epsilon > 0$ can be chosen arbitrarily small, this shows that the infinite-horizon game of this example indeed has a value. This actually turns out to be the case for the general linear-quadratic differential game, under some general conditions, as stated in the following theorem.

Theorem 2.11. *Consider the infinite-horizon time-invariant linear-quadratic differential game, where $Q_f = 0$ and the pair (A, C) is observable, with $C'C = Q$. Then,*

(i) For each fixed t, the solution to (11.36), $Z(t; t_f)$, is nondecreasing in t_f; that is, if (11.36) has no conjugate point in a given interval $[0, t_f]$, $Z(t; t'_f) - Z(t; t''_f) \ge 0$, $t_f > t'_f > t''_f \ge 0$.

(ii) Every nonnegative definite solution of (11.48) is in fact positive definite, and if there exists a positive definite solution there is a minimal such solution, denoted $\bar{Z}^+$. This matrix has the property that $\bar{Z}^+ - Z(t; t_f) \ge 0$ for all $t_f \ge 0$.

(iii) The game has equal upper and lower values if, and only if, the GARE (11.48) admits a positive definite solution, in which case the common value is

$$J^{\infty *} = x_0' \bar{Z}^+ x_0. \tag{11.49}$$

(iv) The upper value is finite if, and only if, the upper and lower values are equal.

(v) If $\bar{Z}^+ > 0$ exists, as given above, the controller $\mu^{1\infty}$ given by (11.47a), with $\bar{Z}$ replaced by $\bar{Z}^+$, attains the finite upper value, in the sense that

$$\sup_{\mu^2 \in \mathcal{M}_{CL}} J^\infty(\mu^{1\infty}, \mu^2) = x_0' \bar{Z}^+ x_0, \tag{11.50}$$

and the maximizing feedback solution above is given by (11.47b), again with $\bar{Z} = \bar{Z}^+$.

(vi) Whenever the upper value is bounded, the two feedback matrices

$$A - (B^1 B^{1'} - \frac{1}{r} B^2 B^{2'}) \bar{Z}^+ \tag{11.51a}$$

and

$$A - B^1 B^{1'} \bar{Z}^+ \tag{11.51b}$$

are stable.

Proof. The proof is similar to that of Theorem 2.6, and follows by constructing natural counterparts of Lemmas 2.2–2.5 in the continuous time. We do not give it here due to page limitations. This result (with the exception of (11.51b)) was first presented in [18] for the case when Q is of full rank, where it was also mentioned that the result is equally valid if the condition $Q > 0$ is replaced by the observability of the pair (A, C). □

Now, given that $\lim_{t_f \to \infty} Z(t; t_f) = \bar{Z}^+$, where $\bar{Z}^+$ exists as the minimal positive definite solution of the GARE (11.48), it is fairly straightforward to obtain the counterpart of Theorem 2.11 under sampled-data measurements. In view of Theorem 2.10, we have, as candidate solutions

$$\mu_\infty^{1s}(t, x(t_k)) = -B^{1'} \bar{Z}^+ \phi(t, t_k) x(t_k), \quad t_k \le t < t_{k+1} \tag{11.52a}$$

$$\mu_\infty^{2s}(t, x(t_k)) = \frac{1}{r} B^{2'} \bar{Z}^+ \phi(t, t_k) x(t_k), \quad t_k \le t < t_{k+1} \tag{11.52b}$$

where ϕ is the state transition function associated with the matrix

$$\bar{F} := A - \left(B^1 B^{1'} - \frac{1}{r} B^2 B^{2'} \right) \bar{Z}^+. \tag{11.52c}$$

In between two sampling points, still the Riccati differential equations (11.46) will have to be used, for existence considerations, with the boundary conditions now being identical, $S(t_k) = \bar{Z}^+$. Being time invariant, the Riccati differential equations themselves are also identical, apart from the length of the time interval on which they are defined. Since these Riccati equations are used only for the determination of existence, the conjugate point condition of the one defined on the longest time interval will be the determining factor – all others will play no role in the solution. Hence, letting

$$\bar{t}_s := \sup_k \,[t_{k+1} - t_k, \quad k = 0, 1, \ldots]$$

the relevant Riccati equation will be

$$\dot{S} + A'S + SA + Q + \frac{1}{r} S B^2 B^{2'} S = 0; \quad S(\bar{t}_s) = \bar{Z}^+ \tag{11.53}$$

which should not have a conjugate point in the interval $[0, \bar{t}_s]$. This now brings us to the following theorem.

Theorem 2.12. *Consider the infinite-horizon time-invariant linear-quadratic differential game, where $Q_f = 0$, the pair (A, C) is observable, with $C'C = Q$, and the information structure is sampled-data. Then,*

(i) The game has equal upper and lower values if, and only if, the GARE (11.48) admits a positive definite solution $\bar{Z}^+$, and the Riccati differential equation (11.53) has no conjugate point in the interval $[0, \bar{t}_s]$, where $\bar{t}_s$ is the length of the longest sampling interval.

(ii) The upper value is finite if, and only if, the upper and lower values are equal, with the common value being (11.49).

(iii) If $\bar{Z}^+ > 0$ exists, and the Riccati equation (11.53) has no conjugate point on the interval $[0, \bar{t}_s]$, the controller μ_∞^{1s} given by (11.52a) attains the finite value.

(iv) If Q is positive definite, the conditions of (i) are satisfied, and the controller (11.52b) is used for u^1, the resulting (hybrid) closed-loop system becomes input-output stable (with u^2 considered to be the input). □

Computation of $\bar{Z}^+$:
We have seen that the minimal positive definite solution $\bar{Z}^+$ of the GARE (11.48) plays an important role under both closed-loop and sampled-data information patterns (in fact under any information pattern) for the infinite horizon LQ zero-sum differential game. A relevant question at this point is whether one can develop an iterative procedure to yield $\bar{Z}^+$, quite parallel to those available for the standard regulator problem (see, for example, [15]). This indeed turns out to be possible, as discussed earlier in [19]. Below we provide the essentials of this iterative procedure. First a useful lemma:

Lemma 2.7. *Let $F := -A'$, $G := C'$, $R := B^1B^{1'} - \frac{1}{r}B^2B^{2'}$, and (A, C) be observable. Then, there exists $\bar{Z}^+ > 0$ which is a minimal positive definite solution to the GARE (11.48) if, and only if, there exists a maximal positive definite solution $\bar{M}^+$ to the ARE*

$$F'\bar{M} + \bar{M}F + R = \bar{M}GG'\bar{M} \tag{11.54}$$

in which case $\bar{M}^+ = (\bar{Z}^+)^{-1}$. □

Note that (11.54) is the ARE associated with the regulator problem

$$J(u) = \int_0^\infty (x'Rx + u'u)dt \rightarrow \text{minimize} \tag{11.55a}$$

subject to

$$\dot{x} = Fx + Gu, \qquad x(0) = x_0 \tag{11.55b}$$

where (F, G) is controllable. This is not a standard regulator problem because R could be indefinite. But nevertheless if a minimizing solution exists, it could be obtained through a policy iteration algorithm, which is also known as "Bellman's approximation in policy space". The following lemma sets the stage for such an algorithm:

Lemma 2.8. *Consider the nonlinear time-invariant infinite-horizon optimal control problem:*

$$J(u,\xi) = \int_0^\infty \ell(x(t), u(t))dt \rightarrow \textit{minimize}$$

subject to

$$\dot{x} = f(x,u); \qquad x(0) = \xi$$

and

$$\lim_{t\rightarrow\infty} x(t) = 0,$$

where $\xi \in \mathbb{R}^n$ *is considered as a parameter. Let* $u(t) = \mu_i(x(t))$ *be any stabilizing controller, defined on the interval* $[0,\infty)$, *and let*

$$V_i(\xi) := J(\mu_i, \xi), \quad V_i(0) = 0$$

with $V_i(x)$ *being continuously differentiable in* x. *Introduce the controller* μ_{i+1} *as*

$$\mu_{i+1}(x) = \arg_u \min \left\{ \ell(x,u) + \frac{\partial V_i(x)}{\partial x} f(x,u) \right\}.$$

If μ_{i+1} *is stabilizing, then*

$$V_{i+1}(\xi) \leq V_i(\xi), \quad \textit{for all } \xi \in \mathbb{R}^n.$$

Proof. The result follows from the inequality

$$V_{i+1}(\xi) = V_i(\xi) + \int_0^\infty [\ell(x, \mu_{i+1}(x)) + \frac{\partial V_i(x)}{\partial x} f(x, \mu_{i+1}(x))]dt$$

by noting that the integrand is nonpositive since

$$\ell(x, \mu_i(x)) + \frac{\partial V_i(x)}{\partial x} f(x, \mu_i(x)) \equiv 0$$

with V_i being the value function corresponding to μ_i. □

If we now identify

$$\ell(x,u) = x'Rx + u'u, \quad f(x,u) = Fx + Gu,$$

Lemma 2.8 suggests the following algorithm for the computation of the maximal positive definite solution of (11.54):

Step 1: Find an M_0 such that the controller $\mu_0(x) = -G'M_0x$ stabilizes the system (11.55b). [If there is no such controller, then this means that (11.54) does not have a positive definite solution, implying that the upper value of the original game is unbounded.]

Step 2: Let $F_i := F - GG'M_i$, $i = 0$, which is stable by hypothesis, and solve for the unique solution of the Lyapunov equation

$$F_i'M + MF_i = -R - M_iGG'M_i,$$

and call it M_{i+1}.

Step 3: Let

$$\mu_{i+1}(x) = -G'M_{i+1}x,$$

and assume that this is stabilizing for (11.55b). Set $i = i + 1$, and go to Step 2.

The above iteration yields

$$V_i(x_0) = x_0'M_{i+1}x_0$$

and

$$\mu_{i+1}(x) = -G'M_{i+1}x = \arg_u \min\left[x'Rx + u'u + 2x'M_{i+1}(Fx + Gu)\right].$$

Hence, if μ_{i+1} is stabilizing, then by Lemma 2.8,

$$M_{i+1} \leq M_i\ ,$$

that is the algorithm generates a monotonically nonincreasing (in the matrix sense) sequence.

To study the conditions under which the algorithm would generate stabilizing solutions at each step, we first note that if (11.54) has a real symmetric solution, then it will also have maximal and minimal solutions, say M^+ and M^-, with the property that

$$\Re\lambda(F - GG'M^+) \leq 0$$

$$\Re\lambda(F - GG'M^-) \geq 0$$

(see [23]). Using this property, the following result can be proven ([17]).

Proposition 2.1. *Let (11.54) have a real symmetric solution, and $M^+ - M^- > 0$. Then,*

(i) The sequence $\mu_i(x)$ generated above is stabilizing for all $i = 1, 2, \ldots$
(ii) $M_0 \geq M_1 \geq \ldots \geq M_i \geq \ldots \geq M^+$ and

$$\lim_{i\to\infty} M_i = M^+ > 0. \qquad \square$$

11.3 Applications in Minimax Controller Design

The results presented in the previous section have important applications in a special class of minimax controller design problems, known as "H^∞-optimal control" problems ([24], [14], [13], to cite just a few representative papers from this voluminous literature). The objective in these design problems is to determine a controller which will lead to a minimum sensitivity from the disturbance to the (controlled) output. Even though both frequency and time domain formulations are possible, the latter is preferable since it also allows for the analysis of transient behavior, and can accommodate time-variant linear systems. In this section we will first present a general formulation of this worst-case design problem, and then show how the theory of the previous section can directly be used for the construction of minimax (H^∞-optimal) controllers. The presentation here will be brief due to page limitations; more detailed analysis, supported by several illustrative examples, can be found in the recent papers [2], [6], [5], [4].

To introduce the "H^∞-optimal control" problem, let us first adopt the discrete-time formulation, with the linear system being

$$x_{k+1} = A_k x_k + B_k u_k + D_k w_k, \quad x_1 = 0 \tag{11.56}$$

where u_k, $k \in \mathcal{K}$, is the control variable, and w_k $k \in \mathcal{K}$, is the disturbance. Let the square of the norm of the mapping from the disturbance to the controlled output, under a given control policy $u_k = \mu_k(\cdot)$, $k \in \mathcal{K}$, be given in terms of a cost function

$$\tilde{L}(u, w) = \sum_{k=1}^{K} \left\{ |x_{k+1}|^2_{Q_{k+1}} + |u_k|^2 \right\}, \quad Q_{k+1} \geq 0 \tag{11.57}$$

as

$$\mathcal{T}_\mu = \sup_w \left\{ \tilde{L}(\mu, w) / \|w\|^2 \right\}. \tag{11.58a}$$

where

$$\|w\|^2 := \sum_{k=1}^{K} |w_k|^2. \tag{11.58b}$$

Note that, apart from a change in notation, (11.56) is the same system equation as (11.1), but with a specific intial condition ($x_1 = 0$). Furthermore, (11.57) is the same as (11.2), with only the negative term missing.

Now, the minimax controller design problem is to determine

(i) the infimum of $\mathcal{T}_\mu$ over $\mu \in \mathcal{M}$, where $\mathcal{M}$ is a given class of controllers, compatible with the underlying information pattern, and
(ii) the controller $\mu^* \in \mathcal{M}$ (if exists) that achieves the infimum. That is,

$$\inf_{\mu \in \mathcal{M}} \mathcal{T}_\mu =: {\gamma^*}^2 = \mathcal{T}_{\mu^*}. \tag{11.59}$$

For the continuous-time problem, we replace (11.56) by

$$\dot{x} = A(t)x + B(t)u(t) + D(t)w(t), \quad x(0) = 0 \tag{11.60}$$

and (11.57) by

$$\tilde{L}(u,w) = |x(t_f)|^2_{Q_f} + \int_0^{t_f} \left\{ |x(t)|^2_{Q(t)} + |u(t)|^2 \right\} dt \tag{11.61}$$

for $Q_f \geq 0$, $Q(\cdot) \geq 0$. The objective is again (11.59), with $\|w\|$ in (11.58a) now defined by

$$\|w\|^2 := \int_0^{t_f} |w(t)|^2 dt. \tag{11.62}$$

Under either formulation, if there exists a minimax (H^∞-optimal) controller μ^*, then (11.59) becomes equivalent to

(i) $\tilde{L}(u^*, w) \leq (\gamma^*)^2 \|w\|^2$, for all permissible w,
and
(ii) there is no other $\mu \in \mathcal{M}$ (say, $\hat{\mu}$), and a corresponding $\hat{\gamma} < \gamma^*$, such that

$$\tilde{L}(\hat{\mu}, w) \leq (\hat{\gamma})^2 \|w\|^2,$$

for all permissible w.

Now, introducing the parameterized (in $\gamma \geq 0$) family of cost functions:

$$L_\gamma(u,w) := \tilde{L}(u,w) - \gamma^2 \|w\|^2, \tag{11.63}$$

(i) and (ii) above become equivalent to finding the smallest value of $\gamma \geq 0$ under which the upper value of the associated game with objective function $L_\gamma(u,w)$ is bounded, *and* finding the corresponding controller that achieves this upper value[1]. Since (11.63) is precisely either (11.2) or (11.9), with r_k or $r(t)$ replaced by constant γ^2, the theory of the previous section is directly applicable here, providing complete answers under various information structures, and for both finite and infinite horizon formulations.

Of particular relevance here are Theorems 2.2, 2.5, 2.6, 2.8, 2.10 and 2.11. Each one corresponds to a different problem formulation, and in each case the performance level γ^* is the smallest value of γ (or the infimum of all such γ) under which the necessary and sufficient conditions of the corresponding theorems are satisfied. In the discrete time this involves the invertibility of some matrices, and in the continuous time the requirement is the nonexistence of conjugate points to some generalized Riccati equations.

As $\gamma \downarrow \gamma^*$, the limiting saddle-point controllers may not always be well-defined, since their gain matrices may become unbounded. In such a case one has to content with ϵ-optimal controllers, which are the saddle-point controllers corresponding to $\gamma = \gamma_\epsilon := \gamma^* + \epsilon$. This indeed turns out to be the case in the continuous-time problem under the closed-loop information pattern ([4]); but under all information patterns for the discrete-time problem ([2], [5]), and under sampled-data information in the continuous-time case ([4]) the limiting (optimal) controllers do exist.

---

1 Here, the disturbance w can be taken to be open-loop, without any loss of generality.

11.4 Nonzero-Sum Dynamic Games

In this section we present a class of coupled generalized Riccati equations that arise in the characterization of noncooperative (Nash) equilibria of linear-quadratic nonzero-sum dynamic games, in both discrete and continuous time, and under various information patterns. The presentation will be brief, and will concern only two-player deterministic games. Details and extensions can be found in ([8], chapter 6).

The state equations are again given by (11.1) and (11.8), in discrete time and continuous time, respectively. The cost functions of the players will now not add up to zero as in Section 2; instead we have

discrete-time:

$$L^i(u^1,u^2) = \sum_{k=1}^{K} \left\{ |x_{k+1}|^2_{Q^i_{k+1}} + |u^i_k|^2_{R^{i1}_k} + |u^2_k|^2_{R^{i2}_k} \right\}$$

$$R^{ii}_k > 0, \quad R^{ij}_k \geq 0, \quad Q^i_{k+1} \geq 0, \quad k \in \mathcal{K}, \quad i \neq j; i,j = 1,2$$

continuous-time:

$$L^i(u^1,u^2) = |x(t_f)|^2_{Q^i_f} + \int_0^{t_f} \left\{ |x(t)|^2_{Q^i(t)} + |u^1(t)|^2_{R^{i1}(t)} + |u^2(t)|^2_{R^{i2}(t)} \right\}$$

$$R^{ii}(\cdot) > 0, \quad R^{ij}(\cdot) \geq 0, \quad Q^i_f \geq 0, \quad Q^i(\cdot) \geq 0, \quad i \neq j; i,j = 1,2.$$

Player i wishes to minimize his own cost function, L^i, by choosing an appropriate policy $\mu^i \in \mathcal{M}^i$, where the policy space $\mathcal{M}^i$ is compatible with the underlying information pattern. Introducing two functions J^1 and J^2, as in (11.6), and the continuous-time counterparts, we have the quadruple $\{J^1, J^2; \mathcal{M}^1, \mathcal{M}^2\}$ constituting the *normal* form of the nonzero-sum dynamic game. This framework now allows us to introduce the notion of a *Nash equilibrium*.

Definition 4.1. Given a nonzero-sum dynamic game $\{J^1, J^2; \mathcal{M}^1, \mathcal{M}^2\}$, a pair of policies $(\mu^{1^*}, \mu^{2^*}) \in \mathcal{M}^1 \times \mathcal{M}^2$ constitutes a *Nash equilibrium* if,

$$J^{1^*} := J^1(\mu^{1^*}, \mu^{2^*}) = \min_{\mu^1 \in \mathcal{M}^1} J^1(\mu^1, \mu^{2^*}) \tag{11.64a}$$

and

$$J^{2^*} := J^2(\mu^{1^*}, \mu^{2^*}) = \min_{\mu^2 \in \mathcal{M}^2} J^2(\mu^{1^*}, \mu^2). \quad \square \tag{11.64b}$$

Property 2.1 has no counterpart here, but the notions of "strong time consistency" and "noise insensitivity" are equally applicable; in Definition 2.2, for example, we simply have to replace the "saddle-point equilibrium" with "Nash equilibrium" (see [3]). These refinements are even more important here, because Nash equilibria under dynamic information are generally nonunique, leading to nonunique equilibrium values for J^{1^*} and J^{2^*}.

11.4.1 The Discrete-time LQ Dynamic Game

11.4.1.1 Open-loop information for both players The procedure for derivation of open-loop Nash equilibrium is similar to that of open-loop saddle-point solution presented in Section 2.1.1. We minimize $L^i(u^1, u^2)$ over u^i for each fixed $u^j, j \neq i$, – the solution (reaction curves) is unique in each case since L^i is a quadratic strictly convex function of u^i – and then solve for u^1 and u^2 from the resulting set of linear equations. The solution may or may not exist, or be nonunique, depending on the common point(s) of intersection of the two reaction curves. This common point of intersection, as well as its existence, is characterized in terms of two coupled Riccati equations as stated in the following counterpart of Theorem 2.1.

Theorem 4.1. *For the discrete-time linear-quadratic nonzero-sum dynamic game with open-loop information, let Λ_k^i, M_{k+1}^i, $k \in \mathcal{K}$, $i = 1, 2$, be matrices generated by*

$$M_k^i = Q_k^i + A_k' M_{k+1}^i \Lambda_k^{-1} A_k; \quad M_{K+1}^i = Q_{K+1}^i \tag{11.65a}$$

$$\Lambda_k = I + \sum_{i=1}^{2} B_k^i (R_k^{ii})^{-1} B_k^{i'} M_{k+1}^i \tag{11.65b}$$

If Λ_k, $k \in \mathcal{K}$, thus recursively defined, are invertible, the game admits a unique open-loop Nash equilibrium solution *given by*

$$u_k^{1^*} = \mu_k^{i*}(x_1) = -(R_k^{ii})^{-1} B_k^{i'} M_{k+1} \Lambda_k^{-1} A_k x_k^*, \quad k \in \mathcal{K}; \quad i = 1, 2 \tag{11.66}$$

where $\{x_{k+1}^*, k \in \mathcal{K}\}$ is the corresponding state trajectory, generated by

$$x_{k+1}^* = \Lambda_k^{-1} A_k x_k^*, \quad x_1^* = x_1. \; \square \tag{11.67}$$

Note the similarity between (11.65a)-(11.65b) and (11.14a)-(11.14b). The differences are that here we have two discrete-time Riccati equations, coupled through Λ_k, and the invertibility of Λ_k is a part of the hypothesis rather than a conclusion.

11.4.1.2 Closed-loop information for both players When the players have access to dynamic information, the Nash equilibrium is (informationally) nonunique, as indicated earlier. The imposition of strong time consistency, however, eliminates informational nonuniqueness and leads to generically unique equilibria, known also as *feedback Nash equilibrium.* Here the procedure is to solve a sequence of static quadratic games in retrograde time, which yields a pair of policies (in Nash equilibrium) that depend only on the current value of the state vector. Before presenting this result in the form of a theorem, let us introduce some notation:

Let P_k^i, $k \in \mathcal{K}$, $i = 1, 2$, be a sequence of matrices satisfying the set of linear matrix equations

$$\left[R_k^{ii} + B_k^{i'} Z_{k+1}^i B_k^i\right] P_k^i + B_k^{i'} Z_{k+1}^i B_k^j P_k^j = B_k^{i'} Z_{k+1}^i A_k, \qquad j \neq i; \quad i, j = 1, 2, \tag{11.68a}$$

where Z_{k+1}^i, $k \in \mathcal{K}$, $i = 1, 2$, are generated by

$$Z_k^i = F_k' Z_{k+1}^i F_k + \sum_{j=1}^{2} P_k^{j'} R_k^{ij} P_k^j + Q_k^i; \quad Z_{K+1}^i = Q_{K+1}^i, \quad i = 1, 2, \tag{11.68b}$$

and

$$F_k := A_k - B_k^1 P_k^1 - B_k^2 P_k^2, \quad k \in \mathcal{K}. \tag{11.68c}$$

Theorem 4.2. *For the discrete-time linear-quadratic nonzero-sum game with closed-loop information structure, there exists a unique feedback Nash equilibrium solution if, and only if, (11.68a) admits a unique solution set $\{P_k^{i^*}, k \in \mathcal{K}, i = 1, 2\}$, in which case the equilibrium strategies are given by*

$$u_k^{i^*} = \mu_k^{i^*}(x_k) = -P_k^{i^*} x_k, \quad k \in \mathcal{K}; i = 1, 2. \tag{11.69}$$

This solution is strongly time consistent and noise insensitive. □

Note that the coupled Riccati equations (11.68b) above are different from the ones characterizing the solution in the open-loop case. The existence condition, which is the solvability of (11.68a) for each $k \in \mathcal{K}$, can be simplified somewhat by solving for P_k^1 from (11.68a) with $i = 1$, substituting it into the same for $i = 2$, and requiring that the coefficient matrix of P_k^2 be invertible. This leads to the single matrix condition:

$$\det\left[R_k^{22} + B_k^{2'} Z_{k+1}^2 B_k^2 - B_k^{2'} Z_{k+1}^2 B_k^1 \left[R_k^{11} + B_k^{1'} Z_{k+1}^1 B_k^1\right]^{-1} B_k^{1'} Z_{k+1}^1 B_k^2\right] \neq 0, \quad k \in \mathcal{K}. \tag{11.70}$$

11.4.2 The Continuous-time LQ NZS Dynamic (Differential) Game

Here again we consider only the open-loop and closed-loop (for both players) cases, and present the continuous-time counterparts of Theorem 4.1 and 4.2.

11.4.2.1 Open-loop information for both players The procedure is similar to the discrete-time case, but here the use of minimum principle results in a simpler derivation ([8], p. 280).

Theorem 4.3. *For the LQ NZS differential game with open-loop information, let there exist a unique solution set* $\{M^i(\cdot), i = 1,2\}$ *to the coupled matrix Riccati differential equations*

$$\dot{M}^i + M^i A + A' M^i + Q^i - M^i \sum_{j=1}^{2} B^j (R^{jj})^{-1} B^{j'} M^j = 0; \qquad (11.71)$$
$$M^i(t_f) = Q^i_f, \quad i = 1,2.$$

Then the differential game admits a unique open-loop Nash equilibrium solution given by

$$u^{i^*}(t) = \mu^{i^*}(t, x_0) = -\left[R^{ii}(t)\right]^{-1} B^i(t)' M^i(t) x^*(t), \quad i = 1,2, \qquad (11.72a)$$

where $x^*(\cdot)$ *denotes the associated state trajectory solving the linear differential equation*

$$\dot{x}^* = \left(A - B^1 (R^{11})^{-1} B^{1'} M^1 - B^2 (R^{22})^{-1} B^{2'} M^2\right) x^*, \qquad (11.72b)$$
$$x^*(0) = x_0. \quad \square$$

The main difference between the zero-sum and nonzero-sum open-loop solutions (i.e., Theorem 2.7 versus Theorem 4.3) is that in the latter a solution may cease to exist even in the absence of a conjugate point. In fact there is no general set of conditions (on the parameters of the problem) that would guarantee the existence of a solution to (11.71), unless the time interval is sufficiently short. One special case, however, deserves further discussion – games with so-called weakly coupled players. For a sufficiently small $\epsilon > 0$, let

$$A = \begin{bmatrix} A_1 & \epsilon A_{12} \\ \epsilon A_{21} & A_2 \end{bmatrix}, \quad B^1 = [B_1', 0]', \quad B^2 = [0, B_2']' \qquad (11.73a)$$

$$Q^1_f = \text{block diag } \{Q^1_{11f}, \epsilon Q^1_{22f}\}; Q^1 = \text{block diag } \{Q^1_{11}, \epsilon Q^1_{22}\} \qquad (11.73b)$$

$$Q^2_f = \text{block diag } \{\epsilon Q^2_{11f}, Q^2_{22f}\}; Q^2 = \text{block diag } \{\epsilon Q^2_{11}, Q^2_{22}\} \qquad (11.73c)$$

$$R^{11} = \text{block diag } \{R^1_1, \epsilon R^1_2\}; R^{22} = \text{block diag } \{\epsilon R^2_1, R^2_2\}, \qquad (11.73d)$$

which implies that for $\epsilon = 0$ the game decomposes into two completely decoupled optimal control problems, one for each player. Now, supposing that $M^i(t;\epsilon)$ admits an expansion in the form

$$M^i(t;\epsilon) = \sum_{\ell=0}^{\infty} \epsilon^\ell M^{i(\ell)}(t); \quad i = 1,2, \qquad (11.74)$$

it readily follows that the individual terms in the expansion above can be obtained from

$$\dot{M}^{i(0)}_{ii} + A_i' M^{i(0)}_{ii} + M^{i(0)}_{ii} A_i - M^{i(0)}_{ii} B_i' (R^i_i)^{-1} M^{i(0)}_{ii} + Q^i_{ii} = 0; \qquad (11.75a)$$
$$M^{i(0)}_{ii}(t_f) = Q^i_{iif}, \quad i = 1,2$$

$$M^{1(0)}(t) = \begin{bmatrix} M_{11}^{1(0)} & 0 \\ 0 & 0 \end{bmatrix}; M^{2(0)}(t) = \begin{bmatrix} 0 & 0 \\ 0 & M_{22}^{2(0)}(t) \end{bmatrix} \tag{11.75b}$$

and for $k \geq 1$,

$$\dot{M}^{i(k)} + M^{i(k)}A^{(0)} + A^{(0)'}M^{i(k)} + A^{(1)'}M^{i(k-1)} + M^{i(k-1)}A^{(1)} + q^{(k)}$$
$$- \sum_{\ell=0}^{k} M^{i(\ell)} \sum_{j=1}^{2} B^j (R_j^j)^{-1} B^{j'} M^{i(k-\ell)} = 0, \tag{11.76}$$
$$M^{i(k)}(t_f) = Q_f^{i(k)}; i = 1,2$$

where

$$A^{(0)} := \text{ block diag } (A_1, A_2), \quad A^{(1)} := A - A^{(0)}$$

$$Q_f^{1(0)} := \text{ block diag } (Q_{11f}^1, 0), \quad Q_f^{2(0)} := \text{ block diag } (0, Q_{22f}^2)$$

$$Q_f^{i(1)} := Q_f^i, -Q_f^{i(0)}; \quad Q_f^{i(k)} = 0, \quad k \geq 2; i = 1,2.$$

Note that (11.75a) is a pair of decoupled Riccati differential equations of the type that arises in LQ optimal control, which admit a nonnegative definite solution for all t, $0 \leq t \leq t_f$. The higher (than *zeroth*) order terms in the expansion (11.74), however, are obtained from even simpler (linear) differential equations (11.76), iteratively in k. Hence, if the expansion (11.74) is valid, then a solution to the coupled Riccati equation (11.71) exists, and can be obtained from the solutions of two regular Riccati and several linear differential equations. Using the implicit function theorem, one can in fact prove that under certain smoothness conditions on the parameters of problem, such an expansion is valid in an open neighborhood of $\epsilon = 0$. For more details see [22].

11.4.2.2 Closed-loop information for both players The feedback Nash equilibrium, whenever it exists, is obtained from the solution of a pair of coupled Hamilton-Jacobi-Bellman equations, which can be expressed in terms of the solutions to some coupled Riccati equations ([8], p. 287). Hence, we have the following counterpart of Theorem 2.8 for LQ NZS games.

Theorem 4.4. *For the LQ NZS differential game with closed-loop information structure, let there exist a set of matrix valued functions, $Z^i(\cdot) \geq 0$; $i = 1, 2$, satisfying the following coupled matrix Riccati differential equations:*

$$\dot{Z}^i + Z^i\tilde{F} + \tilde{F}'Z^i + \sum_{j=1}^{2} Z^j B^j (R^{jj})^{-1} R^{ij} (R^{jj})^{-1} B^{j'} Z^j + Q^i = 0; \tag{11.77}$$
$$Z^i(t_f) = Q_f^i$$

where

$$\tilde{F}(t) := A(t) - \sum_{i=1}^{2} B^i(t) R^{ii}(t)^{-1} B^i(t)' Z^i(t). \tag{11.78}$$

Then, the differential game admits a feedback Nash equilibrium solution given by

$$u^{i^*}(t) = \mu^{i^*}(t,x) = -R^{ii}(t)^{-1}B^i(t)'Z^i(t)x(t), \quad i = 1,2. \tag{11.79}$$

This solution is both strongly time consistent and noise insensitive. □

The set of coupled Riccati equations that characterize the feedback solution (i.e., (11.76)) are generally different from the set that characterizes the open-loop Nash strategies (i.e., (11.71)). Again there is no general set of conditions that would guarantee existence of a solution to (11.77), unless the time interval is short [20]; see also [21] for the infinite-horizon version. For the case of weakly coupled agents (introduced in Section 4.2.1), however, one can again show that, provided that an expansion to $Z^i(t;\epsilon)$ in the form

$$Z^i(t;\epsilon) = \sum_{\ell=0}^{\infty} \epsilon^\ell Z^{i(\ell)}(t); \quad i = 1,2 \tag{11.80}$$

exists, $Z^{i(0)}$ can be obtained by solving a standard Riccati differential equation, for each $i = 1,2$, and the higher-order terms are obtained from the solutions of linear matrix differential equations ([12], [22]). It turns out that $Z^{i(0)}(t) \equiv M^{i(0)}(t)$, even though the higher order terms do not have such a correspondence. Furthermore, one can again prove, under some smoothness conditions, that there is an open neighborhood of $\epsilon = 0$, where the expansion (11.80) is valid, and hence (11.77) admits a unique solution. □

11.5 Conclusion

There are several other classes of games and/or equilibrium solution concepts where some type of Riccati equations play important roles either in the construction of equilibrium policies or in the characterization of conditions of their existence and uniqueness. We have not discussed them here, because their inclusion would have at least doubled the length of this survey/tutorial. Some of these extensions are (i) noise-insensitive Nash equilibria of LQ NZS dynamic games where the information pattern is mixed, such as OL/CL, CL/sampled-data, CL/delayed information patterns, and (ii) Stackelberg equilibria of LQ NZS games, where there is a hierarchy in the roles of the players, and the solution under the OL/OL information pattern is obtained by solving a LQ control problem subject to a constraint characterized in terms of a Riccati equation. For derivations of these solutions and discussions of their properties, the reader is referred to [8] and the references contained therein. Another class would be linear-quadratic-Gaussian (LQG) zero-sum dynamic/differential games where the players have a common noisy measurement. Here, a weak form of the separation result of LQG control theory applies, where the Riccati equation (11.14a) (discrete time) or (11.36) (continuous time) is used in the construction of the pure-strategy saddle-point control gains, but an additional ($2n$-

dimensional) Riccati equation is needed to determine the existence of a saddle point [1]. Finally, it is possible to write down and study properties of infinite-dimensional versions of the Riccati differential equations (11.36) and (11.35), which arise in the closed-loop feedback and open-loop saddle points of LQ zero-sum dynamic games defined on infinite-dimensional Hilbert spaces [9].

Acknowledgement. The writing of this chapter was supported in part by the U.S. Department of Energy under Grant DE-FG-02-88-ER-13939, and in part by the Joint Services Electronics Program under Grant N000-14-90-J-1270.

References

1. T. Başar. On the saddle-point solution of a class of stochastic differential games. *Journal on Optimization Theory and Applications*, 33(4):539–556, April 1981.
2. T. Başar. A dynamic games approach to controller design: Disturbance rejection in discrete time. In *Proceedings of the 29th IEEE Conf. Decision and Control*, pages 407–414, December 13-15, 1989. Tampa, FL.
3. T. Başar. Time consistency and robustness of equilibria in noncooperative dynamic games. In F. Van der Ploeg and A. de Zeeuw, editors, *Dynamic Policy Games in Economics*, pages 9–54. North Holland, 1989.
4. T. Başar. Game theory and H^∞-optimal control: The continuous-time case. In *Proceedings of the 4th International Conference on Differential Games and Applications*. Springer-Verlag, August 1990. Helsinki, Finland.
5. T. Başar. Game theory and H^∞-optimal control: The discrete-time case. in *Communication, Control and Signal Processing*, E. Arikan (edt) (*Proceedings of the 1990 International Conference on New Trends in Communication, Control and Signal Processing*, Ankara, Turkey), pages 669-686, Elsevier, July 1990.
6. T. Başar. Minimax disturbance attenuation in LTV plants in discrete-time. *Proceedings of the Automatic Control Conference*, San Diego, May, 1990.
7. T. Başar. *Theory of Stochastic Dynamic Noncooperative Games*. Basil Blackwell, Oxford, England, 1992. Under preparation.
8. T. Başar and G.J. Olsder. *Dynamic Noncooperative Game Theory*. Academic Press, London/New York, 1982.
9. A. Bensoussan. Saddle points of convex concave functionals. In H.W. Kuhn and G.P. Szegö, editors, *Differential Games and Related Topics*, pages 177–200. North-Holland, 1971. Amsterdam.
10. P. Bernhard. Linear-quadratic two-person zero-sum differential games: necessary and sufficient conditions. *Journal on Optimization Theory and Applications*, 27(1):51–69, January 1979.
11. R. W. Brockett. *Finite Dimensional Linear Systems*. Wiley, New York, 1970.
12. Ü. Özgüner and W.R. Perkins. A series solution to the Nash strategy for large scale interconnected systems. *Automatica*, 13(3):313–315, May 1977.
13. J. Doyle, K. Glover, P. Khargonekar, and B. Francis. State-space solutions to standard H_2 and H_∞ control problems. *IEEE Transactions on Automatic Control*, AC-34(8):831–847, 1989.
14. B.A. Francis. *A Course in H_∞ Control Theory*, volume 88. Lecture Notes in Control and Information Sciences, Springer-Verlag, New York, 1987.
15. D.L. Kleinman. On an iterative technique for Riccati equation computations. *IEEE Transactions on Automatic Control*, AC-13(1):114–115, February 1968.
16. H. Kwakernaak and R. Sivan. *Linear Optimal Control Systems*. Wiley, New York, 1972.
17. E.F. Mageirou. Iterative techniques for Riccati game equations. In *Proc. 15th IEEE Conf. Decision and Control*, 1976.

18. E.F. Mageirou. Values and strategies for infinite duration linear quadratic games. *IEEE Transactions on Automatic Control*, AC-21(4): 547-550, August 1976.
19. E.F. Mageirou and Y.C. Ho. Decentralized stabilization via game theoretic methods. *Automatica*, 13:393–399, 1977.
20. G.P. Papavassilopoulos and J.B. Cruz, Jr. On the existence of solutions to coupled matrix Riccati differential equations in linear quadratic Nash games. *IEEE Transactions on Automatic Control*, AC-24(1), February 1979.
21. G.P. Papavassilopoulos, J.V. Medanic, and J.B. Cruz, Jr. On the existence of Nash strategies and solutions to coupled Riccati equations in linear-quadratic games. *Journal of Optimization Theory and Applications*, 28(1), May 1979.
22. R. Srikant and T. Başar. Iterative computation of noncooperative equilibria in nonzero-sum differential games with weakly coupled players. *Journal on Optimization Theory and Applications*, to appear, 1991; a shorter version will appear in the *Proc. 29th IEEE Conf. Decision and Control*, Honolulu, Hawaii, December 1990.
23. J.C. Willems. On the existence of a non-positive solution to the Riccati equation. *IEEE Transactions Automatic Control*, AC-19(5):592–593, October 1974.
24. G. Zames. Feedback and optimal sensitivity: Model reference transformation, multiplicative seminorms and approximate inverses. *IEEE Transactions on Automatic Control*, AC-26:301–320, 1981.

Subject Index